国家自然科学基金项目（批准号：51408121）
Funded by National Natural Science Foundation of China (No. 51408121)

城垣下的绿谱

南京明城墙绿道空间特征与服务绩效图解

GREEN SPECTRUM ALONG THE CITY WALL

Graphic Analysis of Spatial Characteristics and Service Performance of the Nanjing Ming Dynasty City Wall Greenway

周聪惠 等著
ZHOU Conghui et al

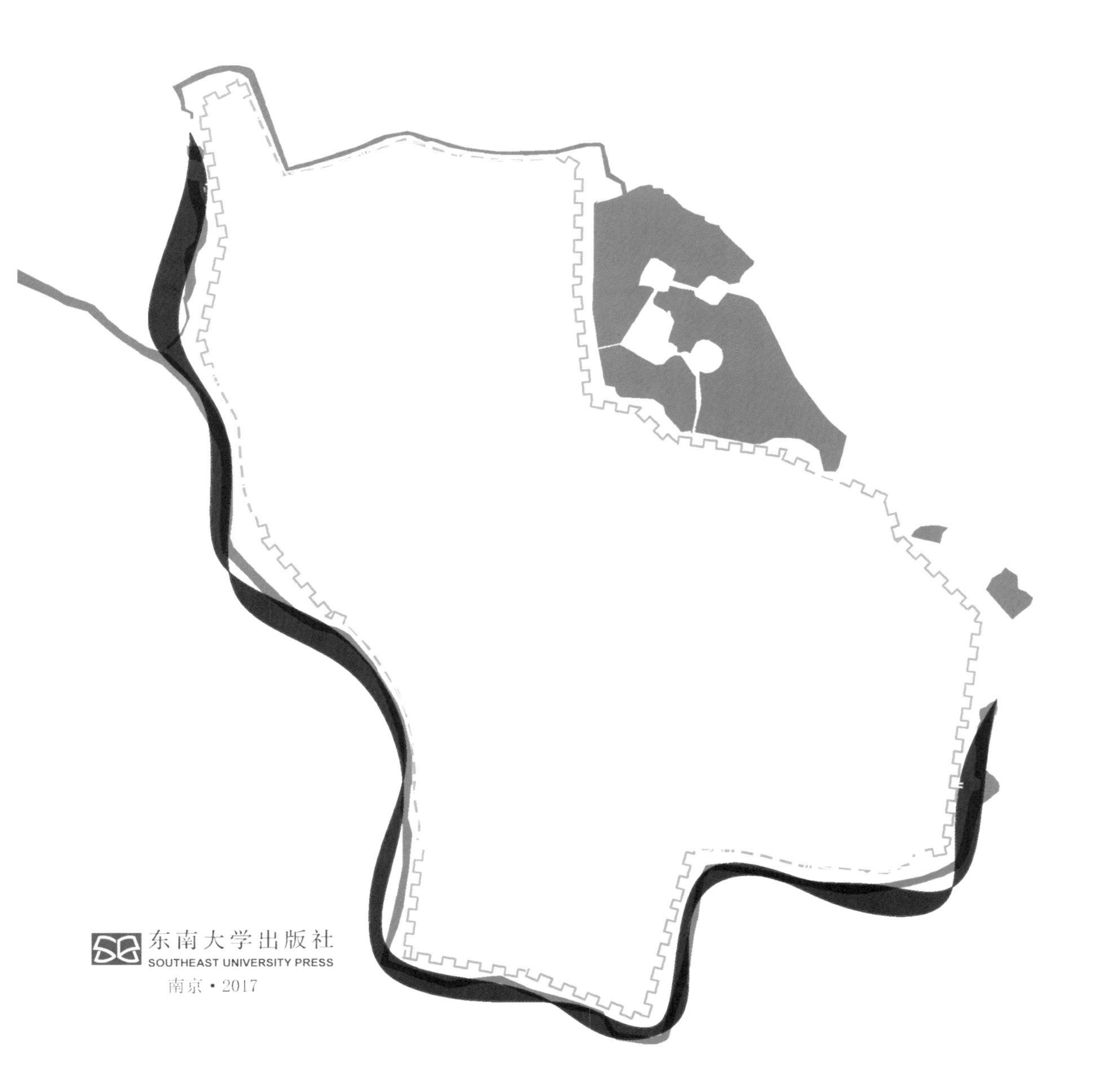

东南大学出版社
SOUTHEAST UNIVERSITY PRESS
南京·2017

内容提要

本书以南京明城墙绿道为研究对象，通过多轮使用人群信息采集、问题分析、信息补充等环节，用定量方法将绿道的服务绩效、内部和外部空间特征进行梳理和描述，并在此基础上将绿道服务绩效指标与其内部和外部空间特征指标进行关联分析，进而尝试从中发现绿道内、外部空间特征对服务绩效的影响方式和作用规律。

全书在表达上依托直观易懂的图解方式来展示研究思路、方法、过程和成果，除了希望其能为相关科研人员和规划设计从业者提供有益的参照信息外，还寄予书中研究内容和成果能被更多非专业人士所认知和理解，并期待激发出广大社会公众的兴趣和积极性，更加主动地关注和投身参与到身边诸如绿道一类城市公共空间规划设计、优化提升和建设使用中来。

图书在版编目（CIP）数据

城垣下的绿谱：南京明城墙绿道空间特征与服务绩效图解 / 周聪惠等著. —南京：东南大学出版社，2017.9

ISBN 978-7-5641-7378-4

Ⅰ. ①城… Ⅱ. ①周… Ⅲ. ①城墙-绿化-研究-南京 Ⅳ. ①TU985.253.1

中国版本图书馆CIP数据核字（2017）第193634号

书　　名：城垣下的绿谱：南京明城墙绿道空间特征与服务绩效图解
著　　者：周聪惠 等
策划编辑：孙惠玉　　责任编辑：徐步政　　邮箱：1821877582@qq.com

出版发行：东南大学出版社　　社址：南京市四牌楼 2 号（210096）
网　　址：http://www.seupress.com
出 版 人：江建中

印　　刷：恒美印务（广州）有限公司
开　　本：889mm×1194mm　1/16　　印张：21　　字数：640 千
版 印 次：2017 年 9 月第 1 版　2017 年 9 月第 1 次印刷
书　　号：ISBN 978-7-5641-7378-4　　定价：139.00 元

经　　销：全国各地新华书店　　发行热线：025-83790519　83791830

编写翻译人员 | AUTHORS AND TRANSLATORS

主要作者

周聪惠（副教授，东南大学建筑学院景观学系）

Main Author

Zhou Conghui (Associate Professor, Department of Landscape Architecture, School of Architecture, Southeast University)

其他作者

吴　韵（研究助理，东南大学建筑学院景观学系）

胡　樱（研究助理，东南大学建筑学院景观学系）

容梓昊（研究助理，东南大学建筑学院景观学系）

戴文嘉（研究助理，东南大学建筑学院景观学系）

Other Authors

Wu Yun (Research Assistant, Department of Landscape Architecture, School of Architecture, Southeast University)

Hu Ying (Research Assistant, Department of Landscape Architecture, School of Architecture, Southeast University)

Rong Zihao (Research Assistant, Department of Landscape Architecture, School of Architecture, Southeast University)

Dai Wenjia (Research Assistant, Department of Landscape Architecture, School of Architecture, Southeast University)

英文翻译

田　恬

English Translator

Tian Tian

英文译审

斯蒂芙·查韦斯　（哥伦比亚大学城市设计实验室）

Edit Proofreader

Steff Chavez (Urban Design Lab at Columbia University)

目录 | CONTENTS

序言 | 绿色基础设施的“道”和“墙”

FOREWORD | Green Infrastructure as Ways and Walls

城市已步入绿色基础设施发展的又一个重要阶段。在全球前所未有的城市化进程和生态风险中，城市成了保障我们未来的关键。鉴于此，重新思考城市基础设施的功能成为当前迫切需求。城市既可能是破坏生态环境的潜在风险，也可成为应对新的生态挑战的灵丹妙药。创新型城市战略必须着力于重塑诸如水和能源等重要资源的分配和消费方式，这也一直是千年以来城市形成的核心所在。而与社会“共融”相关的社会基础设施问题，则对达成公众共识、保障新理念的成功推行同样至关重要。南京明城墙绿道项目就是将古城墙作为新型基础设施利用的再思考。本研究在增进对社会准则的认知理解上做出了重要贡献，而对社会准则的充分认知则是南京明城墙绿道这类有深远影响的高标准项目保持长期成功运行的必然要求。

城市线性绿化景观的规划建设早在现代城市主义潮流中就已开始。长期以来，“绿带”一直是欧美城市发展理论中的重要策略，例如，在公园之间或为引入自然特性来建立联系的相关策略。纽约的 F.L.Olmsted 和伦敦的 E.Howard 留下的遗产逐渐演变成风靡 20 世纪的基础设施概念，如 F.J.Osborn 的《绿带城市》（1946 年版）。作为一个常见的英文术语，“绿道”出现于 1960 年代，并由 W.H.Whyte 在《最后景观》（1968 年版）中将其理论化。在过去半个世纪中，“绿道”一词逐渐成为城市绿色基础设施中最为人熟知的概念，其主要任务是重新利用 19 世纪遗留的基础设施，如已不使用的铁路路段或废弃的工业滨水区。“公园路”名下的绿道则是试图缓解 20 世纪以来城市空间被干道网等交通设施分割所带来的危害。

Cities have arrived at a next moment for green infrastructure. With unprecedented global urbanization and unprecedented global ecological risk, cities are key players in securing our future. Seen in this light, rethinking how our urban infrastructure functions is a priority. Cities are at risk to be casualties; but they also may be panaceas for the new ecological challenges. Innovative urban strategies must of necessity focus on reinventing distribution and consumption of resources such as water and energy, which have been at the heart of city formation for millennia. Also crucial are questions of social infrastructure related to *communitas*; to the kind of public consensus that is needed to insure the success of new concepts. Nanjing's greenway initiative is important for rethinking the function of its ancient urban wall as new infrastructure. And this study is a significant contribution to increased understanding of the social norms required for the long-term success of an ambitious project.

Linear urban green landscapes have a long history of deployment in modernist urbanism. In European and American urban theory the "green belt" has long been a prominent strategy, for example as linkage strategy between parks or for assimilation of natural features. The legacies of Frederick Law Olmsted in New York and Ebenezer Howard in London evolved into prevalent 20th century infrastructural concepts like Frederic J. Osborn's *Greenbelt-Cities* (1946). As a common English-language term "greenway" appears to date from the period of the 1960's and well-theorized by William H. Whyte in *The Last Landscape* (1968). During the last half-century, the term "greenway" evolved as the most ubiquitous concept for urban green infrastructure. Greenways were tasked with repurposing 19th century infrastructure such as unused rail cuts or abandoned industrial waterfronts. The greenway under the rubric of "parkway" would attempt to mitigate the hazards of 20th century urban traffic-related incisions such as highways.

在 1930 年代完成的长达 3500km 的东海岸 Appalachian 游径是北美最宏大的绿道景观。近年来，中国广东省建设了总长达 2372km 的绿道网，其连接了珠江口区域的众多工业遗址。这是目前中国最长的绿道网，它解决了土地再利用尤其是工业废弃地再利用的相关问题。该绿道网最终还将与香港和澳门的绿道系统相连。在北京，目前已建成长达 710km 的绿道，而未来在城市内河沿线还计划修建 134km 的绿道。四川省省会成都市、湖北省省会武汉市以及长三角区域的主要城市均已启动绿道规划建设，并已初见成效，南京市的绿道规划建设就是其中之一。但是，让南京绿道规划建设脱颖而出的则是绿道与明城墙的结合及其在文化层面的彰显。

在世界各地，老城墙已所剩无几。北美城市从未拥有过城墙，在欧洲，目前遗存规模最大的军事防御城墙位于法国卡尔卡索纳（Carcassonne，总人口 4.7 万人）。同等级别的其他“绿墙”案例还有意大利卢卡（Lucca，总人口 8.4 万人）或西班牙阿维拉（Avila，总人口 5.8 万人）。对于欧洲大城市而言，在 20 世纪城墙已消失殆尽，其中包括 1929 年被基本抹去的巴黎市外墙。相反的是，南京（总人口 823 万人）却仍坐拥长达 25.1km 的独特城墙基础设施，并有着与欧洲对应城墙截然不同的城市空间文脉。在中国，唯一可与之媲美的是西安（总人口 847 万人）的明城墙。但两地截然不同的气候条件使得“城墙绿道”战略仅在南京实施起来较为现实。而南京的城墙绿道建构战略以及对城墙沿线土地最优和最高效再利用的评价让人能被本研究所吸引。在城市总人口基本相当的纽约市有一处名

Completed in the 1930's, the 3,500 kilometer long Appalachian Trail on the East Coast is the most ambitious greenway landscape in North America. Recent examples in China include the 2,372 kilometer greenway chain in Guangdong tying together industrial sites along the Pearl River estuary. It is the longest in China and addresses the repurposing question, especially as relates to abandoned industrial sites. Eventually it will connect to greenways in Hong Kong and Macau. Beijing has built 710 kilometers of greenway with plans to build an additional 134 kilometers alongside its aqueducts. Similarly Chengdu, the capital of Sichuan Province; Wuhan, the capital of Hubei province; and several major cities in the Yangtze River Delta started their greenway projects and have reached their initial goals. Nanjing is far from alone in its greenway initiative. Its distinction, however, is its merging with the Ming Dynasty Wall and with an associated cultural dimension.

Little is left of old city walls anywhere. North American cities never had walls. In Europe, the largest remaining fortification walls are at Carcassonne (population 47,000) in France. Other "green wall" examples on that scale include Lucca (population 84,000) in Italy or Avila (population 58,000) in Spain. For larger cities in Europe, the walls disappeared by the 20th Century including the outer wall in Paris which was totally erased by 1929. By contrast Nanjing (population 8.23 million) still enjoys a unique wall infrastructure, at 25.1 kilometers length and within a vastly different context than European counterparts. The only comparable infrastructure in China is the Ming Dynasty Wall at Xi'an (population 8.47 million). But climate differences between the two cities make a "green wall" strategy realistic only at Nanjing. The strategy for "greening" at Nanjing makes this study interesting, together with evaluation of the best and highest re-use. In New York, which shares an equivalent population as Nanjing, perhaps a 20th century equivalent is the High Line elevated rail infrastructure, also recently repurposed as

为“高线”（铁路高架桥基础设施）的地方，在近年也被规划为城市公园，并可作为南京明城墙绿道的对照。虽然“高线”修建于20世纪，较明朝相距甚远，并且其2.33km的总长度在尺度上也相对更小，但并不妨碍两者相互分享和汲取各自的经验教训。

与“高线”一样，南京城墙也与城市的空间和社会肌理紧密交织，并将绩效提升视为绿色基础设施建构的重要策略。促进社会共融是下一代绿色基础设施的重要使命，这也必须与新的社区管理形式结合，因此，有必要从新的社会粒层来理解基础设施的空间文脉。本研究的一个开创性方面正与此问题相关。

鉴于城市在我们不断改变的生态环境中的重要性，共同利益的最大化将成为下一代绿色基础设施的关键问题，因为它对于维持设施运行效率，并在文化维度上激发出全新风貌至关重要。当然，如果在功能或土地使用方面处理失当，绿道将会损耗空间资源或需要过度维护，从而阻碍其正常运行和发展。本研究加大在社会层面的研究力度，也将为各个地区城市绿化战略实施功效的优化提升提供分析和借鉴。

本研究最显著的特征是将远程（无人机）监控与地面调查相结合，采集了新一代的数据进行分析，并成为基础设施研究的关键部分。通过在调查技术和分析手段上的创新，研究能深入掌握绿道的功能运转细节。数据涵盖了南京明城墙绿道沿线的6个典型区段，所有区段均位于公众可达条件下，每个区段以50m为基本单元被划分为20-30个研究“切片”。数据采集过程中，在具有代表性的季节和周天时段进行了超过100次（含预调研与6个区段正式调研合计次数）的调查。数据采集范畴包含了城墙绿道内、外部的公共空间特征以及与绿道使用相关的信息。数据被细分来研究要素之间的相互作用，以解读公共空间的功能共性，进而提取积极的作用因素，并对可能的功能修复策略加以探讨。虽然本研究时段以一年为限，但其发展出

a park, although its origins are obviously far from Ming Dynasty origins, and at 2.33 kilometers a far smaller scale. Yet there are likely lessons to be shared from both initiatives.

Like the High Line, the Nanjing wall is closely intertwined with the city's spatial and social fabric with heightened effectiveness as a green infrastructure strategy. Social propinquity is important to next generation green infrastructure, which must be integrated with new forms of community stewardship. Therefore infrastructural context must be understood at new levels of social granularity. A pioneering aspect of this study has exactly to do with this question.

Given the importance of cities in our changing ecological global context, maximized co-benefits of next generation infrastructure is essential for both efficiencies of maintenance and for encouraging the cultural dimension of a new world outlook. Of course there are obstacles given that greenways are space consumptive and require inordinate maintenance not fully justified by the function; or even justified in terms of land use. The heightened social dimension of this study provides analytics for improved effectiveness of urban greening strategies everywhere.

Most significantly this study deploys a new generation of data analytics to become an essential part of the infrastructure equation; through the deployment of remote (drone) surveillance in combination with ground surveys. A remarkable level of functional detail could be obtained and processed entailing innovation in survey technique and analysis. The data involved six typical segments along greenway, each of which was divided into 20 to 30 fifty-meter "slices" across the entire range of public access conditions. For data collection, more than 100 surveys were completed, distributed over a representative range of time of day and season. Data collected was inclusive of public space characteristics and use conditions inside, outside, and on top of the greenway walls. Data was decomposed into binary relationships between elements as a first step in understanding commonalities in public space function; positive attributes to be promulgated; and possible functional remediation strategies. While the study period was limited to a single year, it has developed techniques that could be adapted for the continuous infrastructure monitoring that will be more and more an essential for distributed infrastructural maintenance in cities everywhere.

了一套能够针对基础设施绩效进行持续监测的技术。值得注意的是，该类持续监测将在各城市基础设施的维护使用中变得日益重要。

在数据可视化方面的创新也是该研究的重要部分，如此一来将能方便社会公众认知和理解整个研究的过程和结果。归根结底，普罗大众才是将南京明城墙转型为下一代绿色基础设施的主要利益相关者。与世界各地城市居民一样，南京市民对城市绿色基础设施规划、设计、优化和重建的兴趣和参与是应对全球生态变化挑战的重要组成部分。

纽约市哥伦比亚大学地球研究院城市设计实验室对该研究尤感兴趣。在精神和方法层面，该研究与我们的基础设施调研有着很多共同点。该研究的发起人周聪惠副教授曾在2010-2011年以访问学者身份到访哥伦比亚大学城市设计实验室，并参与了当时正在进行的纽约布鲁克林区Gowanus运河修复和再利用课题，以及中国四川省泸州市古蔺县新城设施再生与规划设计课题。本次针对南京的研究很大程度上是对全球各城市共同关注问题和类似研究的更进一步探索。

理查德·普朗兹
哥伦比亚大学建筑学教授
哥伦比亚大学地球研究院城市设计实验室主任

（序言作者要感谢哥伦比亚大学城市设计实验室高级研究助理Zhou Yijia在本研究解读时的协助）

Innovation in visualization of data was also an important component of the study, such that the process and results can be easily understood by the general public who are, after all, principal stakeholders in insuring that Nanjing's walls fulfill their mandate as next generation green infrastructure. As with urban dwellers everywhere in the world, in Nanjing citizen interest and participation in infrastructure planning, design, optimization and reconstruction is an essential ingredient in meeting the challenges of global ecological change.

For the Urban Design Lab (UDL) at the Earth Institute at Columbia University in New York, this study is of particular interest. In spirit and method it shares much with our own infrastructural investigations. It is significant that this work has been initiated by Prof. Conghui Zhou, who was Visiting Scholar at the Urban Design Lab during the period 2010—2011. He participated in on-going UDL research on the remediation and repurposing of the Gowanus Canal in Brooklyn; and in explorations of infrastructural innovation for the Gulin New Town in Sichuan Province in China. This study for Nanjing is a significant further exploration of shared concerns - similar explorations by others in cities across the globe.

Richard Plunz
Professor of Architecture, Columbia University
Director, Urban Design Lab, The Earth Institute, Columbia University

（The author wishes to thank Yijia Zhou, Senior Research Assistant in the Columbia Urban Design Lab, for her assistance in researching aspects of this text）

前言 | PREFACE

“绿道”作为官方专业术语首次出现于 1987 年美国总统户外游憩委员会报告中，但其规划思想则可追溯到 19 世纪末美国的公园路规划及后续一系列相关实践探索。经过一百多年的发展，目前的绿道规划设计实践已整合了风景园林学、城乡规划学、生态学、地理学、游憩学等多个学科相关知识，并覆盖了建成环境和自然环境两大空间范畴。对比自然环境中的绿道，建成环境绿道通常是城市慢行系统的重要载体，与居民日常生活联系更加紧密，使用频率和密度需求也更高，但目前世界上对该部分绿道规划设计的相关研究还相对滞后，针对性的规划选线和设计实施技术参照也相当匮乏。

当前城市发展过程中资源与环境问题的日益加剧，迫使城市土地使用由以外延增量为主的粗放低效方式向以内涵存量为主的集约高效方式转变。在许多经济发达但城市人口密集、土地资源紧张的国家和地区，建成环境中的绿道在串联城市空间、整合城市资源、提升城市品质、带动城市更新方面的功能开始逐渐显现。在日本东京，有学者针对建成区中人口密度过高、资源紧张、环境恶化等问题提出了“纤维化”的绿廊建设计划，旨在通过统筹调控道路、遗址和街巷等建成区固有空间和设施来完成绿道精细化布局，达到空间整合、社会服务和生态改善的效果。新加坡则通过在高密度城市环境中建构多级绿道网，对城市用地、现有设施等进行最大整合，提升城市内部的连通度和用地效率。

As an official technical term, “greenway” first appeared in a 1987 report issued by President’s Commission on Americans Outdoors but its planning concept can date back to parkway planning in the United States (US) at the end of the 19th century and a series of following relevant practices and explorations. After more than 100 years of development, the planning, design, and practice of greenways has integrated relevant knowledge of multiple disciplines including landscape architecture, urban and rural planning, ecology, geography, recreation studies, and so on, and covers two spatial categories including built environment and natural environment. Compared to a greenway in a natural environment, the one in a built environment is usually a very important carrier of an urban slow-traffic system. It is more closely linked to residents’ daily life and in high demand with regard to frequency and density of use. However, relevant research on the planning and design of greenways in the world relatively lags behind at present, and the technical reference for targeted planning, design, and implementation are rather scanty.

With the increasing aggravation of issues related to resources and environment in urban development, urban land use has been reduced to transforming from the extensive and inefficient way that centers on extensions and increment to the intensive and highly efficient one that focuses on intension and stock. In many countries and regions that are economically developed but densely populated in urban areas with a shortage of land resources, greenways in a built environment have gradually emerged to serve as a link to urban space, integrate urban resources, improve the quality of urban areas, and drive urban renewal. In Tokyo, Japan, Olmo Hidetoshi has proposed a “fibrillated” greenway construction plan, which is aimed at overpopulation, resource shortage, environmental degradation and other such problems in built-up areas, for the purpose of completing a refined layout for the greenway through adjusting and controlling as a whole the inherent space and facilities in built-up areas including roads, historical relics, streets and alleys, so as to achieve the desired result of spatial integration, social services, and ecological improvement. Singapore has maximized the integration of urban land use, existing facilities and so on by constructing a multilevel greenway network in a highly dense urban built environment, so as to improve connectivity and land use efficiency inside a city.

但在操作层面，通过建成环境绿道合理的规划设计来实现服务绩效的最大化并非易事。一方面，由于绿道外向关联型的功能服务属性，其规划建设的集约程度或用地效率很难通过其自身的单位用地开发强度、土地经济产出效率等工业、商业类用地效率评价指标来进行衡量。另一方面，除满足自身结构合理性需求外，建成环境中绿道的规划设计和服务绩效还将受到城市用地格局、路网水网结构、城市人口分布等外部空间环境的影响，其中影响因素众多，作用方式也较复杂。

为此，从 2013 年开始我通过主持一系列国家和部省级科研课题的机会，围绕建成环境中绿色基础服务设施的规划建构与优化更新问题展开一系列研究，而对绿道服务绩效的界定与衡量、绿道空间布局的影响因素及其作用规律等方面的探讨就是其中的重要方向之一。幸运的是，在研究过程中，我还同时得到负责主持与该研究紧密相关的中心城绿道规划、设计以及调研类课题的机会，从而能及时在规划设计实践中对相关研究成果加以应用检验。南京明城墙绿道服务绩效调查分析及其内外部空间特征研究就是其中的重要板块之一。

依托身在南京的便利条件，研究团队在对南京建成环境多条绿道进行预调研的基础上，锁定位于南京中心城区核心地段、内外部空间特征丰富的明城墙绿道作为研究对象。在南京明城墙绿道已建成并投入使用的西线和南线上选取了 6 个长度在 1000m 至 1500m 之间的典型区段作为调查和分析样本，并通

However, in terms of operation, it is not easy to maximize service performance through feasible planning and design of greenways in a built environment. On the one hand, due to its service functional attribute of external connection, it is very hard to measure the planning and construction intensity or land use efficiency of the greenway through its own industrial and commercial land use efficiency evaluation indicators including development, intensity of unit land use, output efficiency of land economy, and so on. On the other hand, apart from meeting feasible requirements of its own structure, the planning and design of the greenway in a built environment will also be influenced by external space and environment including urban land use pattern, the structures of road system and network of rivers, distribution of urban population and so on, among which there are many influencing factors and complex modes of action.

To this end, since 2013, the author has been conducting research on the planning and regeneration of green infrastructure in built environment, through scientific research projects including funding from the National Science Foundation of China, the China Postdoctoral Science Foundation, and other special funding. In the research, one of the key components is the definition and measurement of service performance of the greenway, influencing factors of spatial distribution of the greenway, and its functions and laws. Fortunately, in the process of research, the author has also had the opportunity to take charge of presiding over the planning, design, and research topics of central city greenways closely related to this research, so as to put into use and test relevant research results during the process of planning, design, and practice in time. One of the important sectors is survey and analysis of service performance of the Nanjing Ming Dynasty City Wall Greenway, and the study on the characteristics of both its interior and exterior spaces.

Relying on the convenience of being in Nanjing, our team focuses on the Nanjing Ming Dynasty City Wall Greenway as its object of study, which is located in the core area of downtown Nanjing and rich in internal and external spatial characteristics, based on

过多轮反复地使用人群信息采集、问题分析、信息补充等环节，用定量方法将绿道的服务绩效、内外部空间特征进行梳理和描述，并在此基础上将绿道服务绩效指标与其内部和外部空间特征指标进行关联分析，进而尝试从中发现绿道内外部空间特征对服务绩效的影响方式和作用规律，希望借此能为众多正在或即将开展的相关规划设计实践做出一些有益探索。

本书在表达上依托直观易懂的图解方式来展示研究思路、方法、过程和成果，除了希望它为相关科研人员和规划从业者提供些许有益的参照信息外，还寄予本书中的研究内容和成果易于被更多非专业人士，尤其是社会公众认知和理解，也期待借此激发广大社会公众对于身边城市公共空间品质的关注和兴趣，并更加主动地参与到所在城市公共空间的规划设计、优化提升和建设使用中来，从而共同推动我们所生活的城市朝着宜居和可持续发展方向不断前行。

a preliminary survey of and review on multiple greenways in the built environment of Nanjing. As samples of study and analysis, our team selects six typical segments (1,000m-1,500m in length) along the west and south lines of the Nanjing Ming Dynasty City Wall Greenway, which have been completed and put into use, and uses quantitative methods to put in order and describe the service performance and interior and exterior spatial characteristics of the greenway through multiple rounds of repeated use of user information acquisition, problem analysis, information supplement, etc. On this basis, our team conducts correlation analysis on the service performance indicators and interior and exterior spatial characteristics of the greenway, so as to attempt to find out how interior and exterior spatial characteristics of the greenway influence service performance, as well as the functions and laws of the greenway, in the hope of making some useful explorations for numerous relevant planning, design, and practice, which are or will be carried out.

In terms of expression, this book relies on a visual, straightforward, graphic method to exhibit study ideas, methods, processes, and results. In addition to hoping that this book can become a convenient bond through which relevant research personnel and planning and design professionals exchange their respective ideas and practices, the author also hopes that the study content and results in this book can be easily perceived and understood by more nonprofessionals, especially the public, by which their interests and concerns towards the quality of urban public space could be partly aroused and may be further attracted to throw themselves into, and participate in the planning, design, optimization, and reconstruction of urban public space like greenways so as to jointly promote the city where we live to move in the direction of sustainability and habitability.

总论

Introduction

什么是绿道

What is a greenway

绿道其实是在被城市快速扩张撕裂成破碎化的自然环境之间建立起来的稳定联系渠道，它通常覆盖建成环境与自然环境，以游憩为主导功能，并兼具生态保护、历史文化资源保护、城乡连接、社区生活品质改善等多种功能。

1987 年，美国总统户外游憩委员会在户外运动报告中首次对绿道进行了官方描述，即"绿道是让我们每一个美国居民便捷地进入自然世界的依托和愿景。绿道是由地方所创造的绿指，它由社区伸出，能够连接并贯穿于美国境内的所有社区。它们

A greenway is a stable channel of connection built between natural environments, which are torn to pieces due to rapid urban expansion. It usually covers built environment and natural environment. With recreation as its dominant function, it also has a variety of functions including ecological protection, historical and cultural resources protection, connection between urban and rural areas, quality improvement of community life and so on.

In 1987, President's Commission on Americans Outdoors described greenways for the first time, "We have a vision for allowing every American easy access to the natural world- Greenways. Greenways are fingers of green that reach out from and around and through communities all across America, created by local action. They will

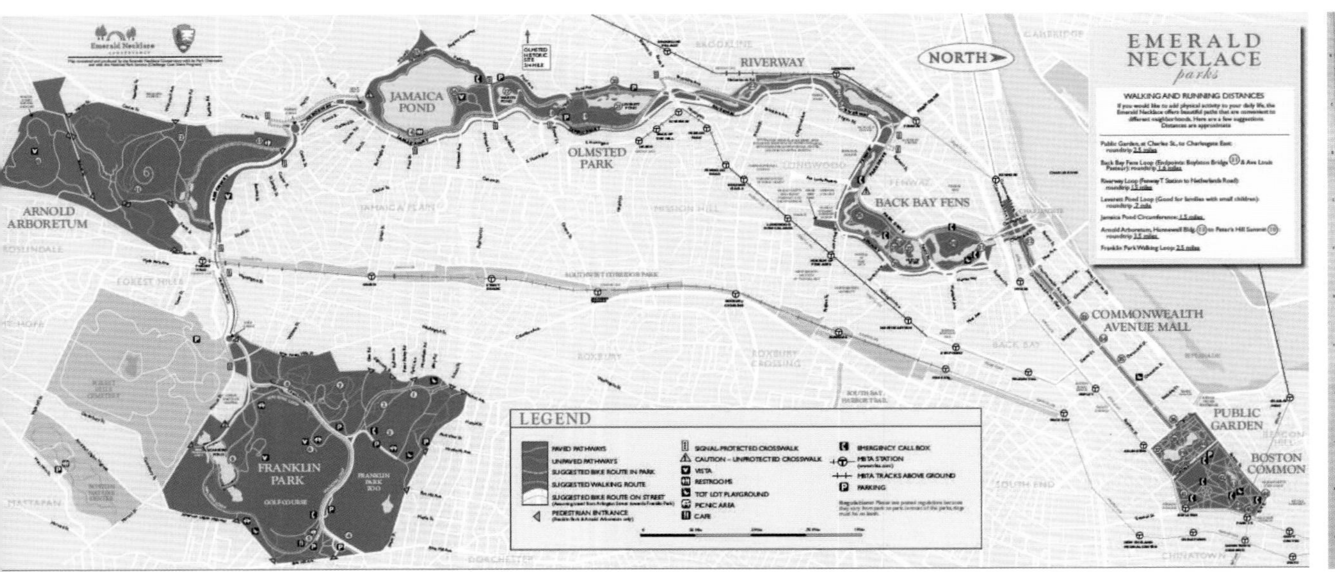

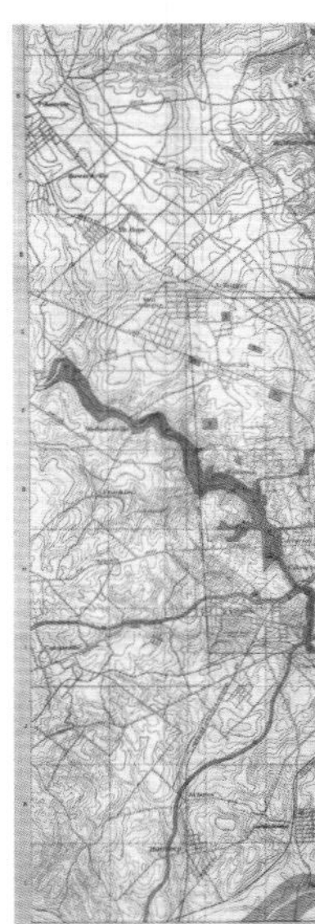

将能够联系公园、公共的或私有的森林和田园乡村，并成为徒步旅行、慢跑、野生动物活动、骑马和骑自行车的游憩通廊”。该描述强调了绿道的游憩服务主导功能和空间连接功能，同时描绘了一幅覆盖“区域—城市—社区”三个层面复合化的绿道空间体系。此后，美国也进入了绿道高速发展期，并带动全世界掀起了一场轰轰烈烈的绿道建设运动。据统计，美国的绿道在1989年约为250条，而到了1993年，绿道总数已突破3000条，在4年时间里增加近10倍。目前美国绿道总长度已超过10万km，其目标是让所有居民都能在15分钟内从家中或工作场所到达最近的绿道。而在亚洲、欧洲和南美等国家和地区，绿道规划建设如火如荼，并在人们生活品质提升、生态和历史文化资源保护、城市形象改善等方面发挥了积极作用。

connect parks, forests, and scenic countrysides, public and private, in recreation corridors for hiking, jogging, horse and bicycle riding, and wildlife movement." The description emphasizes the dominant functions of the recreational services and spatial connection of the greenway, in addition to depicting a complex greenway space system that covers such three layers as "region-city–community." Afterward, US entered the stage of rapid development for greenways and drove the whole world to a vigorous campaign of greenway construction. According to statistics, there were about more than 250 greenways in US in 1989. Until 1993, the number exceeded 3,000, with an addition of nearly 10 times in 4 years. At present, the total length of greenways in US has already exceeded 100,000km, and the aim is to let all residents reach the nearest greenways from their home or workplace within 15 minutes. In countries and regions including Asia, Europe, and South America, the planning and construction of greenways is well under way. They are playing active roles in aspects including quality improvement of people's lifestyle, ecological protection, historical and cultural resources preservation, improvement of cityscape, and so on.

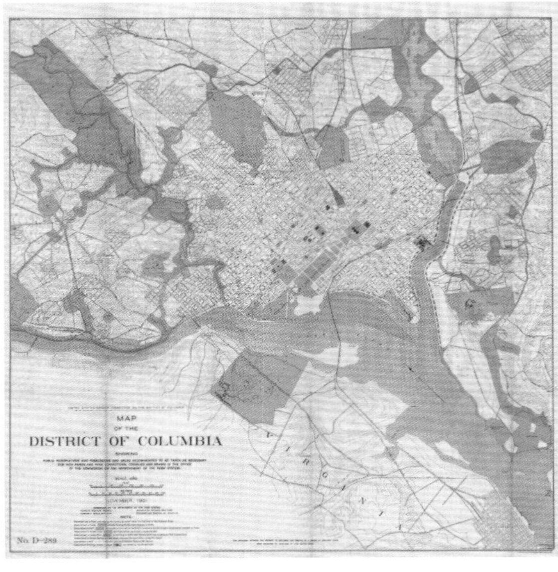

1 | 2 | 3

1　波士顿翡翠项链 1894 年原始规划图
Boston Emerald Necklace Plan (1894)
图片来源：http://www.greeningthegrey.org
Source：http://www.greeningthegrey.org

2　巴尔的摩公园系统平面图
Baltimore Park System Plan (1904)
图片来源：https://www.olmstedmaryland.org/history/
Source：https://www.olmstedmaryland.org/history/

3　华盛顿公园系统平面图
Washington D.C. Park System Plan (1901)
图片来源：http://www.nationalmall.net/resource/mcmillan.html
Source：http://www.nationalmall.net/resource/mcmillan.html

在中国，绿道研究和规划建设起步较晚，中国第一次较为系统地介绍美国绿道是在1992年《国外城市规划》（现《国际城市规划》）刊登的《美国绿道简介》中。其后，随着绿道相关研究的增加，绿道规划的相关思想开始逐渐在我国部分城市绿地系统专项规划中显现。但直到2010年，以《珠江三角洲绿道网总体规划纲要》和《珠三角区域绿道（省立）规划设计技术指引》为代表的独立编制绿道项目和针对性规划指引才由广东省率先编制完成。2016年9月，住建部正式颁布《绿道规划设计导则》（简称《导则》），《导则》对绿道的定义为"以自然要素为依托和构成基础，串联城乡游憩、休闲等绿色开敞空间，以游憩、健身为主，兼具市民绿色出行和生物迁徙等功能的廊道"，同时将绿道空间分解为游径系统、绿化和设施三大组成部分。《导则》还从空间级别上将绿道分为区域、城市和社区三个级别，并又根据区位特征将绿道分为城镇和乡村两个类型。《导则》的颁布是中国官方第一次对绿道做出系统性定义和描述，并开始对全国各地日益风行的绿道规划建设活动进行规范和指导。

In China, the research, planning, and construction of greenways started relatively late. *Introduction of American Greenways* published in *Foreign Urban Planning* (currently *Urban Planning International*) in 1992 marks the first time that American greenways have ever been systematically introduced in China. Afterward, with the increase of study on greenways, relevant ideas about greenway planning has gradually appeared in special plans of green space systems in some of our cities. It was not until 2010 that independently drawn up greenway projects and targeted planning guidelines had been worked out and completed by Guangdong Province for the first time, represented by *Outline of Greenway Network Master Planning of Pearl River Delta Region and Guidelines for Planning and Design of Regional Greenways in Pearl River Delta*. In September 2016, the Ministry of Housing and Urban-Rural Development of the People's Republic of China officially launched *Greenway Planning and Design Guideline* (the *Guideline* for short), which defines a greenway as "a corridor that relies on and is based on natural elements, links together green open spaces for urban and rural recreation and leisure, focuses on recreation and physical exercise, and is used as the corridor for green commuting of residents and migration of wildlife." At the same time, the space of a greenway is broken up into three components, namely, trails system, greening, and facilities. The Guideline also divides greenways into regional, municipal, and community levels in terms of space level, and categorizes greenways into urban greenways and rural greenways according to location characteristics. The launch of the *Guideline* marks the first time that China has ever systematically defined and described greenways and has started to regulate and guide greenway planning, design, and construction that have gradually become increasingly prevalent across China.

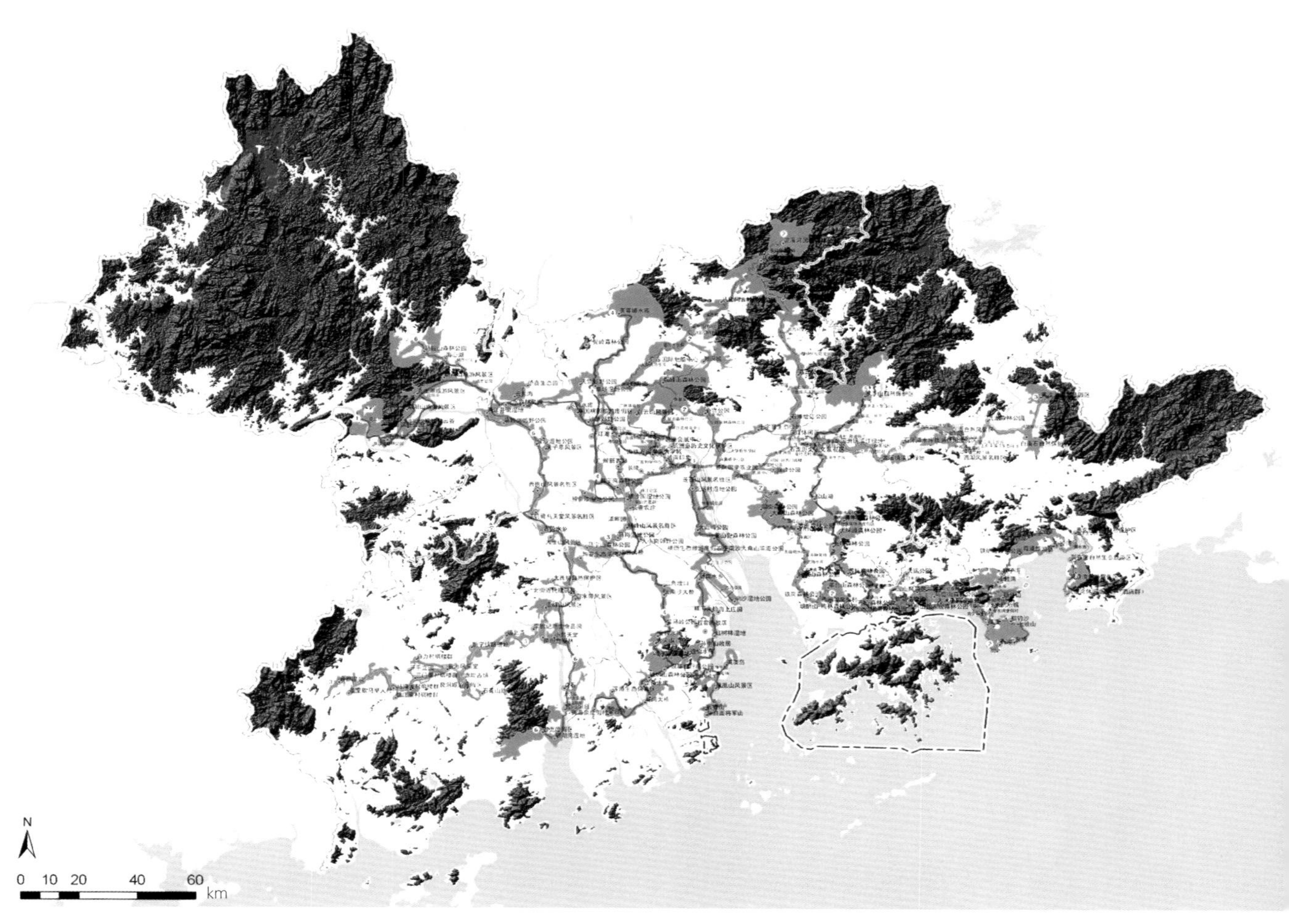

珠江三角洲绿道网总体规划纲要：总体布局图（2010）
Outline of Greenway Network Master Planning of Pearl River Delta Region: Master Plan (2010)

图片来源：广东建设工程信息网
Source：http://gd.jsgc168.com

建成环境中的绿道

Greenways in a built environment

顾名思义，建成环境中的绿道即位于城市建成区中的绿道，其相对概念为自然环境中的绿道。如果对两者进行深入分析和考量，可以发现除了在区位上的差异外，两者在功能、连接对象以及空间载体上也各有侧重。

As the name indicates, a greenway in a built environment is located in a built-up urban area, as opposed to a greenway in a natural environment. If we make in-depth analysis of, examine, and weigh the two kinds, we can find that they not only differ in location but also focus on different functions, connecting objects, and spatial carriers.

建成环境绿道的功能特征

建成环境绿道主要为城市居民的休闲游憩提供服务，并兼具生态提升、历史保护、城市美化等多种功能。与自然环境绿道相比，该类绿道由于位于城市居民的居住地和工作地周边，通常使用频率较高，在空间形态上密度相对较高，但单条绿道长度相对较短，设施类型和密度却较为丰富。

Functional characteristics of the greenway in built environment

A greenway in a built environment mainly provides urban residents with services for recreation, in additional to functions as ecological improvement, the preservation over historic sites, city beautification, and so on. Compared to one in a natural environment, a greenway in a built environment is located where city-dwellers live and work, so it is frequently used and correspondingly planned with a higher density in layout. However, a single greenway is relatively short in length but rich in the types and density of facilities.

建成环境绿道的连接对象

解析建成环境绿道的主导功能，可以发现该类绿道：一方面是通过自身游憩服务能力为城市居民游憩或出行提供服务，即配置跑步、散步、骑行、健身等活动设施以及信息、商业、医疗等相关服务设施；另一方面是能将使用者便捷地引导至游憩资源集中地段，即在游憩服务供需端之间建立空间联系。

Connecting objects of the greenway in a built environment

To analyze the dominant functions of a greenway in built environment, we have to know that on the one hand a greenway provides city-dwellers with services for outdoor recreation and travel through its own recreational service capabilities, in that it is equipped with facilities for jogging, walking, bike riding, fitness, and so on, as well as relevant service facilities for information, commerce, medical treatment, and so on, while on the other hand it can conveniently lead users to a slice where recreational resources are centralized (e.g. parks, museums, stadiums, etc.), which means that spatial linkage is set up between supply and demand sides of recreational services.

根据该功能服务特点，可将建成环境绿道的连接对象分为两类：一类为游憩服务供给端，主要是公园绿地、近郊生态型绿地（含林地、湿地、农田等）、风景区、旅游区等绿地生态型游憩资源聚集区以及城市广场、历史街区、商业文化、娱乐体育中心等公共人文型游憩资源聚集区；另一类为游憩服务需求端，主要指具有游憩需求的被连接对象，在建成环境内主要是以城市社区为核心的城市居民聚集地区。

建成环境绿道的空间载体

绿道并非独立的城市建设用地类型，因此在绿道规划建设时，须依附各类用地作为规划建设实施的空间载体。在建成环境中，主要被用作绿道规划建设空间载体的有城市道路、河道水系、线性历史遗址（如城墙、古运河等）、废弃铁路等线性要素及其沿线绿地（如带状公园、道路防护绿地等）。

其中，最常见被建成环境绿道用来作为空间载体的是城市道路及其沿线绿地。一方面因为城市道路是建成环境中最常见的线性开放空间；另一方面也因为在建成环境中河道、线性历史遗址等空间要素沿线通常会建有城市道路。两者的走向通常相互叠合。

According to the characteristics of their functions and services, the connecting objects of a greenway in a built environment can be divided into two types. One is the supply side of recreational services, mainly including green and eco-friendly recreational resources clusters such as public parks, suburban ecological green lands (including forest lands, wetlands, farmlands and so on), scenic spots, tourist attractions and so on, as well as public or cultural recreational resources clusters such as municipal squares, historic districts, commercial and cultural centers, and entertainment and sports centers, and so on. The other is the demand side of recreational services, including connected objects in need of recreation, such as urban settlements with urban communities as core in a built environment.

Spatial carriers of a greenway in a built environment

A greenway is not a type of independent urban development land. Therefore, during greenway planning and construction, we must depend on all types of land for use as spatial carriers. In a built environment, spatial carriers used for planning and construction of greenways mainly include urban roads, river systems, linear historic sites (such as city walls, ancient canals, and so on), abandoned railways and other such linear elements (belt-shaped parks, green buffers, and so on).

Among them, urban roads, together with the linear green space along the way, are the most common spatial carriers used by the greenway in a built environment. On the one hand, it is because urban roads are the most common linear open spaces in built environments. On the other hand, in a built environment, spatial elements such as watercourses and linear historic sites are usually built with urban roads. Their directions are often mutually superimposed.

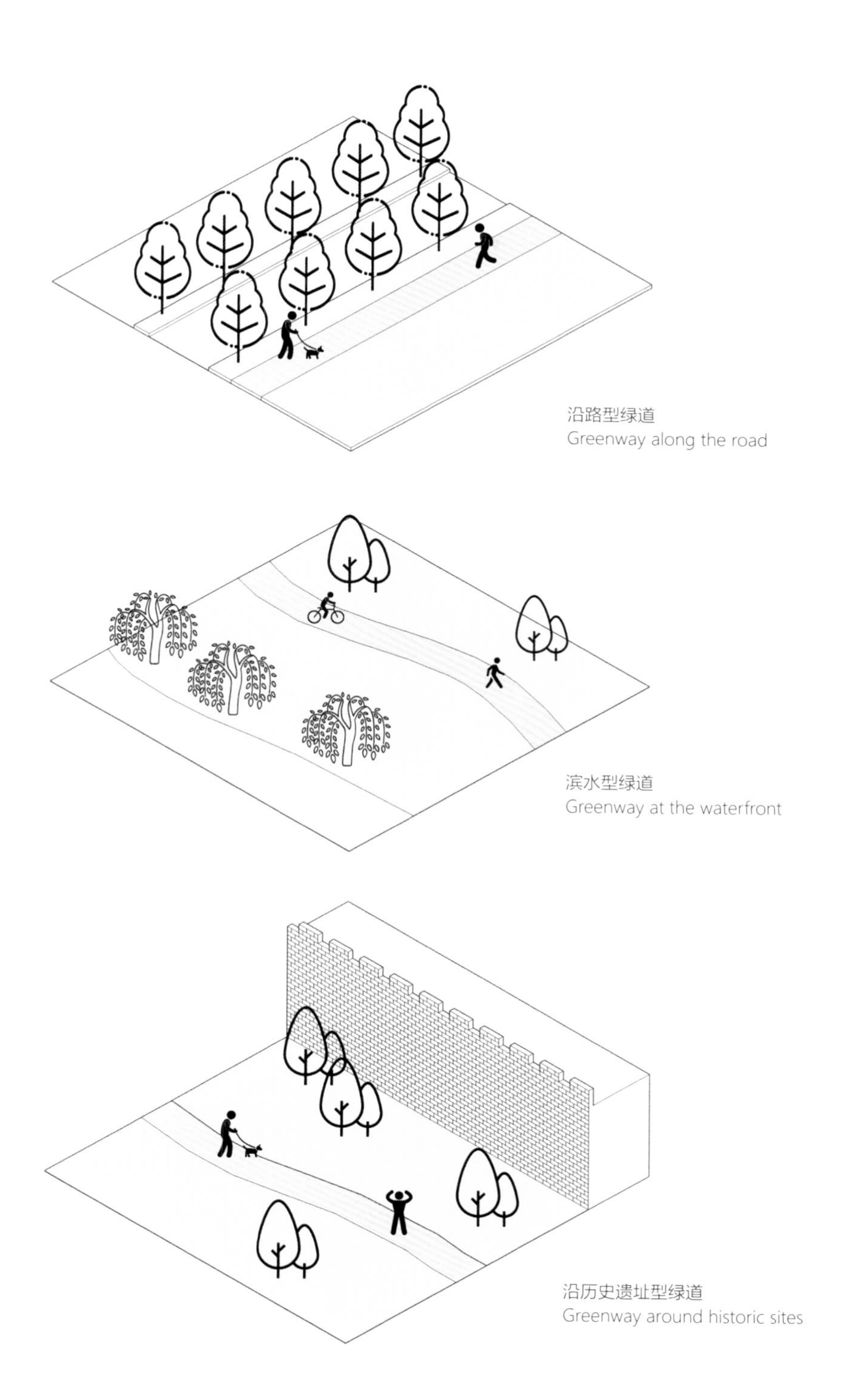

沿路型绿道
Greenway along the road

滨水型绿道
Greenway at the waterfront

沿历史遗址型绿道
Greenway around historic sites

为什么要关注建成环境中绿道的服务绩效与空间特征问题

Why should importance be attached to the spatial characteristics and service performance

综观绿道规划建设实践的发展历程，其实践的空间范畴重心经历了起源于建成环境、拓展于自然环境、回归于建成环境三个阶段。如把奥姆斯特德（F.L.Olmsted）设计的公园路视为绿道的初始原型，绿道缘起于19世纪末对美国城市建成环境中一系列环境、卫生及社会问题的思考。至20世纪中前期，随着汽车的普及以及生态学的兴起，绿道功能开始走向多元化，该阶段大量的规划建设实践使得其在国土层面及自然环境中得以广泛拓展。至20世纪后期，随着紧凑城市、新城市主义以及景观都市主义等规划思潮的兴起，建成环境中绿道的功能开始被重新审视，并被视为激活旧城活力、优化固有结构、带动社区发展的重要工具。但目前绿道的相关研究稍显滞后，所用方法仍主要以绿道自身为中心，侧重从土地建设适宜性、景观生态学等角度探讨绿道选线或评价问题。这类方法更适用于在自然环境中指导绿道布局选线和生态绩效评价，因为与城市空间环境关联较弱，对于使用人群的关注也不够，所以对于建成环境绿道的规划设计指导价值有限。因此，开展相关研究来为当前如火如荼开展的建成环境绿道规划设计提供针对性指导成为迫切需求。

After making a comprehensive survey of the development history of the planning, construction, and implementation of a greenway, we have found out that the center of spatial category of its implementation originates from a built environment, expands from a natural environment, and returns to the built environment. For example, if we take parkway raised by F. L. Olmsted as the initial prototype of a greenway, we find out that the greenway originates from reflections on a series of environmental, sanitation, and social problems in the built environments of American cities at the end of the 19th century. Until the early and mid 20th century, with the popularity of motor vehicles and the emergence of ecology, the functions of a greenway had moved toward diversification. During that time, a great deal of planning, construction, and implementation helped its wide expansion in national territory and natural environment. Until the late 20th century, with the emergence of *compact cities, new urbanism, landscape urbanism* and other such new ideas of planning, the functions of the greenway in built environments had started to be re-examined and had been considered a very important tool to activate the vitality of old cities, optimize city inherent structure, and drive community development. However, relevant study on greenways at present lags behind a little bit, and most methods mainly focus on the greenway itself, which puts extra emphasis on the suitability of ecology, land development, construction, and other such angles to explore greenway route selection or evaluation. All these approaches are better suited for guiding greenway layout establishment, route selection, and ecological performance evaluation in a natural environment. Since it is weakly correlated to urban space and the elements in built environments with a lack of attention to users, it is not well applicable in the planning and design of a greenway in a built environment. As a result, we need to carry out relevant study to provide targeted tools for the planning and design of greenways in a built environment, which is now in full swing.

在中国，绿道规划和建设的初始重心主要放在以自然环境为主的区域绿道上，对于建成环境绿道的关注也同样不足。直到近年，随着国家对城市外扩型发展模式的严格限制以及对集约发展、精明增长、紧凑城市等规划理念的大力倡导，许多城市规划的关注点开始转移到城市内部固有空间结构的优化。在此背景下，一方面，建成环境绿道的规划建设作为“城市双修”（生态修复、城市修补）的重要渠道开始逐渐成为焦点。但另一方面，新的问题也开始伴随目前的实际规划建设热潮开始显现，其中最突出的就是许多新建绿道大量闲置、无人使用，造成土地、人力和物力资源浪费，而大量城市居民则在抱怨无法享受绿道的规划建设成果。这种供需错位的问题广泛存在于当前我国在建成环境中进行绿道规划建设的大中城市里，也使决策者和从业人员不得不重新审视如何通过合理规划设计来提升建成环境绿道的服务绩效问题。

通过对建成环境绿道服务绩效的调查与空间特征加以分析，我们就能以绿道上使用人群的状态特征为线索，有效梳理居民对绿道的服务需求，揭示其在绿道上的活动规律，同时将其与绿道的内部空间结构和外部空间环境进行匹配和关联分析，在此基础上对影响绿道服务绩效的内外部空间因素及其作用规律做出针对性研究，并能为绿道规划选线、结构布局和空间设计过程中缝合绿道服务供需错位、提升绿道服务绩效提供一些思路。

In China, the initial focus of greenway planning and construction is mainly on regional greenways that centers on natural environment, with an equal lack of attention to greenways in built environments. Till recent years, with the close restraint of urban sprawl development pattern and strong advocacy of the planning concepts of intensive development, smart growth, compact cities and so on, the focus of planning in a lot of cities has started to shift to the optimization of inherent space structures inside those cities. Against such a backdrop, as a very important way of “dual repairs in the city”(a recent initiative raised by Chinese central government centering on ecological restoration and city renovation), the planning and construction of greenways in built environments has gradually become a focus of attention. Yet, on the other hand, new problems have begun to appear accompanying the current hot trend of completed planning and construction practice, among which the most prominent is that a number of newly built greenways are idle and out of use, resulting in a waste of land, labor, and material resources. On the other hand, a lot of city-dwellers complain that they cannot easily enjoy and access the results of greenway planning and construction. The problem of unmatched supply and demand widely exists in large and medium sized cities where greenway planning and construction are well underway in built environments in China, so that decision-makers and professionals have no choice but to re-examine how to improve service performance of greenways in built environments through delicate planning and design.

Through the survey of service performance and the analysis of spatial characteristics of a greenway in a built environment, we can refer to the conditions and characteristics of users on the greenway, effectively put in order the service requirements of residents on the greenway, bring to light their patterns on the greenway, and match and conduct correlation analysis of these residents and the interior spatial structure and exterior space environment of the greenway. Based on these data, we can conduct targeted studies on internal and external spatial factors, which influence the service performance of the greenway, and functions and laws, and can provide some ideas for closing the gap between unmatched supply and demand, and improving the service performance of the greenway during the process of greenway route selection, structural layout, and space arrangement.

研究的困难与挑战

Study difficulties and challenges

难点一：
如何界定建成环境绿道的服务绩效

工业、居住、商业等类型城市用地的功效可以用建筑容积率、建筑密度等开发建设强度指标来评价，也可以用投入强度、产出强度、亩均税收等经济投入产出强度指标来评价。但绿道以提供游憩服务为主导功能，既无法进行高强度开发，也无法直接产生经济效益，因而其功效很难直接用开发建设强度或经济投入产出强度指标来衡量，因此与其功能特征相符的服务绩效指标常被用来衡量其实际功效。但鉴于绿道上的服务人群及其使用状态一直处于动态变化过程中，如何统计和衡量其服务人群的使用状态也并非易事。鉴于此，如何根据绿道自身功能服务属性来界定和衡量其服务绩效是该研究所要解决的首要问题。

Difficulty 1:
How to define the service performance of a greenway in a built environment

The functions of industrial, residential, and commercial lands can be evaluated through indicators related to development and construction intensities including plot ratio, building density and so on, as well as indicators related to economic input-output intensities including investment strength, yield strength, tax revenue per mu and so on. However, with the provision of recreational services as its dominant function, a greenway cannot endure high-strength exploitation, nor can it directly bring about economic benefits. Therefore, the service performance indicators that match its functional characteristics are often used for measuring its actual effects. However, since the groups of people that a greenway serves and its working conditions are always in dynamic change, it is not very easy to add up and measure it accurately. In view of this, how to define and measure the service performance of a greenway based on its own functional characteristics is the primary problem to be resolved at the beginning of this study.

难点二：
如何判别和选取典型的调研对象和样本

研究团队所在地——南京城内已有的建成绿道，总长度超过100km。由于人力和物力条件所限，研究团队无法对南京城内所有的建成绿道展开调查和分析，只能选取特征鲜明的典型区段进行取样调查和分析研究。要确保选取样本的典型性和研究价值，则需对目前南京城内的建成绿道进行全面了解，并在此基础上进行一系列的预调研和预分析，进而才有可能锁定合适的研究对象。

Difficulty 2:
How to differentiate and select typical study samples

The total length of the existing greenways in the city of Nanjing where our team is located is more than 100km. Due to limited labor and material conditions, the team cannot conduct a study on and make analysis of all the built greenways inside Nanjing. It can only select distinctive typical segments as samples for survey and analysis. To guarantee the representativeness and study value of the samples, the team has to gain a full understanding of all the built greenways in the city of Nanjing, based on which the team conducts a series of preliminary surveys and makes a series of preliminary analyses so as to be able to lock the ideal and suitable objects of study.

4G

难点三：
如何制定合理的信息采集方案

在调查研究中，需要采集的信息内容主要涉及静态要素和动态要素两个方面，其中静态要素主要包含绿道内部空间要素（如游径系统、服务设施、绿地、硬地等）和绿道外部空间要素（如周边用地类型、交通设施、开发强度、居民密度等）；动态要素则主要是绿道使用者特征，其中包含使用人群数量、使用活动类型、年龄结构等。两者对比而言，静态要素的调查难度相对较低，而动态要素的调查难度较大。因为，一方面，绿道作为公共游憩空间，其使用情况将会受到气温高低、天气状况、人群出行特点等多方面影响，并且使用人群特征将始终处于动态变化中。如何划定调查范围、判断调查内容、选取调查工具、确定调查条件等均将直接影响信息采集的客观性和准确性。另一方面，如果采用传统的单点场地调查方式，将很难反映出作为线性空间绿道的其他区段的使用状态，而多点调查方式则对人力条件要求过高，同时绿道人群一直在流动，如何规避对象的重复调查又将是一个巨大难题。因此，制定相对合理的信息采集方案对研究的顺利进行至关重要。

Difficulty 3:
How to make a reasonable proposal for information collection

During the survey, the information that the team needs to collect mainly involves static elements and dynamic elements. The former mainly includes internal space elements (such as trail systems, service facilities, greenbelts, hard sites and so on) and external space elements (peripheral land types, transport facilities, development intensity, density of settlements, and so on) of the greenway, while the latter mainly includes the characteristics of greenway users, which include the number of users, the types of activities, age structure and so on. In contrast, the difficulty in surveying static elements is not too great, while that of surveying dynamic elements is the opposite. The reason is that the greenway functions as a public space of recreation; its use is influenced by many aspects including temperature, weather, and so on, and that the characteristics of its users are always in dynamic change. How to define the scope of survey, how to determine the content of the survey, how to select survey tools, and how to identify survey conditions will directly influence the objectivity and accuracy of the collected information. On the other hand, if the traditional method of single-point site for survey is adopted, it will have difficulty reflecting the working conditions of other slices of the greenway as a linear space. Since the method of multi-point site for survey expects too much from labor, and the crowds on the greenway are always moving about, how to avoid repeated survey of the study object will also be a very difficult problem. Therefore, it is of great importance to make a relatively reasonable plan for information collection so that our study can go smoothly.

难点四：
如何梳理影响绿道服务绩效的内外部空间因素及其作用规律

建成环境绿道的使用和服务状态不仅取决于内部空间结构特征，也在很大程度上受制于外部空间环境状态，例如用地结构、开发强度、用地布局紧凑度、交通可达性等因素。由于影响绿道服务绩效的内外部空间因素众多且作用方式复杂，导致在目前的规划实践中尚难以对其进行有效把控。如何在研究中梳理和分析这些空间影响因素，并最终分析揭示其对绿道服务绩效的作用方式将成为本研究的另一大挑战。

Difficulty 4:
How to put in order internal and external spatial factors and laws that influence the service performance of the greenway

Since the use and service condition of greenways in a built environment not only depend on the structural characteristics of the interior space but are also subject to the environmental state of their exterior space to a great extent, such as land-use structure and development intensity, the compactness of layout of land use, traffic accessibility, and so on. Due to many internal and external spatial elements that influence the service performance of the greenway and the complexity of its modes of action, we have yet to effectively control them in the current planning and implementation. How to put in order and analyze these spatial influencing factors so as to analyze and bring to light their modes of action to the service performance of the greenway at the end will become another great challenge for this study.

怎样开展本研究

How to carry out this study

研究目标

目标一：

探索建成环境绿道服务绩效的衡量和评价方法。

目标二：

研究建成环境绿道内部空间要素对其服务绩效的作用方式。

目标三：

研究建成环境绿道外部空间要素对其服务绩效的影响规律。

Study goals

Goal 1:
Exploring the measuring and evaluation methods of the service performance of the greenway in its built environment.

Goal 2:
Studying the modes of elements action in interior space of the greenway in its built environment for the service performance.

Goal 3:
Studying the influences from the elements in exterior space of the greenway in its built environment on its service performance.

方法技术

整个调查和研究的技术路线可以分为"四个阶段、两条线索"，其中四个阶段分别为研究对象确定、研究信息采集、信息整理分析、讨论与启示。两条线索则分别为绿道内部空间特征与服务绩效关联分析、绿道外部空间特征与服务绩效关联分析。以此为框架来探索绿道服务绩效的内外部空间影响因素及其作用规律，并为建成环境绿道的规划选线、设计布局以及改造提升提供一些新的思路。

Methods and techniques

The technical route of the entire survey and study can be divided into "four phases, two clues." The four phases are the locking of study object, study information collection, information compilation and analysis, reflection and discussion of planning and design, respectively. The two clues are correlation analysis of internal spatial characteristics and service performance of the greenway, as well as correlation analysis of external spatial characteristics and service performance of the greenway, respectively. Based on this framework, we explore the influencing factors of the greenway service performance both lying in interior space and exterior space of the greenway, as well as their modes of action so as to provide some new ideas for planning, design, and improvement of the greenway in its built environment.

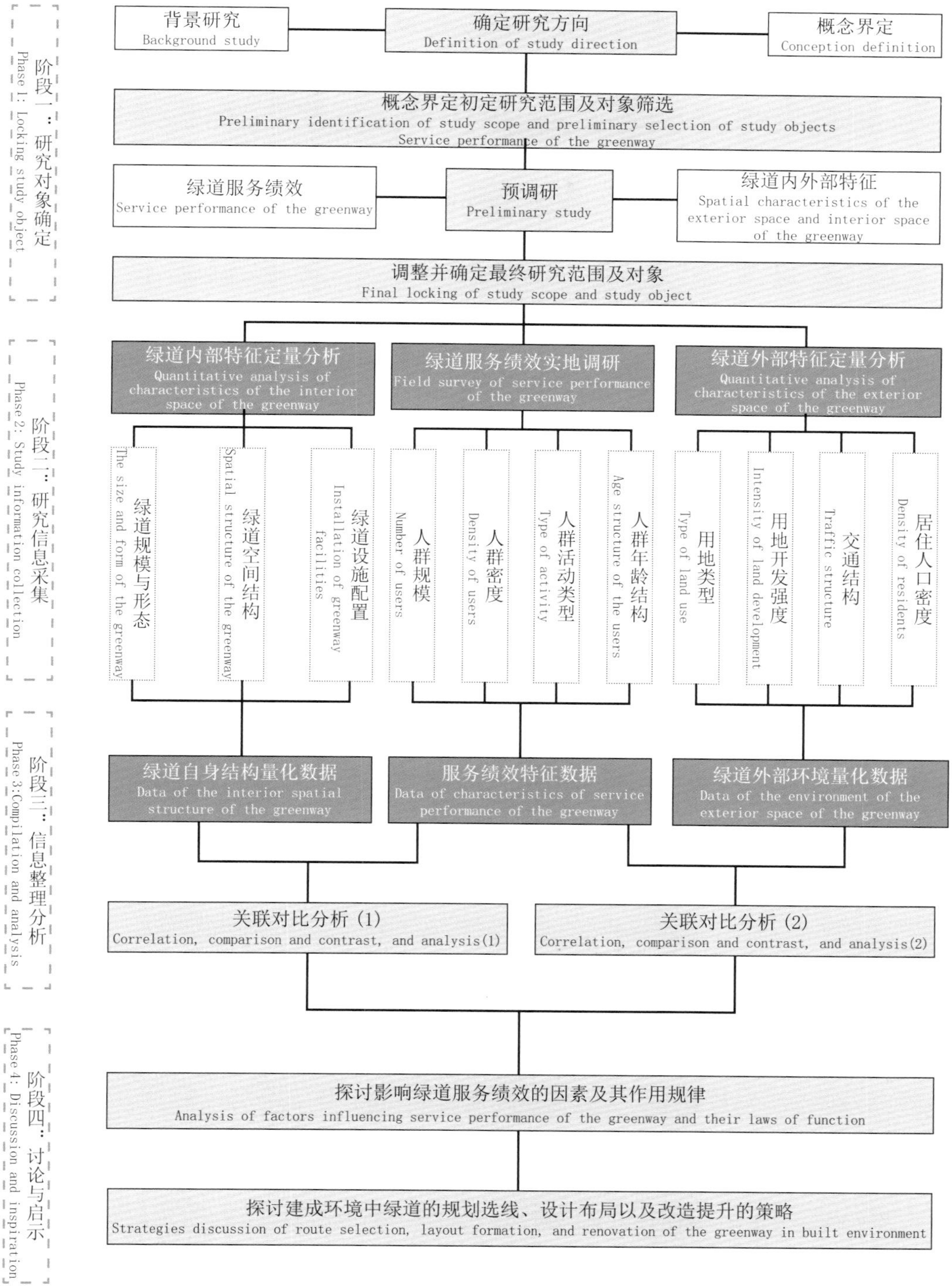

阶段一：研究对象确定
Phase 1: Locking study object

背景研究
Background study

确定研究方向
Definition of study direction

概念界定
Conception definition

概念界定初定研究范围及对象筛选
Preliminary identification of study scope and preliminary selection of study objects
Service performance of the greenway

绿道服务绩效
Service performance of the greenway

预调研
Preliminary study

绿道内外部特征
Spatial characteristics of the exterior space and interior space of the greenway

调整并确定最终研究范围及对象
Final locking of study scope and study object

阶段二：研究信息采集
Phase 2: Study information collection

绿道内部特征定量分析
Quantitative analysis of characteristics of the interior space of the greenway

- 绿道规模与形态 The size and form of the greenway
- 绿道空间结构 Spatial structure of the greenway
- 绿道设施配置 Installation of greenway facilities

绿道服务绩效实地调研
Field survey of service performance of the greenway

- 人群规模 Number of users
- 人群密度 Density of users
- 人群活动类型 Type of activity
- 人群年龄结构 Age structure of the users

绿道外部特征定量分析
Quantitative analysis of characteristics of the exterior space of the greenway

- 用地类型 Type of land use
- 用地开发强度 Intensity of land development
- 交通结构 Traffic structure
- 居住人口密度 Density of residents

阶段三：信息整理分析
Phase 3:Compilation and analysis

绿道自身结构量化数据
Data of the interior spatial structure of the greenway

服务绩效特征数据
Data of characteristics of service performance of the greenway

绿道外部环境量化数据
Data of the environment of the exterior space of the greenway

关联对比分析（1）
Correlation, comparison and contrast, and analysis(1)

关联对比分析（2）
Correlation, comparison and contrast, and analysis(2)

阶段四：讨论与启示
Phase 4: Discussion and inspiration

探讨影响绿道服务绩效的因素及其作用规律
Analysis of factors influencing service performance of the greenway and their laws of function

探讨建成环境中绿道的规划选线、设计布局以及改造提升的策略
Strategies discussion of route selection, layout formation, and renovation of the greenway in built environment

挹江门 Yijiang Gate

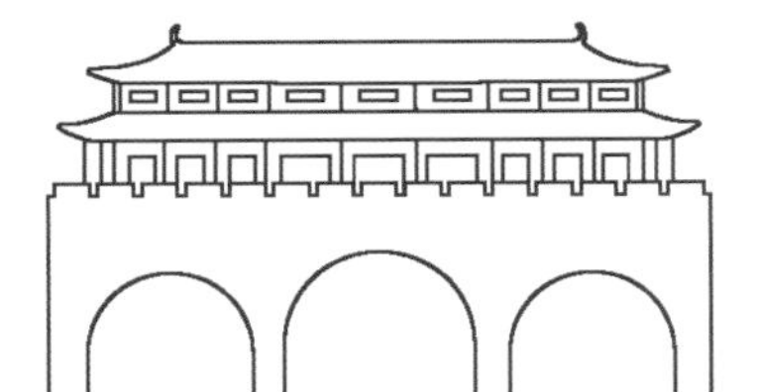

华严岗门 Huayangang Gate

清凉门 Qingliang Gate

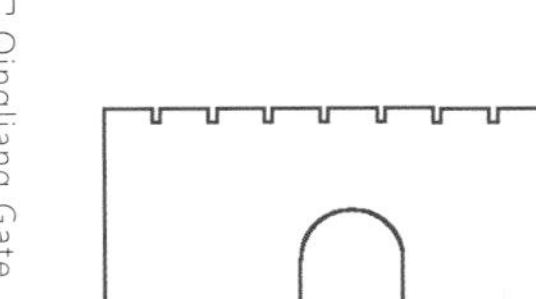

研究对象的确定

Establishment of Study Object

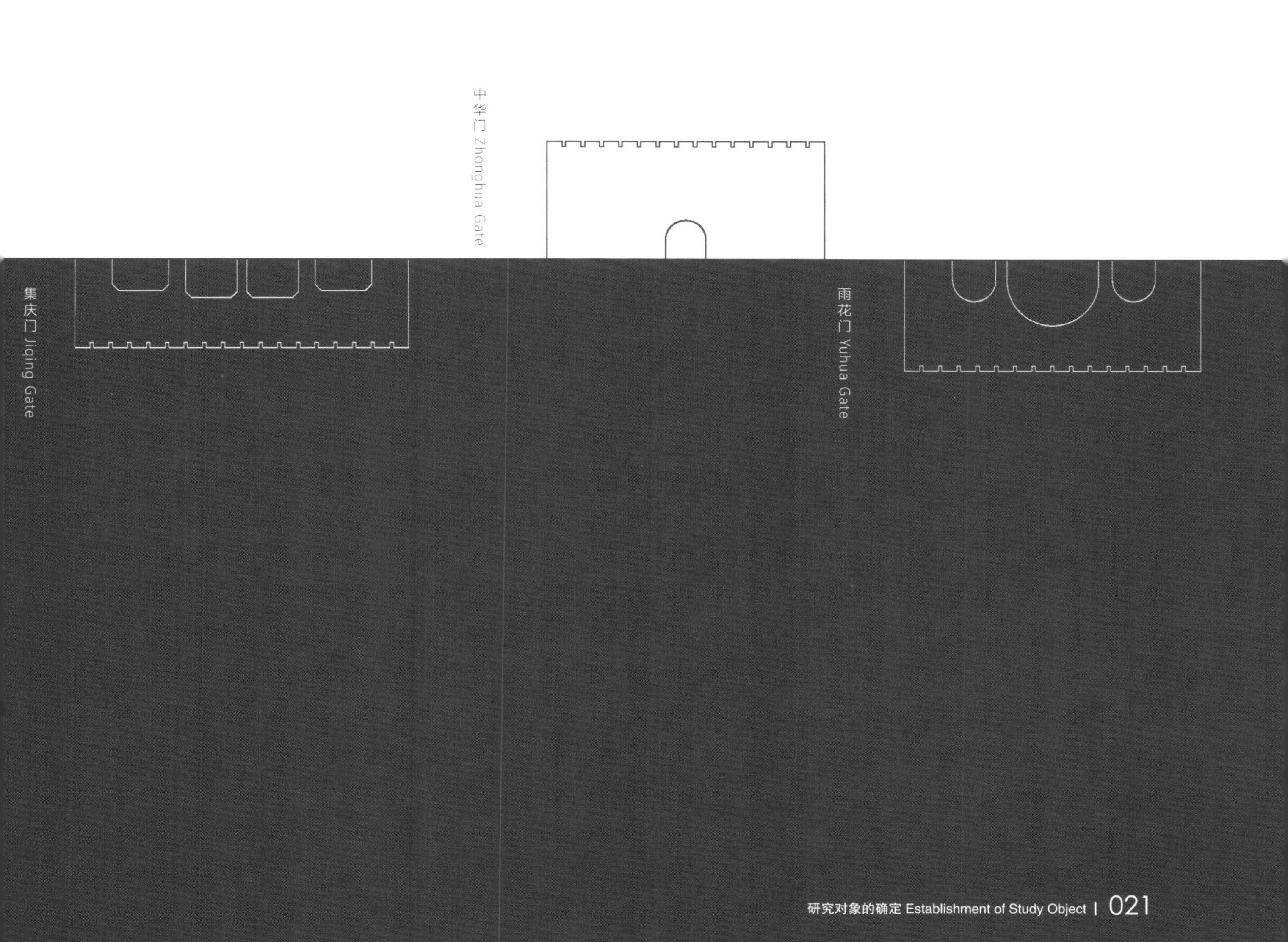

为什么选择南京明城墙绿道
Why choose the Nanjing Ming Dynasty City Wall Greenway

为什么选择南京

2012年底，南京完成了《南京市绿道规划暨三年行动计划》，计划在长江以南城区形成“一江三环五带”的绿道结构，在长江以北形成“一江一环两带”的绿道结构。截至2015年底本研究启动初期，南京已完成绿道建设200km有余。仔细审视南京绿道规划建设现状和规划结构，不难发现自然环境中的绿道无论在数量还是规模上仍占据主导，建成环境中已建成或部分建成的绿道目前仅有明城墙绿道、滨江绿道等。

一方面，与中国其他城市一样，南京建成环境绿道同样存在使用不均、部分绿道服务绩效低下等问题。另一方面，南京建成区具有用地紧凑、人口密集等特点，能被绿道利用作为选线空间载体的要素（如水系、道路及明城墙等）较丰富，在我国目前正在开展建成环境绿道规划建设的城市中具有较强的代表性和研究价值。

Why choose Nanjing

At the end of 2012, Nanjing completed *Nanjing Greenway Planning and Three-Year Action Plan*, which planned to establish a greenway structure of “the Yangtze Riverside-Three Rings-Five Greenways” in the city proper on the south of the Yangtze River, as well as a greenway structure of “the Yangtze Riverside-One Ring-Two Greenways” on the north of the Yangtze River. By the end of 2015 during the initial stage of our study, the city of Nanjing had already completed the construction of more than 200km of greenways. When we carefully review the current situation of the planning and construction of greenways in Nanjing and the planning structure, it is not too hard for us to find that the greenways in natural environments hold a dominant position in terms of quantity and scale while those in built environments, which are already completed or partially completed, only include the Nanjing Ming Dynasty City Wall Greenway and the Yangtze Riverside Greenway, etc.

Similar to the other cities in China, Nanjing also faces problems concerning the greenways in built environments, including uneven use, low service performance of some greenways and so on. On the other hand, the built-up areas in Nanjing are characteristic of land compactness, dense population, etc. and rich in elements (such as river system, roads, the Ming Dynasty City Wall, etc.) that can be used by greenways as spatial carriers of route selection, so that they are very representative with very high values for study among the cities where China is currently carrying out the planning and construction of greenways in built environments.

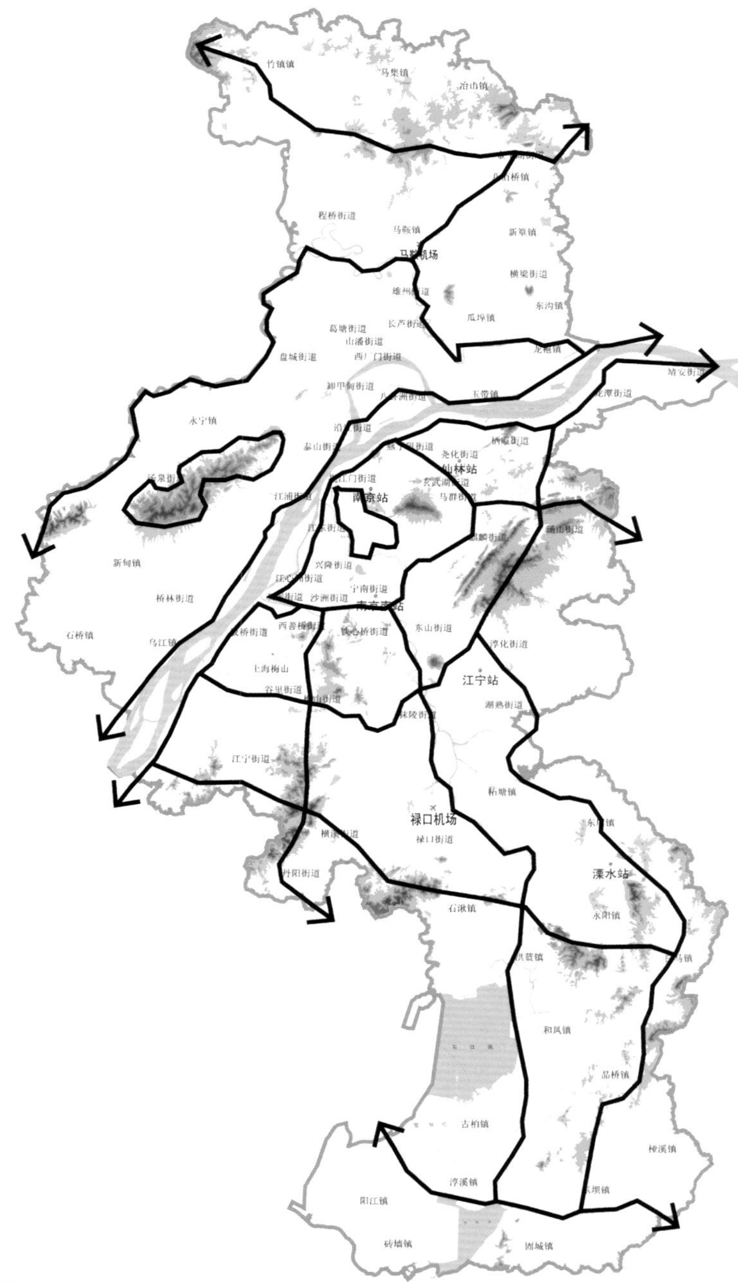

南京市域休闲绿道规划图（2013—2020）
Recreational greenway planning chart of administrative region of Nanjing (2013—2020)

图片来源：《南京市绿地系统规划（2013—2020）年》
Source: Nanjing Greenway System Planning (2013—2020)

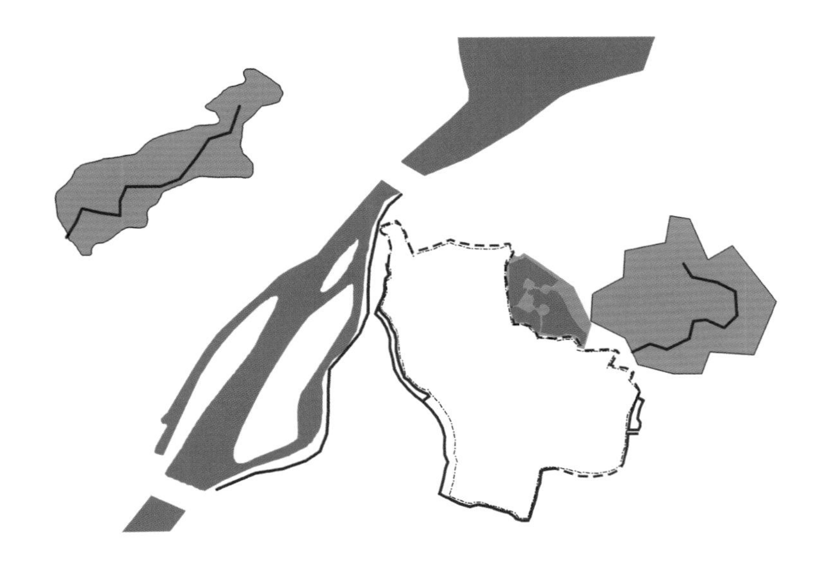

为什么选择南京明城墙绿道

南京目前建成的绿道涉及建成环境和自然环境两个区域，由于本研究主要聚焦于建成环境绿道，因此，在最初绿道筛选时就排除了老山绿道、紫金山绿道等自然环境中的绿道。

Why Nanjing Ming Dynasty City Wall Greenway is chosen

At present, the greenways, which are already built in Nanjing, involve built and natural environments. Since this study mainly focuses on the greenways in built environments, those in natural environments have already been excluded from the process of selection at the beginning, which included Laoshan Greenway, Purple Mountain Greenway, etc.

①

在 2015 年研究起始阶段，南京建成环境绿道中建成水平较高并投入使用的主要为滨江绿道和明城墙绿道两条，因此研究团队以滨江绿道及明城墙绿道为预调研对象，并初步在两条绿道中选取了 12 个区段（各区段长为 1000m 至 1500m），在 2015 年 11 月至 2016 年 5 月约半年的时间中，对各绿道区段进行了多次实地调研。

During the initial phase of our study in 2015, the Yangtze Riverside Greenway and the Nanjing Ming Dynasty City Wall Greenway were the two greenways that were highly built and put into use among all the greenways in built environments in Nanjing. Therefore, our team took the Yangtze Riverside Greenway and the Nanjing Ming Dynasty City Wall Greenway as the objects of preliminary study and preliminarily selected 12 segments (1,000-1,500m per segment) from the two greenways. From November 2015 to May 2016, in about half a year, our team conducted on-the-spot surveys on those greenway segments multiple times.

预调研信息记录表 Log sheet of preliminary study information

次数 Times	调研日期 Date	周末 / 工作日 Weekend/Workday	时间段 Time Interval	天气 Weather	调研区段 Segment under preliminary study
01	2015-11-22	周末 /Weekend	14:00-16:00	晴 /Sunny	绿博园—奥体中心区段 /Green Expo-Olympic Sports Center
02	2015-11-29	周末 /Weekend	14:00-16:00	阴 /Cloudy	定淮门—汉中门区段 /DHG-HZG
03	2015-12-24	工作日 /Workday	14:00-16:00	晴 /Sunny	挹江门—华严岗门区段 /YJG-HYGG
04	2015-12-19	周末 /Weekend	08:00-09:00	晴 /Sunny	月牙湖公园区段 /Yueya Lake Park
05	2015-12-20	周末 /Weekend	14:00-16:00	晴 /Sunny	方家营—建宁路区段 /FJR-JNR
06	2015-12-27	周末 /Weekend	08:00-09:00	晴 /Sunny	草场门大街—清凉门大街区段 /CCG Street-QLG Street
07	2016-02-27	周末 /Weekend	08:00-09:00	晴 /Sunny	清凉门—汉中门区段 /QLG-HZG
08	2016-03-05	周末 /Weekend	08:00-09:00	阴 /Cloudy	定淮门—草场门区段 /DHG-CCG
09	2016-03-11	工作日 /Workday	14:00-16:00	晴 /Sunny	定淮门—草场门区段 /DHG-CCG
10	2016-03-12	周末 /Weekend	18:00-19:00	阴 /Cloudy	挹江门—华严岗门区段 /YJG-HYGG
11	2016-03-19	周末 /Weekend	14:00-16:00	阴 /Cloudy	水西门—集庆门区段 /SXG-JQG
12	2016-03-20	周末 /Weekend	14:00-16:00	晴 /Sunny	东水关—武定门区段 /DSG-WDG
13	2016-03-26	周末 /Weekend	14:00-16:00	晴 /Sunny	凤台路—雨花门区段 /FTR-YHG
14	2016-03-27	周末 /Weekend	18:00-19:00	晴 /Sunny	凤台路—雨花门区段 /FTR-YHG
15	2016-04-04	周末 /Weekend	14:00-16:00	晴 /Sunny	草场门大街—清凉门大街区段 /CCG Street-QLG Street
16	2016-05-11	工作日 /Workday	14:00-16:00	晴 /Sunny	凤台路—雨花门区段 /FTR-YHG
17	2016-05-14	周末 /Weekend	18:00-19:00	晴 /Sunny	草场门—清凉门区段 /CCG-QLG
18	2016-05-17	工作日 /Workday	18:00-19:00	晴 /Sunny	挹江门—华严岗门区段 /YJG-HYGG

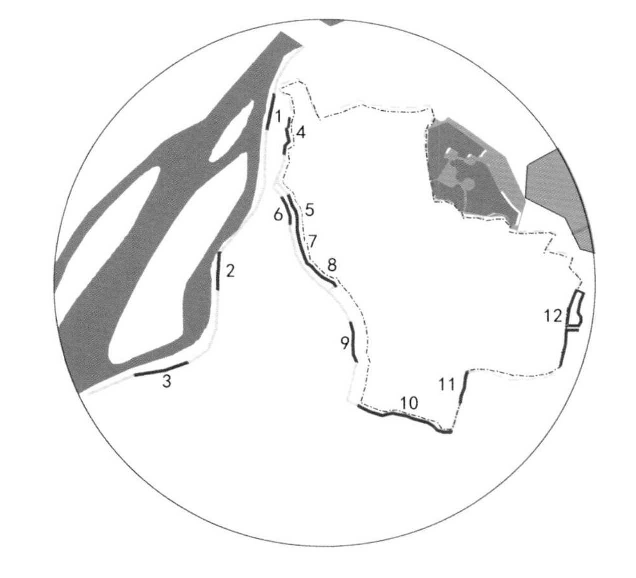

1 方家营—建宁路区段 FJY-JNR
2 草场门大街—清凉门大街区段 CCG Street-QLG Street
3 绿博园—奥体中心区段 Green Expo-Olympic Sports Center
4 挹江门—华严岗门区段 YJG-HYGG
5 定淮门—草场门东区段 DHG-CCG
6 定淮门—草场门西区段 DHG-CCG
7 草场门—清凉门区段 CCG-QLG
8 清凉门—汉中门区段 QLG-HZG
9 水西门—集庆门区段 SXG-JQG
10 凤台路—雨花门区段 FTR-YHG
11 东水关—武定门区段 DSG-WDG
12 月牙湖公园区段 Yueya Lake Park

游径宽度达到6m的滨江绿道是滨江风光带规划建设中的重要一环，骑行或漫步其中，可充分领略南京的大江风貌与山水城林特色。但尽管拥有极佳的景观资源、良好的设施配套，滨江绿道的利用率却一直较低。

The Yangtze Riverside Greenway, the width of whose trails is 6 meters, is a very important part of the planning and construction of Yangtze Riverside Scenic Belt. When riding or walking on it, people can fully experience the style and characteristics of the great riverscape in Nanjing, incorporating mountains, rivers, the city, and forests. Although with excellent landscape resources and high-standard supporting facilities, the utilization ratio of Yangtze Riverside Greenway is always low.

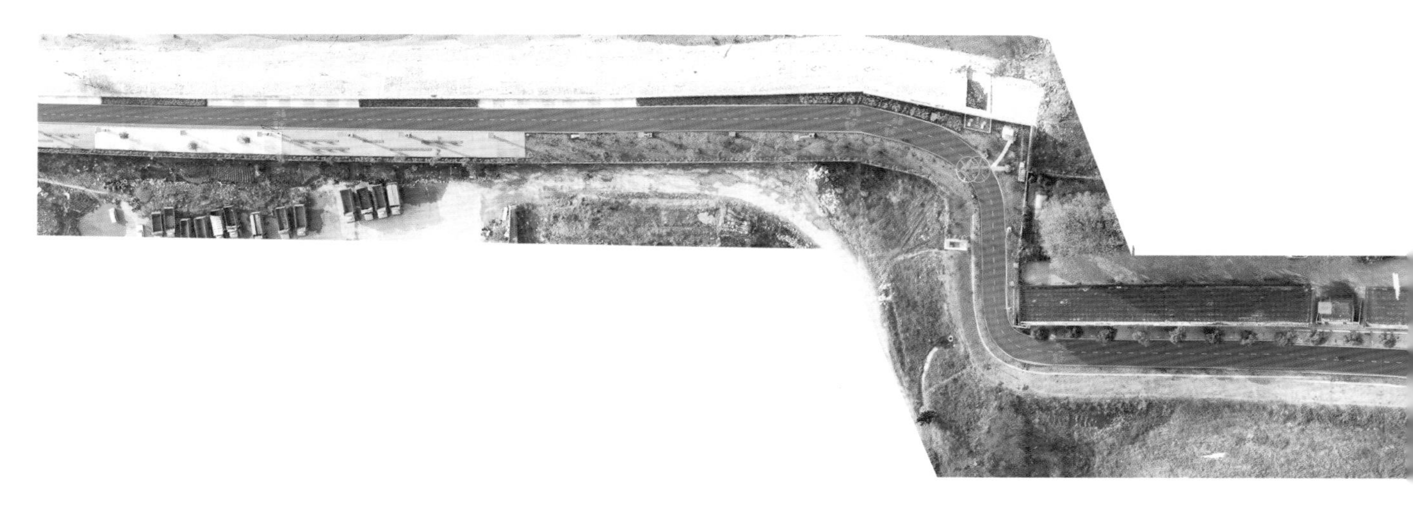

通过预调研对比明城墙绿道与滨江绿道，我们可以发现滨江绿道在空间载体、内部空间以及周边环境的丰富度上均不及明城墙绿道，同时与主城居住人口密集地段距离较远，当前利用率也较低，难以为后期数据分析提供充分的样本支撑。

明城墙绿道位于南京主城，依附明城墙与护城河（含秦淮河一部分区段）而建，是一条集日常游憩服务、旅游观光、生态保护、历史资源整合等功能于一体的综合性城市绿道。它在空间上串联了明城墙沿线各类绿色开敞空间，周边用地类型丰富，居民集中，与南京城市各类功能区关联紧密，具有典型的建成环境特征和较高的研究价值。综合比对，选取在建成环境中更具典型性的明城墙绿道作为研究对象。

After contrasting the Nanjing Ming Dynasty City Wall Greenway and the Yangtze Riverside Greenway through the preliminary study, we can find that the Yangtze Riverside Greenway is not as good as the Nanjing Ming Dynasty City Wall Greenway in terms of the richness of spatial carriers, the interior space, and surroundings, while it is far away from densely-populated areas in the main city, with a relatively low utilization ratio at present, so it is of low feasibility in providing enough and diverse data support for analysis at a later period.

Located in the main city of Nanjing, the Nanjing Ming Dynasty City Wall Greenway attaches itself to the Ming Dynasty City Wall and the city moat (including a large segment of the Qinhuai River). It is a comprehensive urban greenway that incorporates daily recreational services, tourism, ecological protection, historical resources preservation, and so on. In terms of space, it connects in series a variety of open spaces along the Ming Dynasty City Wall. The surrounding areas of the greenway are rich in land use types, densely-populated by residents, and closely link to all kinds of functional districts in Nanjing, so it is characteristic of a typical built environment and provides very high values for study. As a result, the Nanjing Ming Dynasty City Wall Greenway is finally locked as our object of study.

明城墙绿道与滨江绿道的预调研比较 Comparison and contrast of the Nanjing Ming Dynasty City Wall Greenway and the Yangtze Riverside Greenway in preliminary study

	明城墙绿道 Nanjing Ming Dynasty City Wall Greenway	滨江绿道 Yangtze Riverside Greenway
空间载体 Spatial carriers	河道、道路、历史遗址 River courses, roads, historical sites	河道、道路 River courses, roads
内部空间 Internal space	空间变化丰富 Richness of spatial variation	空间开敞宽阔，但较均质，变化较少 Wide-open space, with homogeneity and fewer changes
周边环境 Surroundings	开发成熟，用地类型多元，环境要素丰富 Mature development, diversity of land types, richness of environmental elements	周边许多用地尚未开发，用地类型和环境要素较单一 A lot of land in the surrounding areas is not yet developed, with single land types and environmental elements
与主城关系 Relationship with the main city	较紧密 In core area of the main city	较弱 On the edge of the main city
使用情况 Service condition	利用率有高低变化，样本容量能支持后期数据分析 Utilization ratio ßuctuates; sample size can support data analysis in later period	利用率普遍较低，样本容量过小，将为后期数据分析带来较大偶然性变化 Utilization ratio is generally low; sample size is too small to support data analysis in later period
典型性 Representativeness	强 Strong	弱 Weak

怎样选取研究区段

How to select study segments

至2015年底研究启动时，南京明城墙绿道西线与南线基本建成并投入使用，总长度约为15 km。为保障调研的可操作性，并兼顾预调研过程中对绿道内外部空间特征、使用情况、空间载体等多个因素的梳理和比对，研究团队在绿道西线与南线上均匀选取了6个典型区段作为调查和研究样本，每个区段长度在1000m至1500m之间，区段首末端均为明城墙绿道的重要出入口。

其中，南京明城墙绿道西线总长约10 km，研究选取4个调研区段，分别为区段1（挹江门—华严岗门）、区段2（定淮门—草场门）、区段3（草场门—清凉门）、区段4（水西门—集庆门）；南线总长约5 km，对应选取2个调研区段，分别为区段5（凤台路—中华门）、区段6（中华门—雨花门）。

By the end of 2015 when the study was initiated, both the west and south lines of the Nanjing Ming Dynasty City Wall Greenway loop had already been completed and put into use, with a total length of about 15km. To guarantee the operability of the survey and take into account the classification and comparison of multiple elements including the internal and external spatial characteristics of the greenway, current greenway service conditions, spatial carriers, and so on, during the course of preliminary study, our team evenly selects six typical segments as samples for survey and study along the built up west and south lines, with the length of each segment ranging from 1,000m to 1,500m and both ends of each segment being key greenway entrances-exits, most of which are the main gates of the city wall.

Between the two lines, the west line of the Nanjing Ming Dynasty City Wall Greenway is about 10km in total length. Four segments are chosen for the study, namely, Segment 1 Yijiang Gate-Huayangang Gate (S1 YJM-HYGG), Segment 2 Dinghuai Gate-Caochang Gate (S2 DHG-CCG), Segment 3 Caochang Gate-Qingliang Gate (S3 CCG-QLG), and Segment 4 Shuixi Gate-Jiqing Gate (S4 SXG-JQG). The south line is about 5km in total length. Correspondingly, two segments are chosen, namely, Segment 5 Fengtai Road-Zhonghua Gate (S5 FTR-ZHG) and Segment 6 Zhonghua Gate-Yuhua Gate (S6 SZG-YHG).

南京明城墙沿线的主要绿色空间 Main parks and open spaces along the Nanjing Ming Dynasty City Wall

序号 / 公园或开放空间 Order / park or open space		
1 阅江楼景区 Yuejianglou Scenic Area	9 莫愁湖公园 Mochouhu Park	17 明故宫 Ming Palace
2 绣球公园 Hydrangea Park	10 南湖公园 South Lake Park	18 钟山风景区 Zhongshan Mountain
3 小桃园 Xiao Taoyuan	11 朝天宫 Chaotian Palace	19 白马公园 White Horse Park
4 古林公园 Gulin Park	12 愚园 A Fool Garden	20 玄武湖公园 Xuanwu Lake Park
5 石头城遗址公园 Stone City Ruins Park	13 白鹭洲公园 Bailuzhou Park	21 九华山公园 Jiuhua Mountain Park
6 清凉山公园 Qingliangshan Park	14 东水关遗址公园 Dongshuiguan Ruins Park	22 和平公园 Peace Park
7 乌龙潭公园 Wulongtan Park	15 月牙湖公园 Yueyahu Park	23 北极阁公园 Arctic Pavilion Park
8 汉中门广场 Hanzhongmen Square	16 午朝门公园 Wuchaomen Park	24 神策门公园 Shencemen Park

最终调研区段
Greenway segments in the final study

研究区段基本概况

Basic introduction of study segments

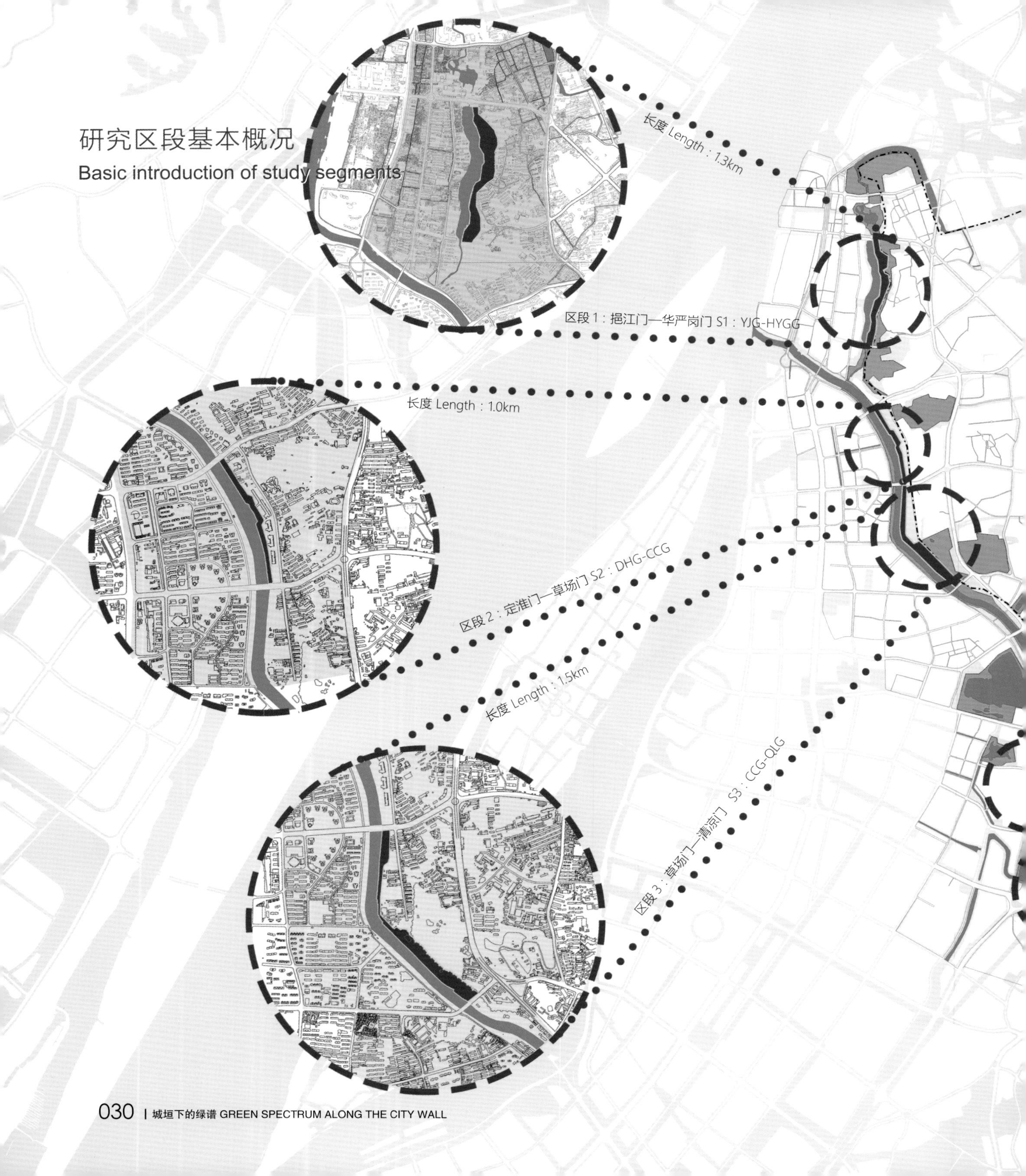

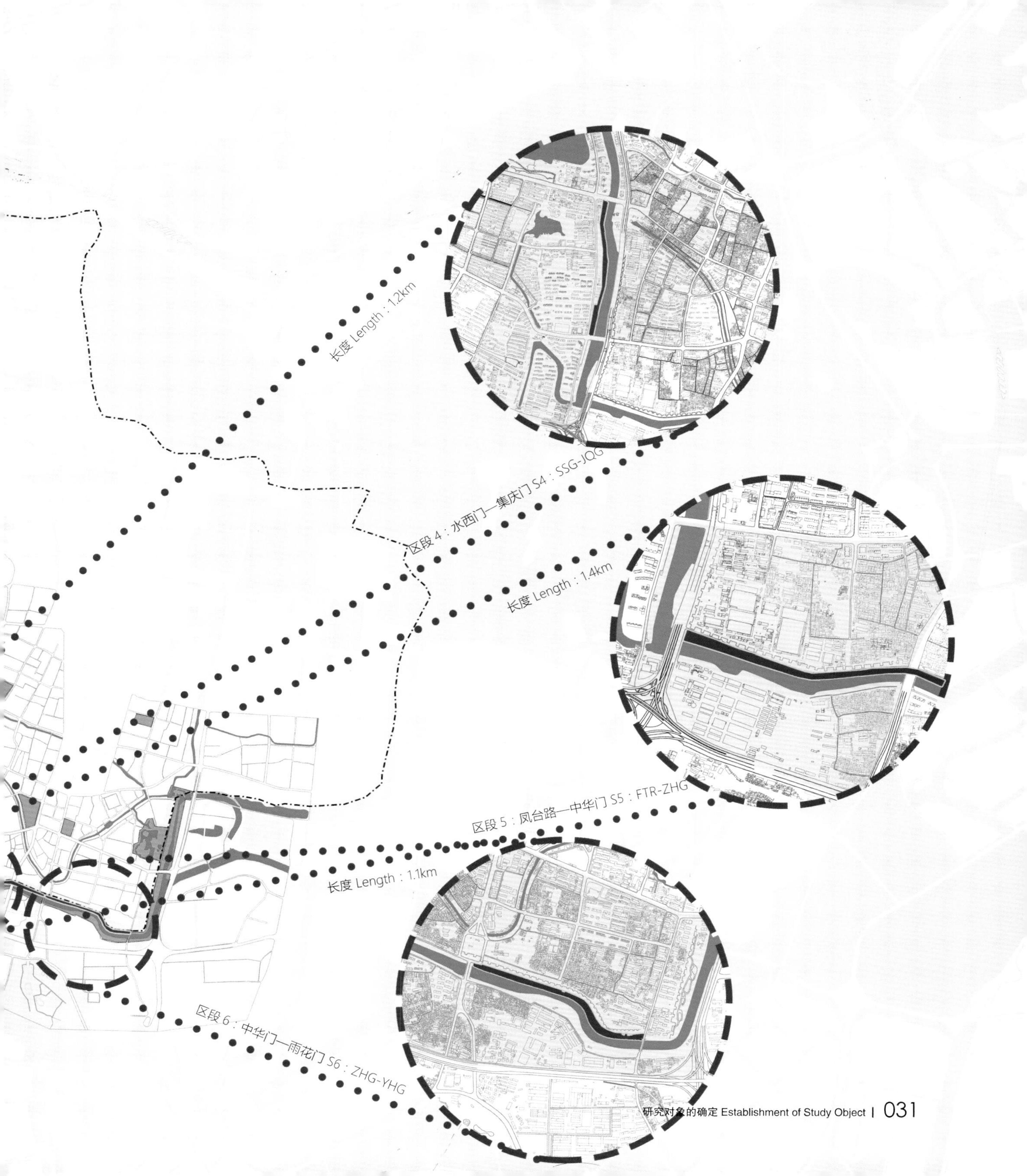
长度 Length：1.2km
区段 4：水西门—集庆门 S4：SSG-JQG
长度 Length：1.4km
区段 5：凤台路—中华门 S5：FTR-ZHG
长度 Length：1.1km
区段 6：中华门—雨花门 S6：ZHG-YHG

区段 1：挹江门—华严岗门
Segment 1: Yijiang Gate-Huayangang Gate

区段 1（挹江门—华严岗门）是南京明城墙绿道已建成部分的最北段，东侧依靠明城墙，西侧毗邻护城河（秦淮河支流）。这里的风景优美、清雅静谧，以种植数千株桃树闻名，人称"小桃园"。该区段绿道连接着多个公园绿地和城市景区，其中规模较大的有北侧的绣球公园和东侧的八字山公园。绿道所在区域树木茂盛，距离市中心有一定距离，是市民避开城市喧嚣、享受静谧美景的理想游憩目的地。

S1 YJM-HYGG is the northernmost segment of the Nanjing Ming Dynasty City Wall Greenway. Its east side is supported by the Ming Dynasty City Wall, while its west side is next to the city moat (a tributary of the Qinhuai River). It owns an elegant and peaceful scenery and is famous for thousands of peach trees, hence the name Little Peach Orchard. The greenway of this segment connects multiple urban public parks and scenic spots, among which Xiuqiu Park on the north side and Bazishan Park on the east are relatively large. The region where the greenway is located is well wooded with a certain distance from downtown Nanjing, making it an ideal destination of recreation for city-dwellers to get away from the noisy area and enjoy the beautiful scenery.

绿道基本信息
Basic information of the greenway

绿道长度	Length	1.3km
绿道总宽度	Total width	18-89m
自行车道宽度	Width of bikeways	4.5m
步行道宽度	Width of footpaths	4.5m
用地面积	Area of land	5.6hm²
硬地面积	Area of hard sites	1.2hm²
硬地率	Percentage of hard sites	21.4 %
绿地面积	Area of green space	4.0hm²
绿地率	Percentage of green space	71.4 %

区段 2：定淮门—草场门

Segment 2: Dinghuai Gate-Caochang Gate

区段 2（定淮门—草场门）北起定淮门桥，南至草场门大桥，紧邻护城河（秦淮河）。明城墙遗址在本区段中断，因此该区段绿道无法像其他区段一样依墙傍水。绿道东侧有高耸的江苏电视台电视塔，周边最突出的绿地资源为东侧的古林公园，占地面积约为 27hm^2。该区段空间较狭窄，空间变化单一，中部有游船码头，但很少有游船停靠，东侧紧邻古玩市场及南京艺术学院。

S2 DHG-CCG starts from Dinghuaimen Bridge in the north, spreads to Caochangmen Bridge in the south, and stands next to the city moat (the Qinhuai River). The historic site of the Ming Dynasty City Wall is broken off at this segment, so the greenway here cannot stand both near the city wall and by the river like the other segments. On the east side of the greenway stands the television tower of Jiangsu TV. The most significant green space in the surrounding area is Gulin Park on the east, with an area of about 27 hectares. The pattern of space of this segment is narrow and homogeneous. Although there is a cruise terminal in the middle, sightseeing-boats seldom berth at this dock. The east side of the segment is close to an antique market and Nanjing University of the Arts.

绿道基本信息
Basic information of the greenway

绿道长度	Length	1.0km
绿道总宽度	Total width	33-50m
自行车道宽度	Width of bikeways	6.0m
步行道宽度	Width of footpaths	5.0m
用地面积	Area of land	3.6hm^2
硬地面积	Area of hard sites	1.7hm^2
硬地率	Percentage of hard sites	47.2%
绿地面积	Area of green space	1.7hm^2
绿地率	Percentage of green space	47.2%

区段 3：草场门—清凉门
Segment 3: Caochang Gate-Qingliang Gate

区段 3（草场门—清凉门）空间和利用情况呈现“两极分化”。从北入口草场门大桥至中部国防园人行桥部分，基本承接区段 2（定淮门—草场门）单一均质的空间特征，利用率较低；但人行桥南侧区段则与石头城公园紧密关联，沿明城墙南下空间大幅敞开，环境品质也得以显著提升，其中游步道众多、硬质场地丰富，使用人群也显著增加。尤其在周末下午，区段南部以“鬼脸照镜”为代表的集中活动场地成为家长陪伴小孩和聚集游玩的理想场所。

The space pattern and utilization condition at the northern and southern parts of S3 CCG-QLG are “polarized.” The part from Caochang Gate Bridge at the north entrance to the foot bridge of Guofang Park in the middle basically carries on the spatial characteristic of Dinghuai Gate-Caochangmen Segment, which is single and homogeneous, with a very low utilization ratio. However, the segment on the south of the foot bridge is closely linked to Stone City Wall Park and the space along the City Wall is wide open with substantial improvement of the environmental quality, which is rich in footpaths and open sites so that the number of visitors noticeably increases. Especially on a weekend afternoon, with “Funny Face Mirrors” as its central activity yard, the southern part of the segment is the ideal place where parents can gather around and play with their children.

绿道基本信息
Basic information of the greenway

绿道长度	Length	1.5km
绿道总宽度	Total width	26-120m
自行车道宽度	Width of bikeways	5.0m
步行道宽度	Width of footpaths	5.0m
用地面积	Area of land	9.7hm^2
硬地面积	Area of hard sites	4.4hm^2
硬地率	Percentage of hard sites	45.4 %
绿地面积	Area of green space	5.1hm^2
绿地率	Percentage of green space	52.5 %

区段 4：水西门—集庆门

Segment 4: Shuixi Gate-Jiqing Gate

区段 4（水西门—集庆门）是 6 个区段中空间最局促、人气最低的一段。该区段是 6 个区段中唯一位于护城河外侧的绿道区段，主要是由于该地段明城墙与护城河之间的空间所剩无几，无法再支持新建绿道所致。该区段绿道的整体空间结构较单一，内部空间主体要素为一条狭窄的综合慢行道主游径，游径沿线会有少量周边居民进行垂钓、健身等活动。

S4 SXG-JQG is the most cramped and the least popular among the six segments. It is the only segment of the greenway that is located outside the city moat among the six segments, mainly because there is not much space between the part of the Ming Dynasty City Wall and the city moat, which can no longer support the newly built greenway. The entire spatial structure of this segment is relatively single, and the main body of the interior space is a narrow main trail of a comprehensive slow lane, along which few surrounding residents go fishing and do physical exercise.

绿道基本信息
Basic information of the greenway

绿道基本信息	Basic information of the greenway	
绿道长度	Length	1.2km
绿道总宽度	Total width	12-35m
自行车道宽度	Width of bikeways	2.6m
步行道宽度	Width of footpaths	2.0m
用地面积	Area of land	2.7hm^2
硬地面积	Area of hard sites	1.1hm^2
硬地率	Percentage of hard sites	40.7 %
绿地面积	Area of green space	1.5hm^2
绿地率	Percentage of green space	55.6 %

区段 5：凤台路—中华门
Segment 5: Fengtai Road-Zhonghua Gate

区段 5（凤台路—中华门）位于南京明城墙的西南部，西起凤台路，东至中华门瓮城。这里也是南京老城西南角，周边建筑物低矮密集。绿道和饮马桥相交，并从桥底通过，该区段在周末午后时段，通常会聚集大量下棋、打牌的中老年人，这些市民大都自备板凳，以公共座椅为桌，聚集打牌和围观，这已成为老城特色一景。

S5 FTR-ZHG is located in the southwest of the Nanjing Ming Dynasty City Wall. It starts from Fengtai Road to the west and spreads to the barbican entrance to Zhonghua Gate to the east. It is also the southwest corner of the old city of Nanjing near which buildings are low, short, and densely distributed. The greenway intersects Yinma Bridge and passes under the bridge. On the afternoon of a weekend, a lot of middle-aged and senior people gather around this segment to play Chinese chess or card. Most of them bring their own stools, use public benches as tables, and gather around playing or watching, which has become unique scenery of the city.

绿道基本信息
Basic information of the greenway

绿道基本信息	Basic information of the greenway	
绿道长度	Length	1.4km
绿道总宽度	Total width	18-40m
自行车道宽度	Width of bikeways	4.2m
步行道宽度	Width of footpaths	4.2m
用地面积	Area of land	4.1hm²
硬地面积	Area of hard sites	1.2hm²
硬地率	Percentage of hard sites	29.3 %
绿地面积	Area of green space	2.8hm²
绿地率	Percentage of green space	68.3 %

区段 6：中华门—雨花门
Segment 6: Zhonghua Gate-Yuhua Gate

区段 6（中华门—雨花门）人气较高，该区段所在区域经常举行南京传统节庆活动。区段西连中华门瓮城，且与大报恩寺遗址公园隔河相望，景观环境优良。该区段北面是老门东街区低矮密集的南京老建筑群。与区段 5（凤台路—中华门）相似，该区段下午也聚集着打牌、下棋以及围观的中老年人，一到节庆日则人山人海，老城特色浓郁。

S6 SZG-YHG is very popular because traditional Nanjing festivals and events are often hosted in the region where this segment is located. This segment connects the barbican entrance to Zhonghua Gate on the west and is across the river from Dabao'en Temple, with very good scenery around. The north side of this segment is Nanjing's old architectural complex in Laomendong Block, which is low, short, and densely distributed. Similar to S5 FTR-ZHG, this segment also attracts a lot of middle-aged and senior people who gather around to play cards or Chinese chess or just watch in the afternoon. On the occasion of a festival or during holidays, this segment is full of people with strong unique atmosphere of the old city.

绿道基本信息
Basic information of the greenway

绿道长度	Length	1.1km
绿道总宽度	Total width	32-50m
自行车道宽度	Width of bikeways	4.0m
步行道宽度	Width of footpaths	1.2m
用地面积	Area of land	4.4hm^2
硬地面积	Area of hard sites	1.4hm^2
硬地率	Percentage of hard sites	31.8 %
绿地面积	Area of green space	2.9hm^2
绿地率	Percentage of green space	65.9 %

绿道服务绩效调查

Survey of Service Performance of the Greenway

怎样开展调查

How to conduct the survey

绿道的使用状态会随季节与气候、工作日与周末以及一天内各时段的差异产生较大变化，鉴于本书主要是探讨绿道内外部空间特征对绿道服务绩效的影响，为了将除空间特征以外的其他外部因素对绿道使用及服务状态所产生的影响减至最小，研究团队经过多次预调研和分析后，决定选取游憩出行外部条件最理想、居民游憩需求最旺盛的时段，即绿道使用和服务需求最强的时段来对绿道的人群使用瞬时状态进行调研，并将其作为本书绿道服务绩效的主要衡量途径。

结合预调研结果以及对南京市民游憩出行特征的研究，我们对每次调研气候、时间和时段等一系列条件进行了限定，从而基本确保每次采集的绿道人群使用信息均是在游憩出行外部条件较佳、居民游憩需求最旺盛的时段采集的。为了降低信息采集过程的偶然性，每个绿道区段共进行 12 次信息采集，时间跨越春秋两季，并在游憩需求最旺盛的周末下午和日常傍晚两个时段各采集 6 次数据。

The working condition of the greenway tends to change drastically according to seasons and climate, change of workdays and weekends, as well as the differences in different time periods of a day. Since this study mainly discusses the influence of both internal and external spatial characteristics of the greenway on its service performance, to minimize the influence of the other external factors but spatial characteristics on the use and the service status of the greenway, our team has undergone multiple preliminary studies and analyses to decide and pick up the time period when the external conditions for recreation and travel are the most ideal and residents are in greatest demand of recreation. It means that we conduct surveys on the instantaneous state of the use of the greenway during the time period when the greenway is in greatest demand. The information and data collected during this time period is taken as the main indicator to measure the service performance of the greenway in this study.

Combining the result of the preliminary study and the review on the outdoor recreational characteristics of Nanjing residents, we set a limit to a series of criteria including climate, time point, time period, and so on, for each survey, so as to guarantee that the user information of the greenway is collected during the time period when the external conditions for outdoor recreation are the most ideal and residents are in greatest demand of recreation. To reduce the contingency of the information collection process, we collect information twelve times for each segment of the greenway, which spans spring and fall, and collect data six times both on the afternoon of a weekend and every day at nightfall, when the recreational demand of the residents in Nanjing is greatest.

调查条件设定
Survey condition set up

	下午时段 Afternoon	傍晚时段 Nightfall
次数 Times	6	6
气温 Temperature	13℃-29℃	13℃-29℃
天气 Weather	晴或阴 Sunny or overcast	晴或阴 Sunny or overcast
季节 Season	春季、秋季 Spring, Fall	春季、秋季 Spring, Fall
时段 Time Interval	14:00-16:00	18:00-20:00
时间 Time	周末 Weekend	工作日和周末 Everyday

注：预调研发现在工作日和周末傍晚建成环境绿道的使用均以周边居民日常游憩为主，其间并未有显著差别，因此在正式调查的傍晚时段中没有区别对待周末和工作日，在书中统一用"日常傍晚"来表示该调查时段。

Note: Through preliminary survey, we find out that surrounding residents use the greenway in built environmentmainly for the purpose of daily recreation at nightfall of a workday and a weekend without any noticeable differences, so we do not make a difference between weekend and workday at nightfall during formal survey. Therefore, we uniformly use "every day at nightfall" to mean this time period of the survey in this book.

怎样采集人群使用信息

How to collect information on greenway utilization

为了方便有效地对绿道上的人群使用瞬时信息进行采集，我们主要通过航拍飞行器与地面调查相结合的方法来进行。利用航拍飞行器的高空垂直拍摄一方面可获得精度较高的绿道空间平面，另一方面由于航拍器具有高机动性与灵活性，能够快速直接地获取地面人群在绿道中的瞬时使用信息，能有效避免和修正由于绿道人群流动所造成的重复统计。而辅以地面调查则能有效地对航拍飞行器信息采集过程中的盲点进行补充和修正。

由于航拍飞行器的飞行速度和距离有一定限制，因此将绿道调研区段的长度定为1000—1500m，在确保采集到的绿道使用人群信息具有瞬时特征的同时，也能充分体现绿道的线性空间特点。

In order to effectively collect instantaneous information of the status of greenway utilization, we mainly rely on a method that combines drones and ground surveys. Due to their high maneuverability and flexibility, drones can quickly and directly collect the instantaneous information of the people on the greenway, so as to effectively avoid and modify repeated statistics arising out of crowd flow on the greenway. Supplementary ground surveys can effectively replenish and modify the blind spots during the process of information collection by drones.

Due to the limitation of flight speeds and the flying distances of the drones, we set the length of the study segments as 1,000-1,500m so as to fully demonstrate the linear spatial characteristics of the greenway, in addition to guaranteeing that the collection information on the greenway's utilization status is instantaneous.

调查方案

Survey plan

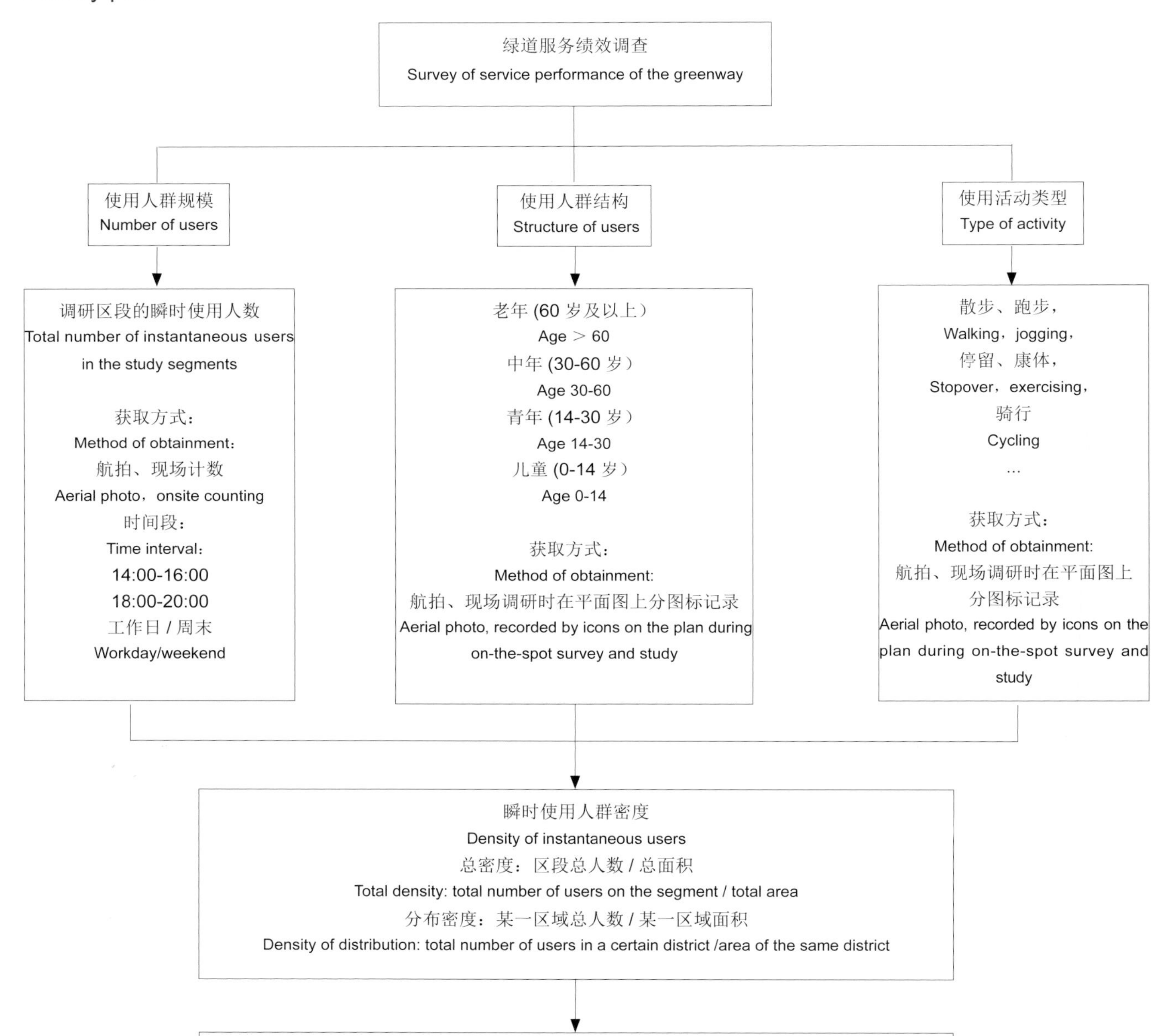

调查结果

Survey result

各区段的总体平均数据显示，周末下午使用人数最多的区段为区段 3（草场门—清凉门）和区段 1（挹江门—华严岗门），该两个区段使用人数相对接近。排在第 3 的区段 6（中华门—雨花门）使用人数不到区段 3 人数的一半。使用人数最少的为区段 4（水西门—集庆门）甚至不到区段 3 人数的一成。

区段 3（草场门—清凉门）和区段 1（挹江门—华严岗门）两个区段各年龄结构人数相对均衡，差距不大，区段 3 中年人明显偏多，而区段 1 青年人略多。其他 4 个区段使用人群年龄结构则明显以老年人和中年人为主，其他年龄结构人群占比较低。

6 个区段活动类型结构相似，虽然散步比重均较高，但除了区段 1（挹江门—华严岗门）和区段 2（定淮门—草场门）外，其余 4 个区段散步活动占比均未过半。该时段进行康体和停留休息的人数在各区段中的占比也相对较大，但骑车和跑步人群在 6 个区段中的人数和占比均相对较小。

According to the average data of each segment, S3 CCG-QLG and S1 YJM-HYGG attract the biggest number of people using the greenway segment on a weekend afternoon. On both of the segments, the number of people is relatively the same. The third is S6 SZG-YHG where the number of people is less than half of that on S3 CCG-QLG. S4 SXG-JQG has the smallest number of people, which is less than 10 percent of that on S3 CCG-QLG.

The age structures of both S3 CCG-QLG and S1 YJM–HYGG are relatively balanced without any huge differences in each age group. S3 CCG-QLG has more middle-aged people, while S1 YJM-HYGG has slightly more young people. The rest of the four segments evidently attract more middle-aged and senior people, and the percentages of the other age groups are rather small.

Among the types of activities in all six segments, the structures of activity types are similar, although walking covers a great deal. However, none of the percentages of walking on the rest of the four segments exceed 50%, except for on S1 YJM–HYGG and S2 DHG-CCG. During the time period, the numbers of people who exercise and who stop for conversation or a short break cover more among all the segments, but the numbers and percentages of people who ride bicycles and jog are rather small among all six segments.

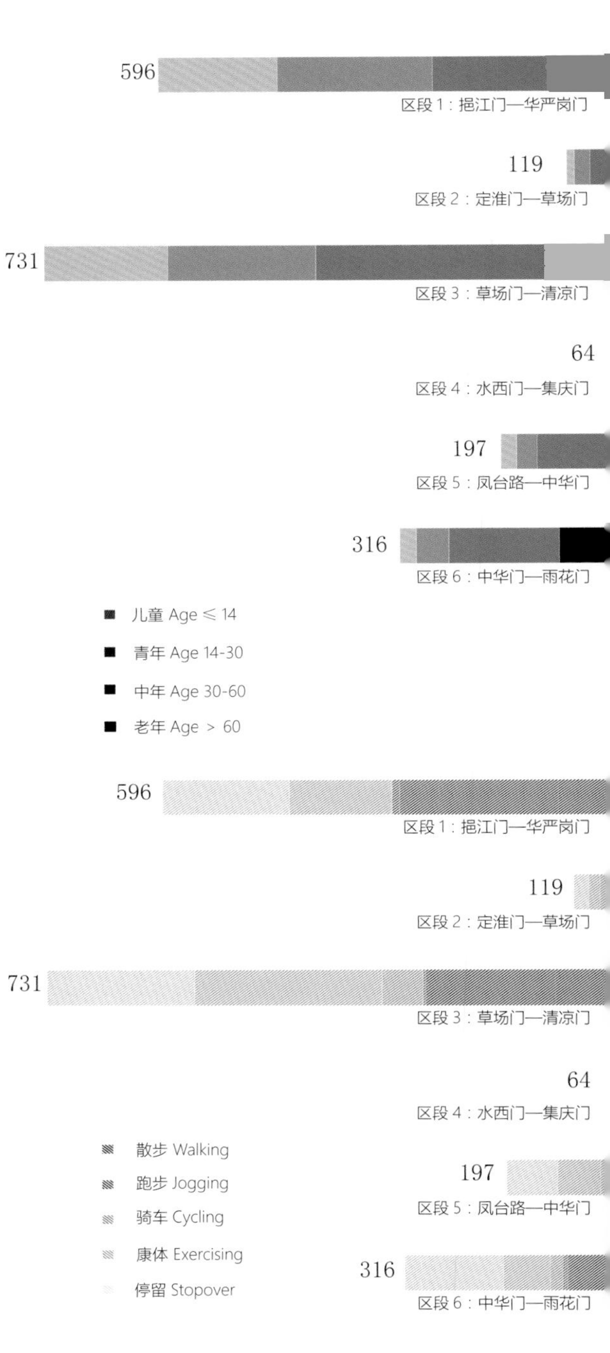

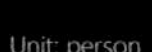

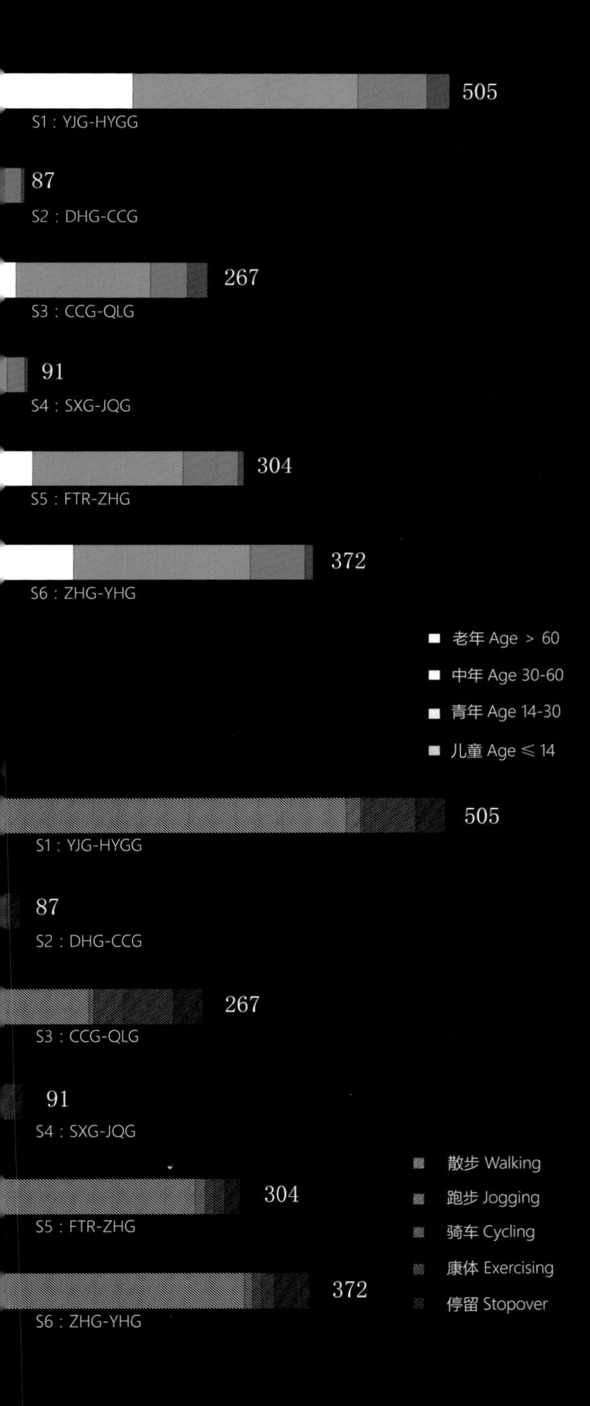

与周末下午不同，日常傍晚使用人数最多的区段变为区段 1（挹江门—华严岗门），区段 6（中华门—雨花门）和区段 5（凤台路—中华门）排位分列第 2 位和第 3 位，而周末下午使用人数最多的区段 3（草场门—清凉门）在日常傍晚仅列第 4 位，区段 4（水西门—集庆门）和区段 2（定淮门—草场门）则排在最后 2 位。

与周末下午相比，该时段使用人群年龄结构中的中老年人占比明显增加，各区段两者比重总计均超过七成，成为各区段的主导使用群体。

鉴于中老年人群成为主体使用人群，该时段内使用活动类型中的散步活动占比也大幅上升，除区段 3（草场门—清凉门）占比约为六成外，其他区段占比均接近或超过七成，成为该时段各个区段的主导活动类型。

Different from weekend afternoons, the segment which has the biggest number of people every day at nightfall is S1 YJM-HYGG, while S6 SZG-YHG and S5 FTR-ZHG rank 2nd and the 3rd, respectively. S3 CCG-QLG, which has the biggest number of users on weekend afternoons, only ranks the 4th every day at nightfall, while S4 SXG-JQG and S2 DHG-CCG rank 5th and 6th.

Compared to weekend afternoons, the number of middle-aged and senior people who use the segment every day at nightfall is noticeably increased. The total proportions of the two age groups on all segments both exceed 70%. Therefore, they become the dominant user groups on all segments.

In light of the fact that middle-aged and senior people are the dominant user groups, the percentage of walking as an activity type during this time period has increased drastically. The percentages of walking on the other segments all come close to or exceed 70%-except for S3 CCG-QLG, which covers about 60%-thus turning walking into a dominant activity on all the segments during this time period.

区段 1（挹江门—华严岗门）实景图
Site view of S1 (YJG–HYGG)

区段 1：挹江门—华严岗门

调查结果显示，周末下午绿道使用人群规模变化较大，在 360 人至 800 人之间变化，人群年龄结构较均匀，开展散步、康体和停留休息活动较多。

Segment 1: Yijiang Gate-Huayangang Gate

The result shows that the number of greenway users varies within a large range on the afternoon of a weekend. The data ranges from 360 to 800 people. The age structure of the crowd is relatively even. They usually walk, exercise, or stop over.

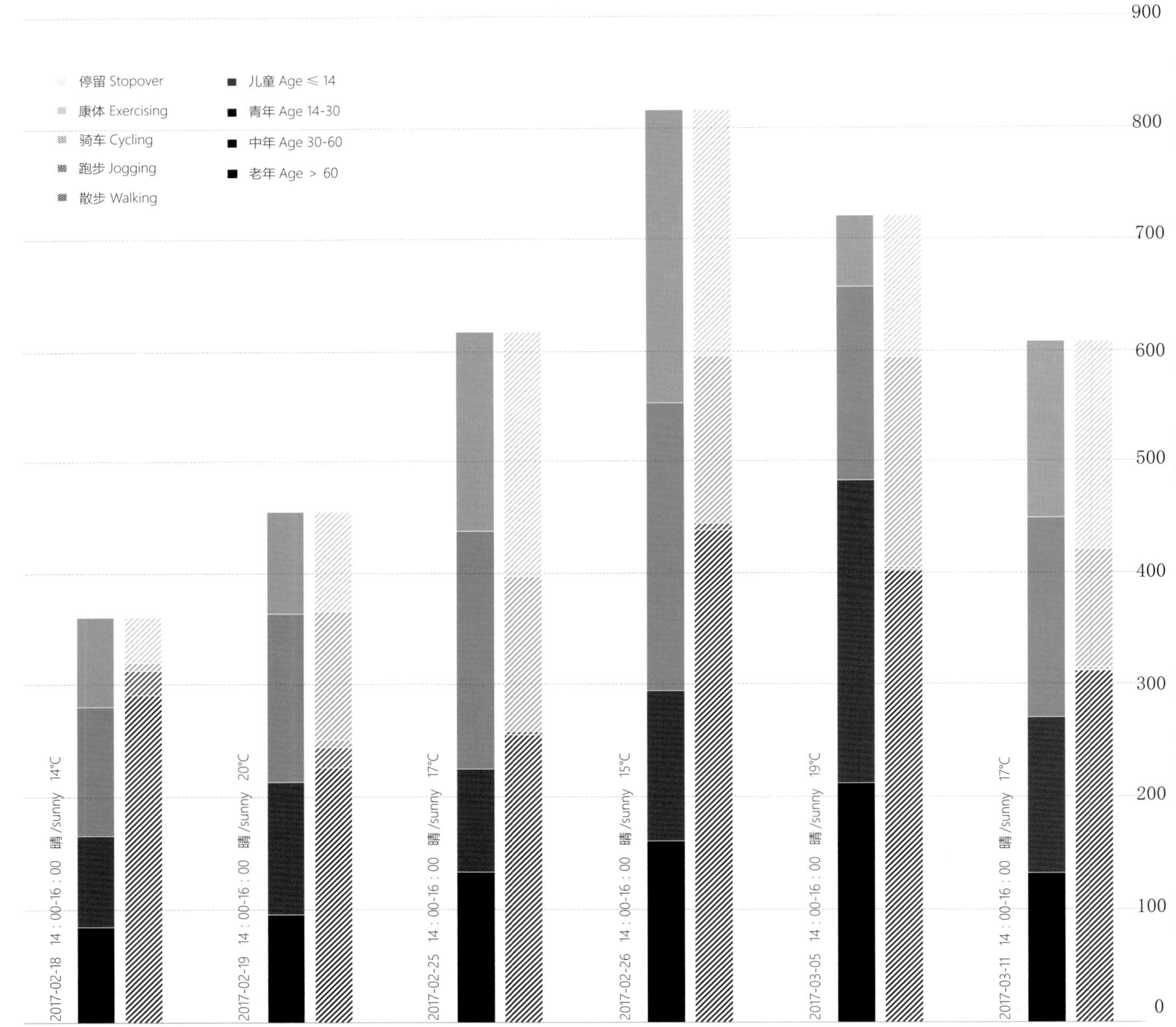

注：本书中提到的使用人群规模和使用人群密度数据均为绿道调查采集的瞬时数据。

Note：The data of number of users and density of users mentioned in the book are both instantaneous, which are collected during the survey of greenways.

单位：人

日常傍晚绿道使用人群规模在 380 人至 670 人之间变化，人群中、老年人居多，以开展散步活动为主，部分调研中也有一定数量人群开展广场舞等康体活动。

Every day at nightfall, the number of users ranges from 380 to 670; most of them are middle-aged and senior people. They mainly walk. In some of the surveys, a certain number of people also square dance or other such physical exercises.

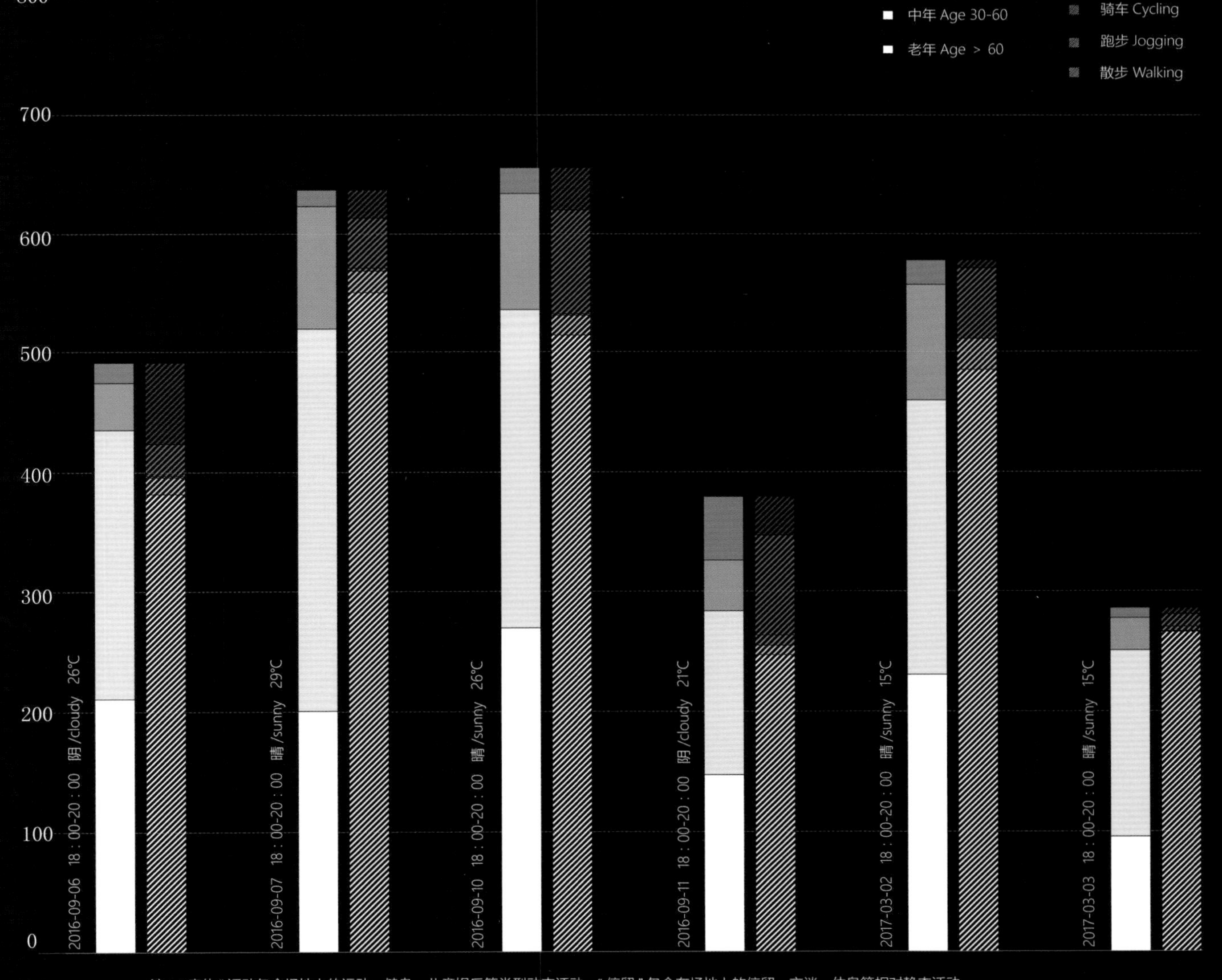

注：“康体”活动包含场地上的运动、健身、儿童娱乐等类型动态活动，“停留”包含在场地上的停留、交谈、休息等相对静态活动。

Note: “Exercising” includes types of sports on the ground, fitness, entertainment for small children and other such dynamic activities, while “staying” includes staying on the ground, chatting, resting, and other such static activities.

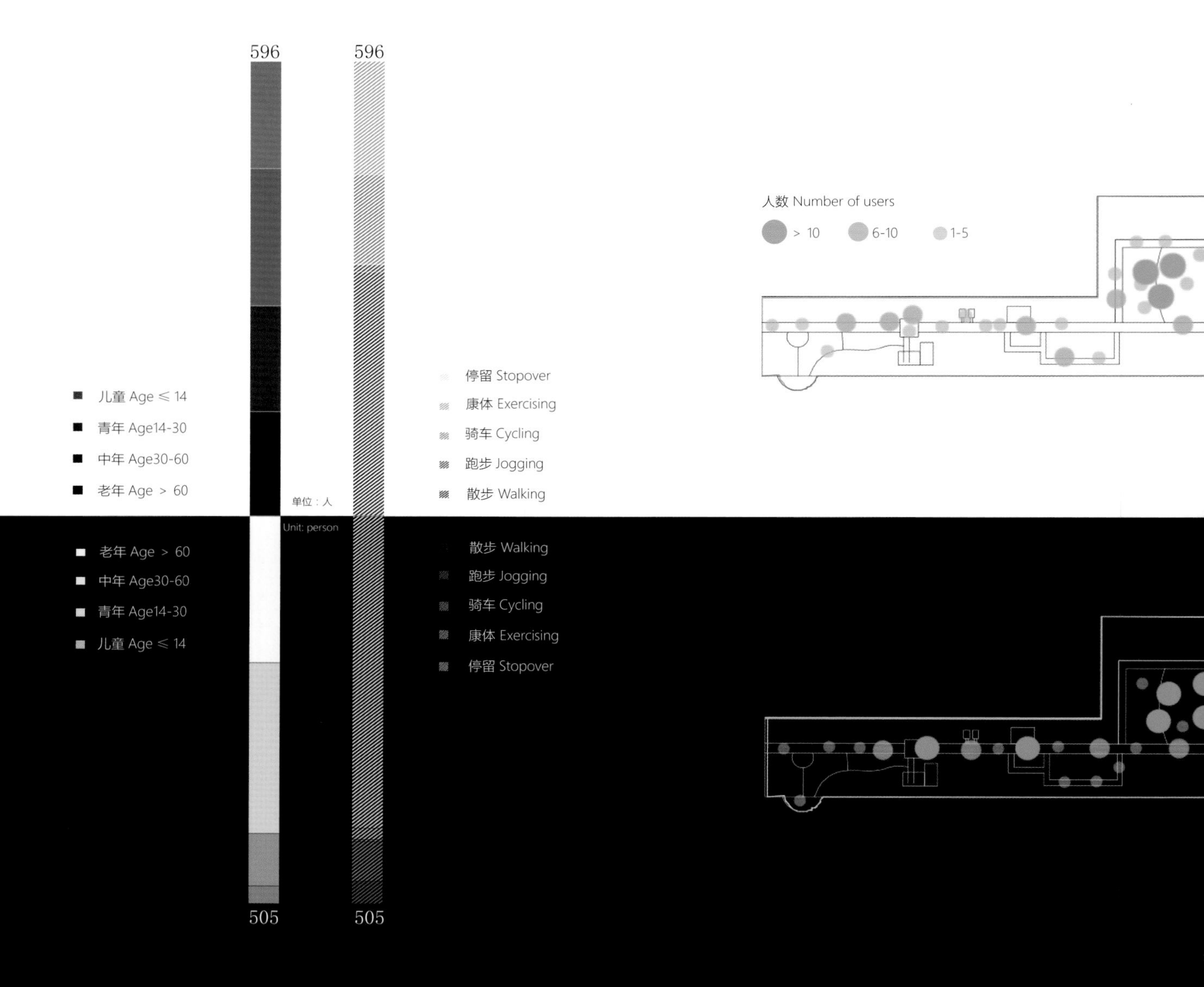

596
596
人数 Number of users
> 10
6-10
1-5
停留 Stopover
康体 Exercising
骑车 Cycling
跑步 Jogging
散步 Walking
儿童 Age ≤ 14
青年 Age14-30
中年 Age30-60
老年 Age > 60
单位：人
Unit: person
老年 Age > 60
中年 Age30-60
青年 Age14-30
儿童 Age ≤ 14
散步 Walking
跑步 Jogging
骑车 Cycling
康体 Exercising
停留 Stopover
505
505

从平均数据来看，周末下午的使用人数要略高于日常傍晚，并且年龄组成更为均衡、使用活动类型更多样，日常傍晚主要以中老年人散步为主。

Judging from the average data rate, we found that the number of users on a weekend afternoon is slightly greater than that of every day at nightfall, with a more balanced age composition, so that the types of activities are more diverse. Every day at nightfall, most users of the greenway are middle-aged and seniors who mainly walk.

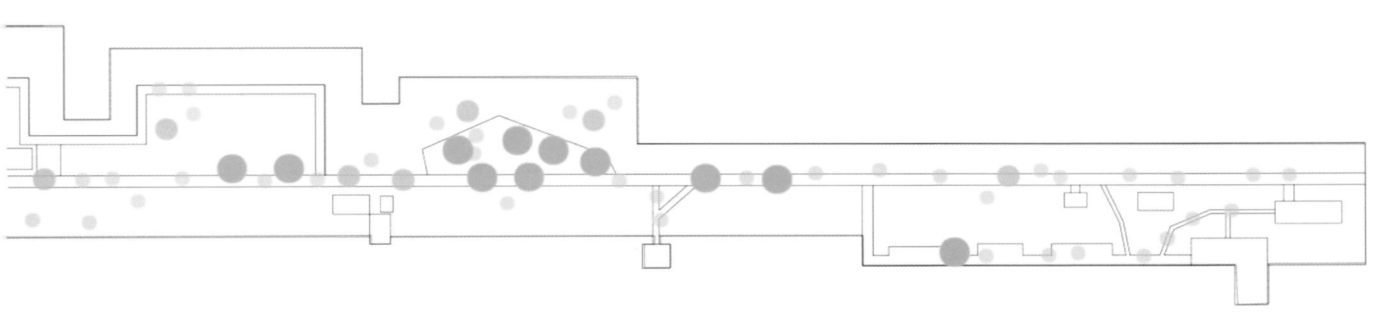

14:00-16:00 人群典型分布图
Typical distribution map of users

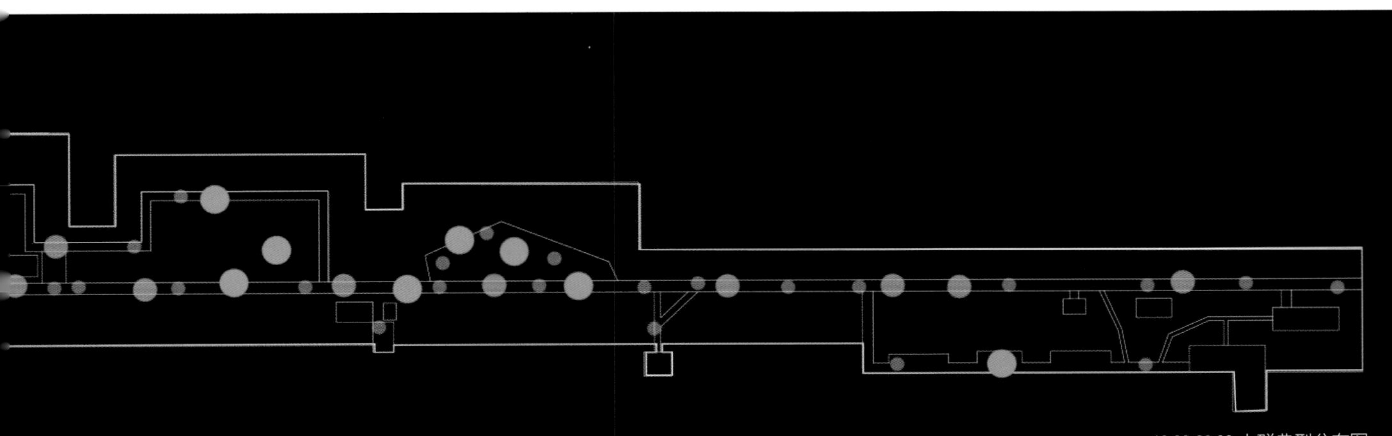

18:00-20:00 人群典型分布图
Typical distribution map of users

在人群空间分布上，该区段的几个集中场地在周末下午和平常傍晚的利用率均较高。在周末下午，人群分布更为分散，其中在绿道绿地以及滨水区域均有一定使用人群；而在日常傍晚，人群则主要聚集在游径及两侧主要空间。

In terms of spatial distribution of population, several central places on this segment have a high utilization ratio on weekend afternoons and every day at nightfall. On a weekend afternoon, user distribution in space is more dispersed. The green space and waterfront area of the greenway attract a certain number of users. Every day at nightfall, the crowd mainly gathers on the trails and in the main space on both sides.

区段 2（定淮门—草场门）实景图
Site view of S2（DHG–CCG）

区段 2：定淮门—草场门

该区段周末下午绿道使用人群规模大部分在75人至160人之间变化，仅有一次使用人群规模明显较大并接近300人。该时段使用人群年龄结构较均匀，以中老年人为主，使用活动类型多为散步。

Segment 2: Dinghuai Gate-Caochang Gate

On a weekend afternoon, the number of people using this segment ranges from 75 to 160, and the number neared 300 only once. The age structure of the crowd is relatively even, most of who are middle-aged and senior people. They mainly walk.

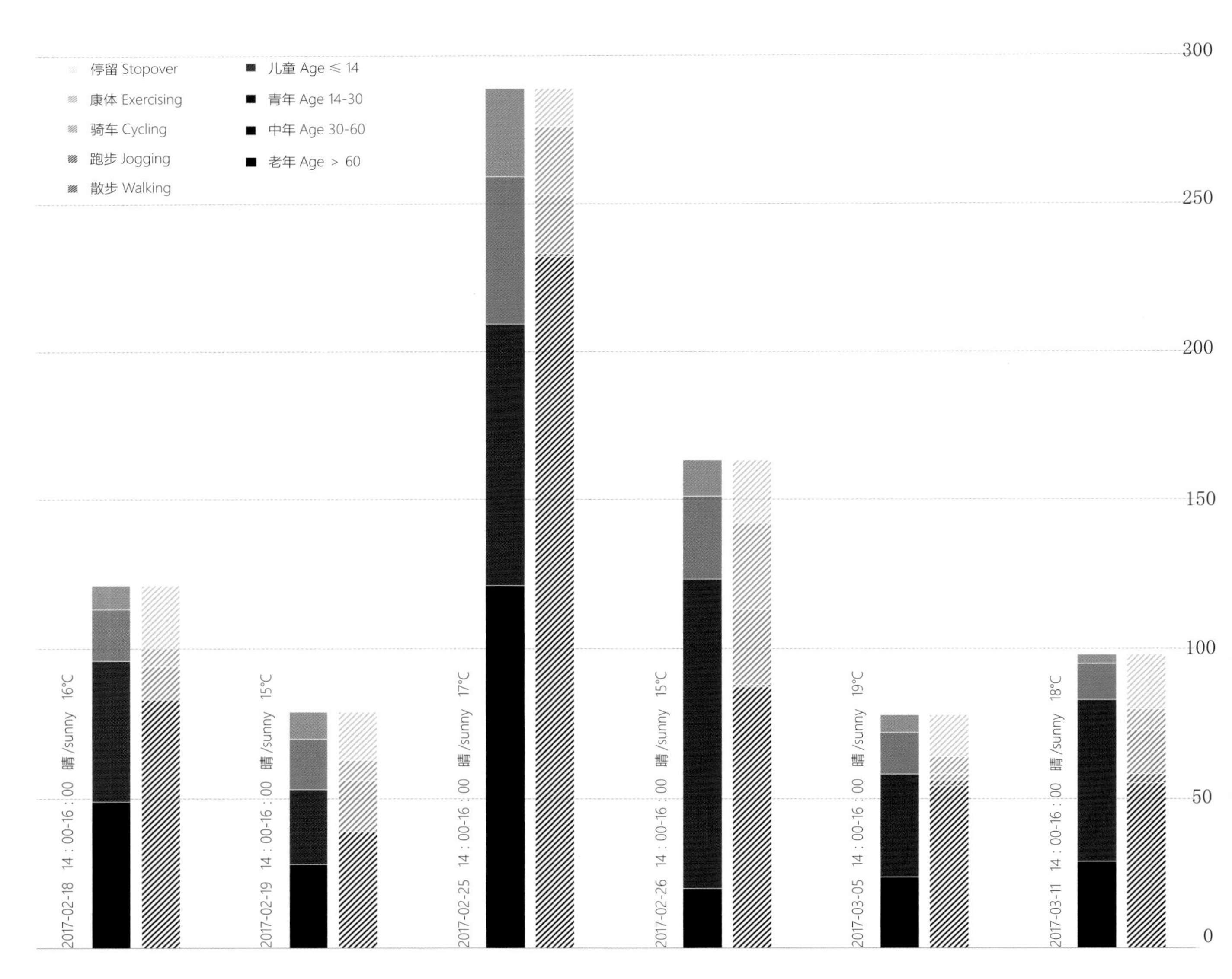

日常傍晚除了一次调查的使用人群规模超过 150 人外，其余 5 次使用人群规模均在 100 人以下。该时段使用人群以中老年人为主，使用活动类型以散步为主。

Every day at nightfall, except once when the number of greenway users exceeded 150 during the survey, that number goes below 100 for the other five time periods. During this time period, most users are middle-aged and senior people who mainly walk.

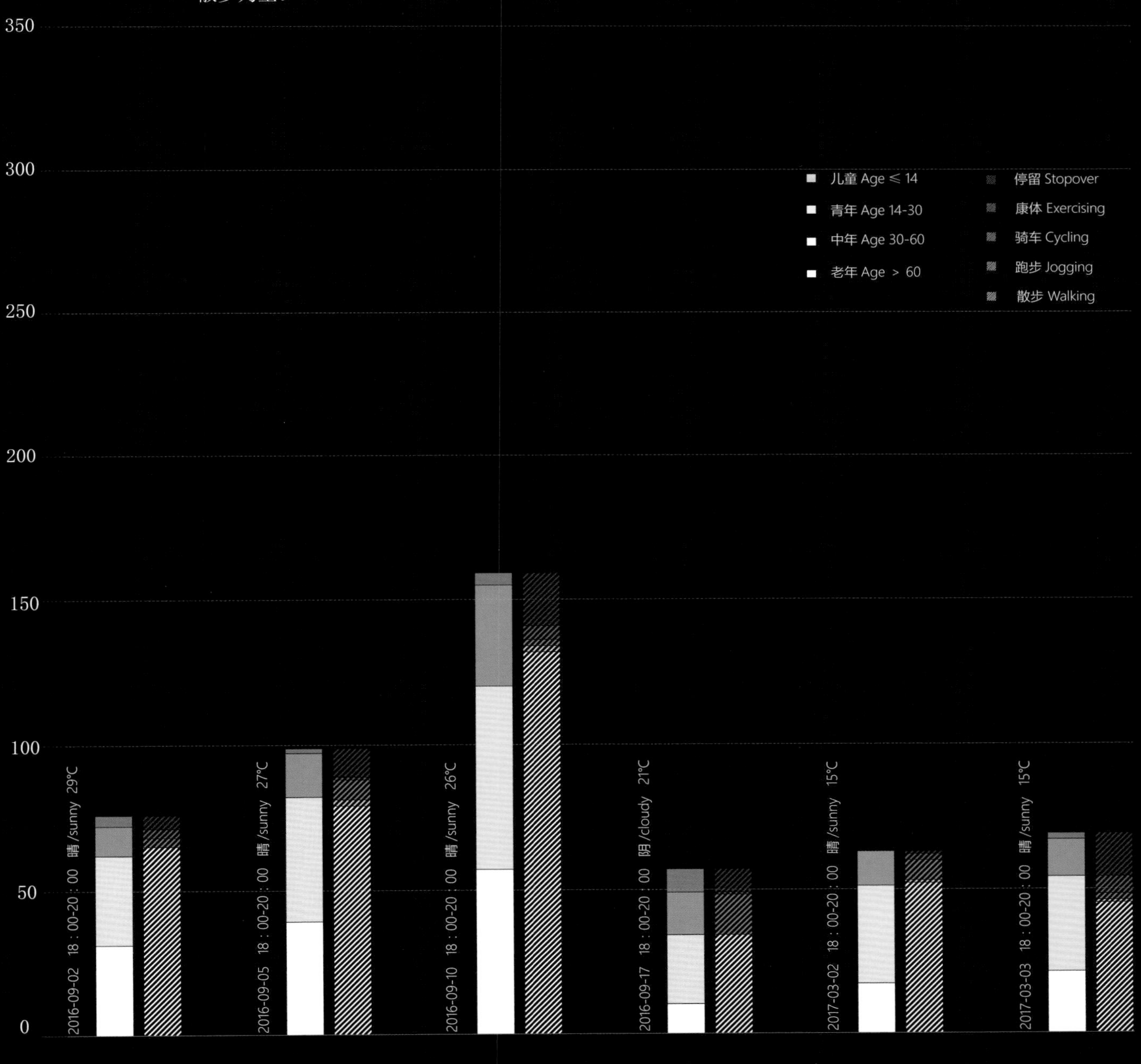

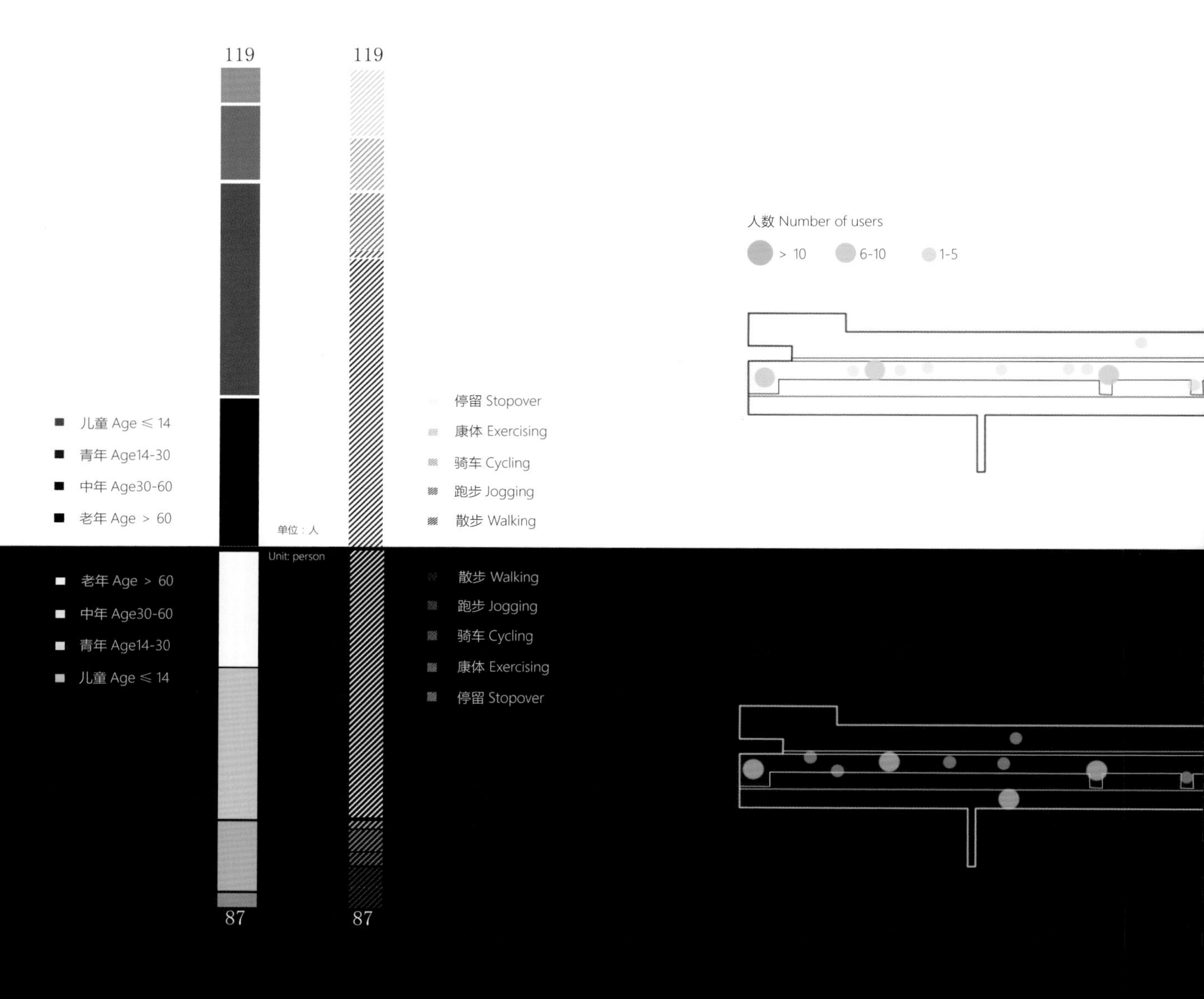

119
119
人数 Number of users
> 10
6-10
1-5
停留 Stopover
康体 Exercising
骑车 Cycling
跑步 Jogging
散步 Walking
儿童 Age ≤ 14
青年 Age14-30
中年 Age30-60
老年 Age > 60
单位：人
Unit: person
老年 Age > 60
中年 Age30-60
青年 Age14-30
儿童 Age ≤ 14
散步 Walking
跑步 Jogging
骑车 Cycling
康体 Exercising
停留 Stopover
87
87

审视平均数据，周末下午的使用人群规模要小于日常傍晚，但人群年龄结构类似，中老年使用群体占据绝对主导，因此呈现的使用活动类型结构也以散步为主。

Looking at the average data, we've found out that the number of greenway users on a weekend afternoon is smaller than every day at nightfall, with a similar age structure. Middle-aged and senior people dominate the crowd, so the structure of the types of activities mainly involves walking.

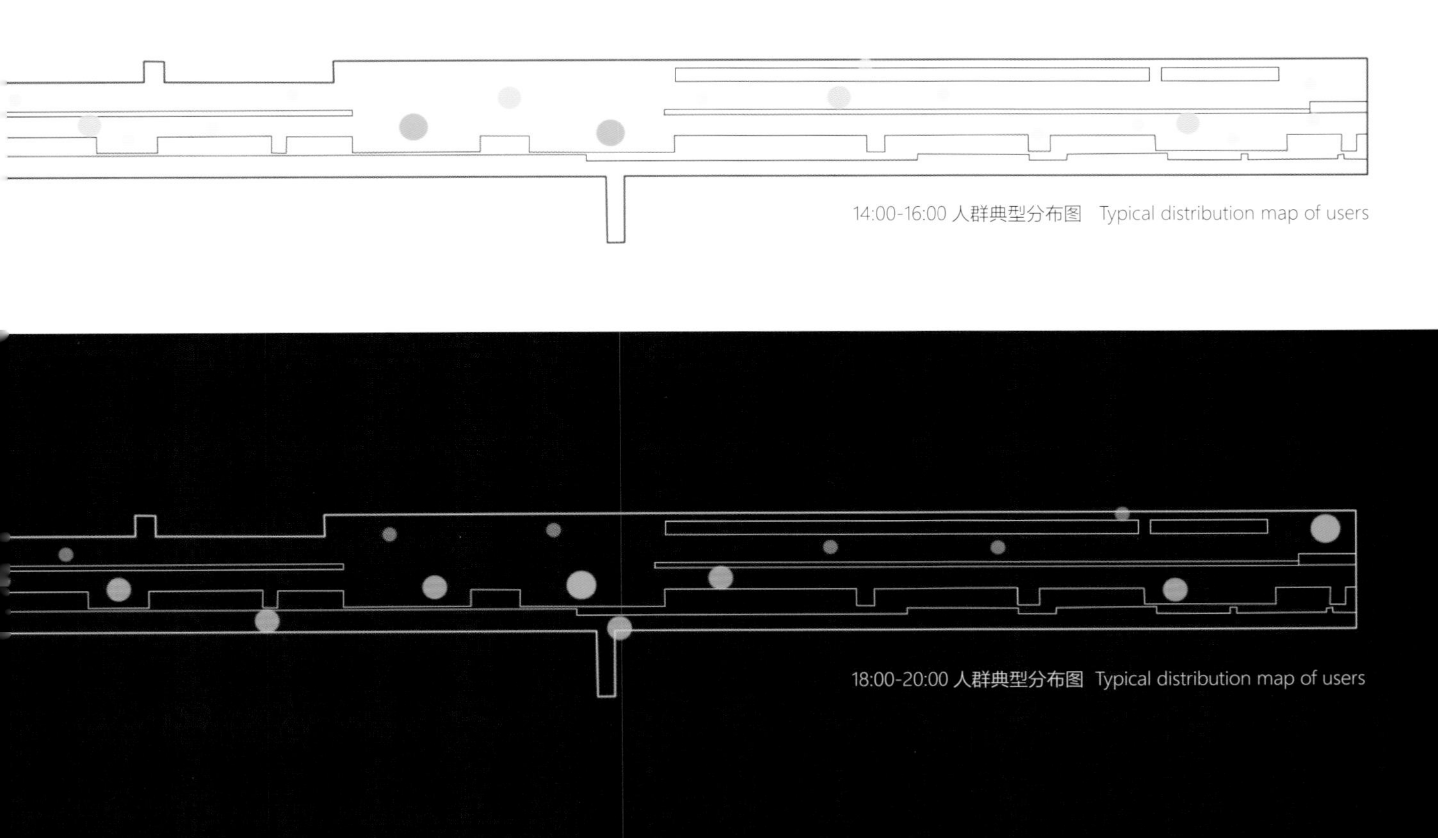

14:00-16:00 人群典型分布图　Typical distribution map of users

18:00-20:00 人群典型分布图　Typical distribution map of users

在人群空间分布上，无论周末下午还是日常傍晚，该区段的人群分布都较为分散，由于使用活动类型以散步为主，人群主要分布在游径上，较少有人群的聚集。

In terms of spatial distribution of the users, whether it is in the afternoon of a weekend or every day at nightfall, the user distribution of this segment is relatively disperse. Since the dominant activity here is walking, the crowd is mainly on the trails rather than gathering together in one place.

区段 3（草场门—清凉门）实景图
Site view of S3 (CCG–QLG)

区段 3：草场门—清凉门

该区段周末下午使用人群规模较大，在 590 人至 1000 人之间。该时段使用人群年龄结构较均匀，中老年略高于青幼年，活动类型也较丰富，进行散步、康体和停留休息的使用人群分布也较均衡。

Segment 3: Caochang Gate-Qingliang Gate

On the afternoon of a weekend, the number of greenway users is relatively big, with the data ranging from 590 to 1,000. During this time period, the age structure is even, with the number of middle-aged and senior people being slightly larger than that of young people and small children. The types of activities are richer, and the distribution of different groups of people who walk, exercise, or stop over are relatively balanced.

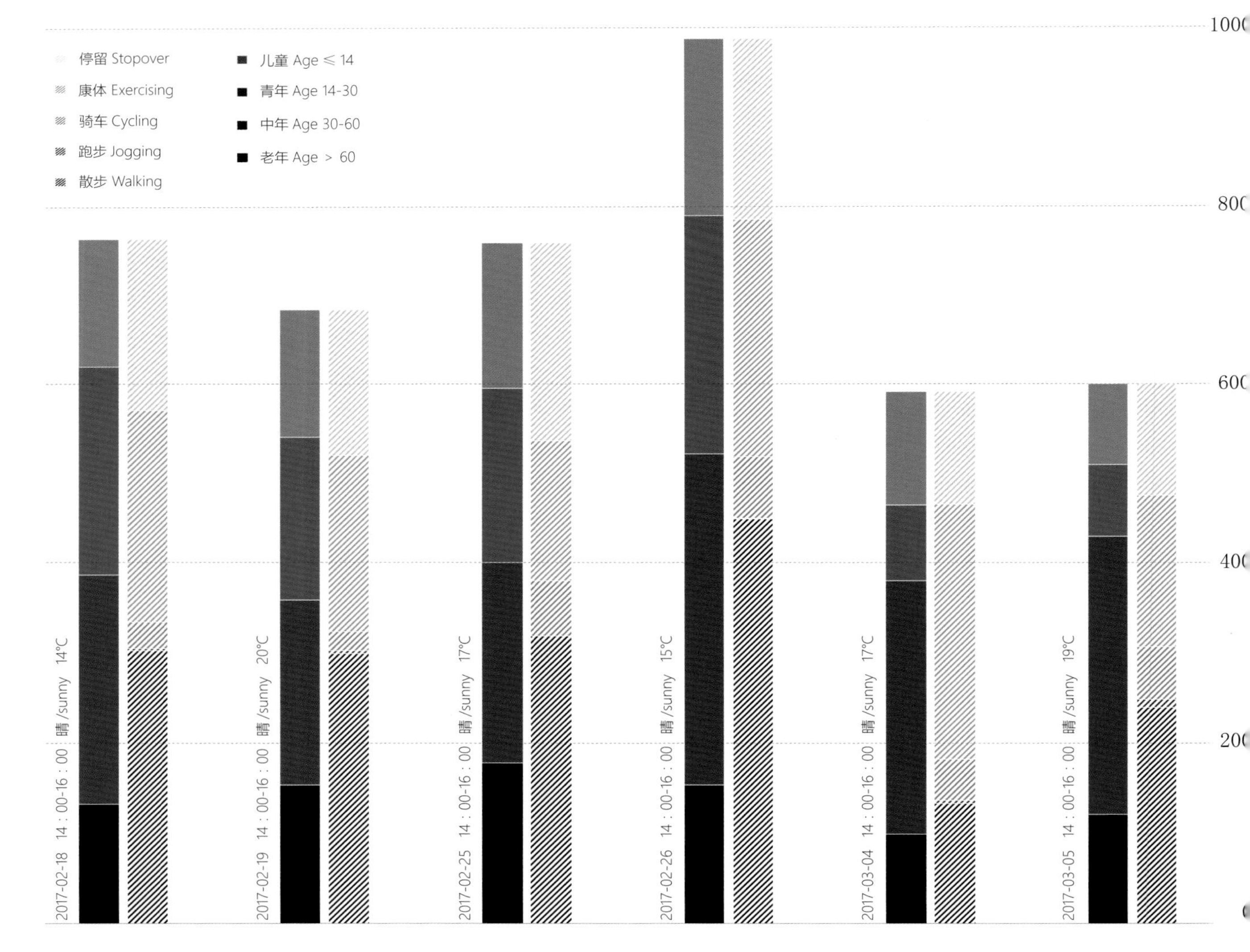

日常傍晚使用人群规模下降明显，在 160 人至 430 人之间。人群以中老年人为主，使用活动类型以散步为主，在调查中偶然也能碰到一定数量的人群进行骑车、康体等活动，但占比不高。

Every day at nightfall, the number of users suffers a noticeable decrease, with the data ranging from 160 to 430. Most of the people are middle-aged and senior people who mainly walk. During the surveys, we often came across a certain number of people who go bike riding and exercise, but the proportion is not large.

1200

1000

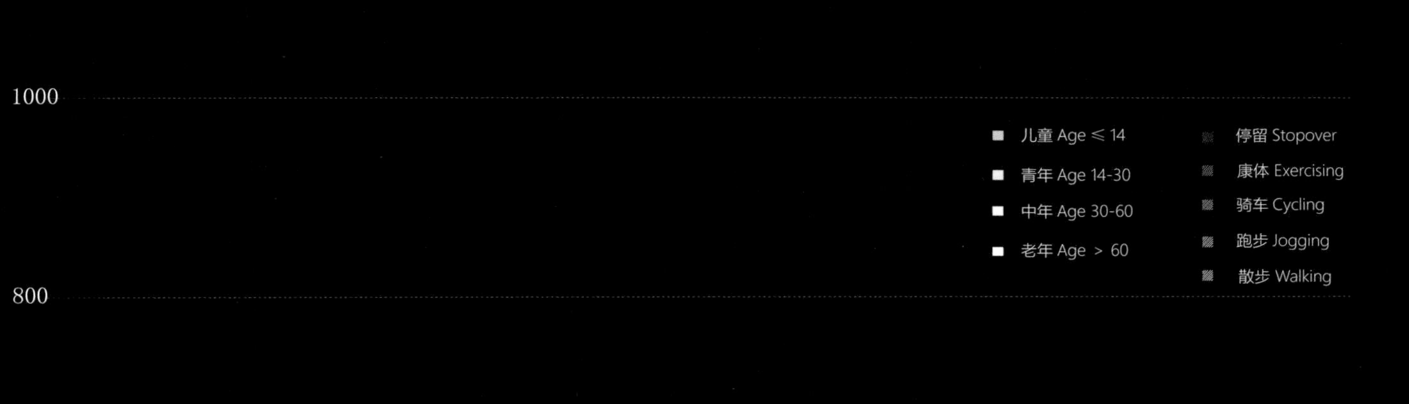

800

600

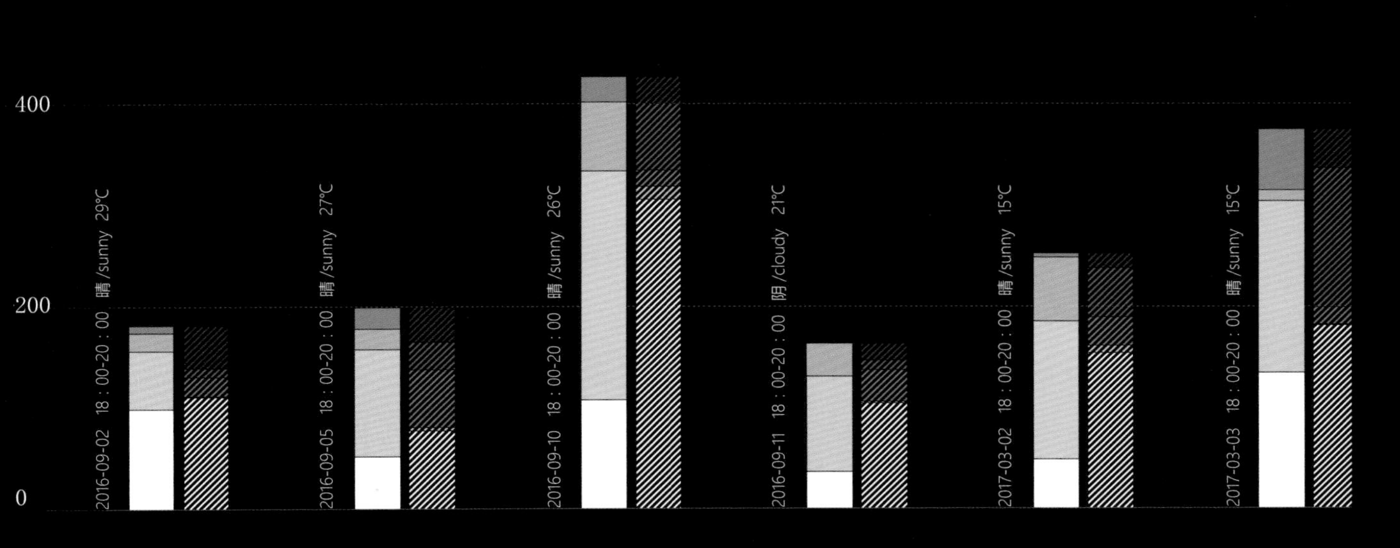

Unit: person

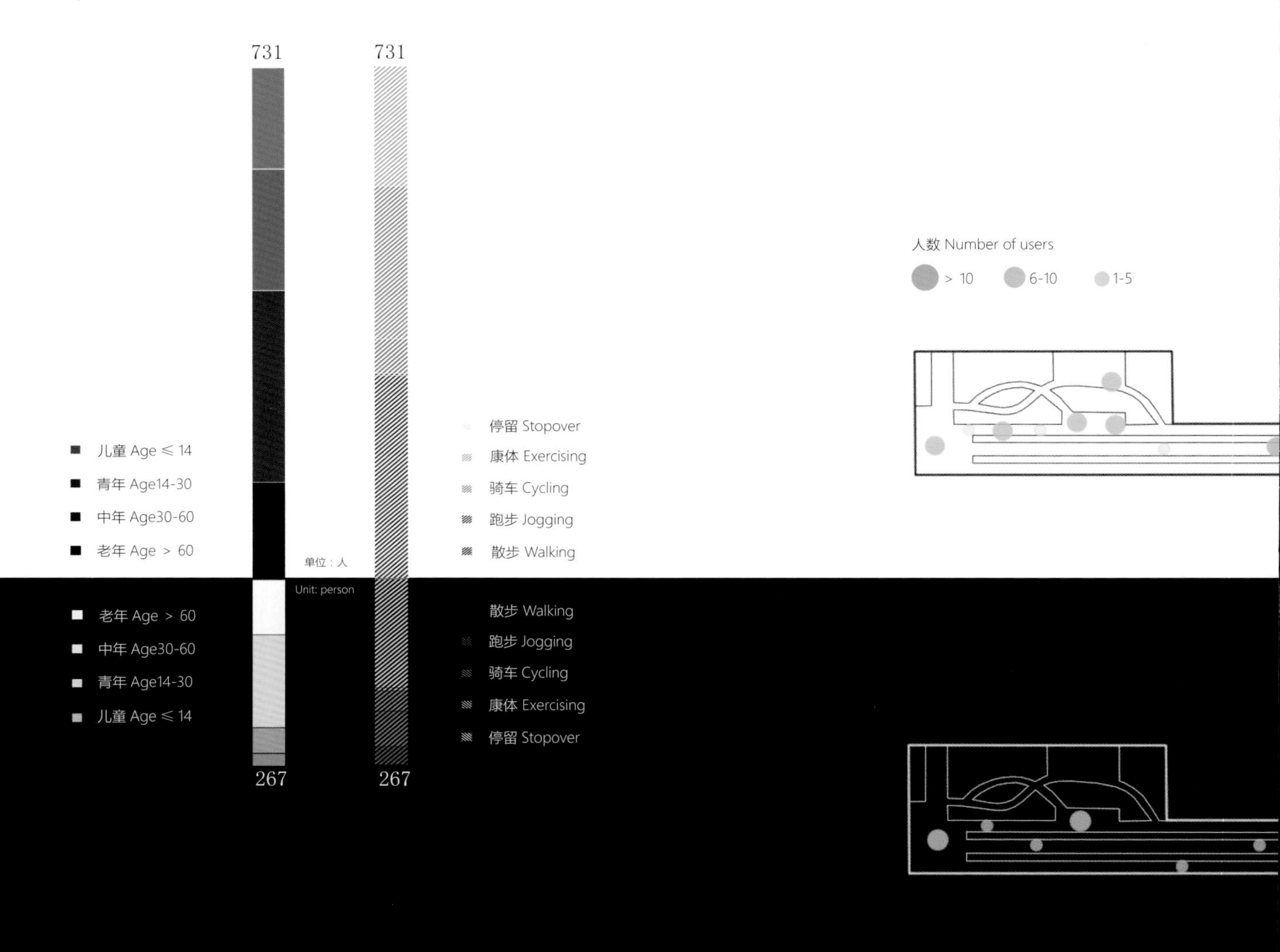
731
731
267
267
单位：人
Unit: person
儿童 Age ≤ 14
青年 Age14-30
中年 Age30-60
老年 Age ＞ 60
老年 Age ＞ 60
中年 Age30-60
青年 Age14-30
儿童 Age ≤ 14
停留 Stopover
康体 Exercising
骑车 Cycling
跑步 Jogging
散步 Walking
散步 Walking
跑步 Jogging
骑车 Cycling
康体 Exercising
停留 Stopover
人数 Number of users
＞ 10
6-10
1-5

平均数据显示，周末下午使用人群的总体规模约是日常傍晚的近 3 倍，并且年龄构成更为均衡，使用活动类型也更多样，有较多使用人群开展康体与停留休息活动，日常傍晚则以中老年人散步为主。

According to average data, the total number of users on a weekend afternoon is nearly three times that of every day at nightfall, the age composition is more balanced, and the types of activities are more diverse. Many people exercise and stop over, while middle-aged and senior people walk every day at nightfall.

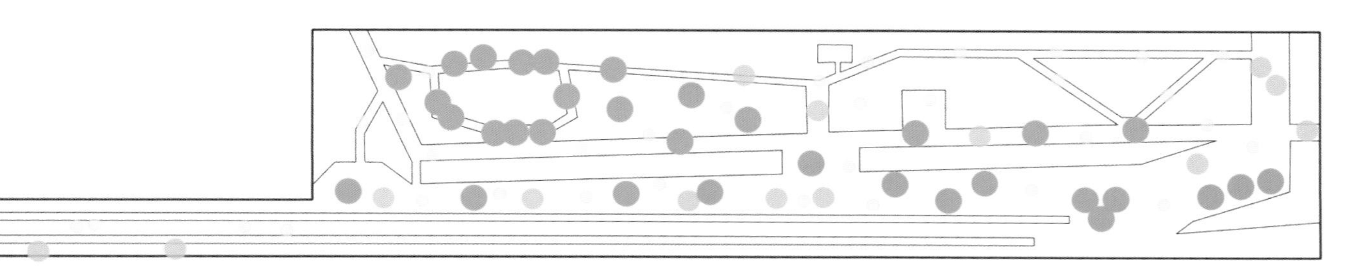

14:00-16:00 人群典型分布图
Typical distribution map of users

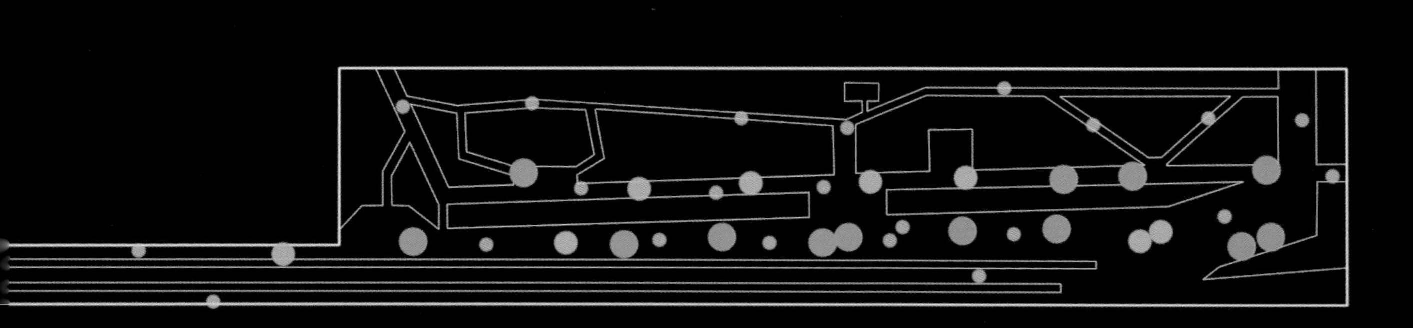

18:00-20:00 人群典型分布图
Typical distribution map of users

在人群空间分布上，该区段在周末下午和日常傍晚的人群分布都呈现出南多北少的现象。周末下午几个集中硬质场地的利用率较高，其中“鬼脸照镜”池边场地是人群最主要的聚集点；而日常傍晚人群则主要集中在游径上。

In terms of spatial distribution of users, the same condition of utilization on a weekend afternoon and every day at nightfall is that the number of people using the segment on the southern part is larger than that on the northern part. On the afternoon of a weekend, the utilization ratios of some large open sites are higher, among which the pool side of “Funny Face Mirrors” is the major rallying point, while the crowd often goes to the trails every day at nightfall.

区段 4（水西门—集庆门）实景图
Site view of S4（SXG–JQG）

区段 4：水西门—集庆门

该区段周末下午使用人群规模总体较小，仅在50人至90人之间，使用人群以中老年人为主，活动以散步为主，也有部分人群开展钓鱼、健身等康体活动。

Segment 4: Shuixi Gate-Jiqing Gate

On the afternoon of a weekend, the number of greenway users of the greenway is relatively small, with the data ranging from 50 to 90. Most users are middle-aged and senior people who mainly walk. Some of them also fish and exercise.

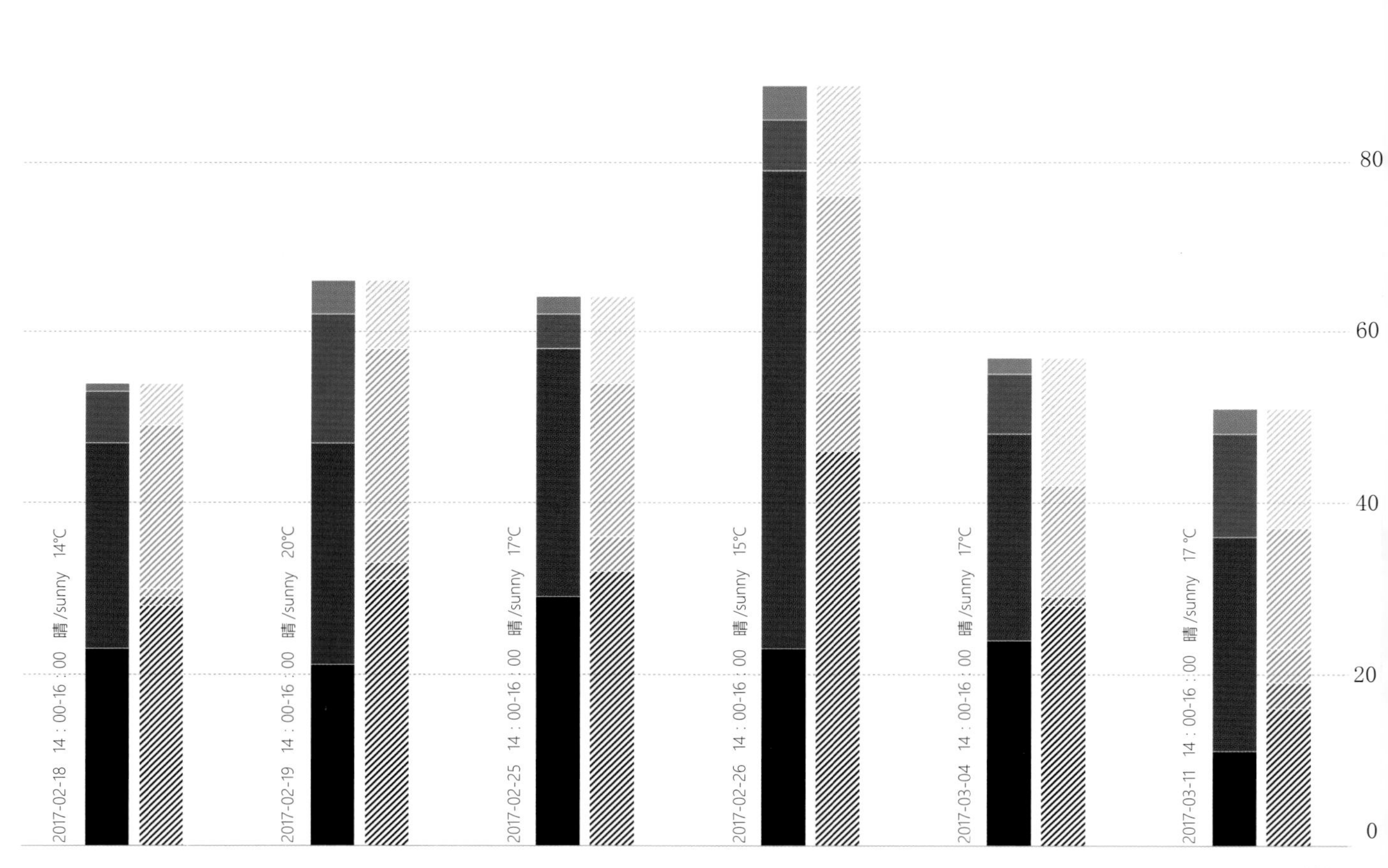

日常傍晚使用人群规模在 55 人至 120 人之间变化，略高于周末下午，但总体规模仍然较小，仍旧以中老年人散步为主。

Every day at nightfall, the number of users ranges from 55 to 120, slightly higher than that on weekend afternoons, but the total number remains relatively small. Most users are middle-aged and senior people who mainly walk.

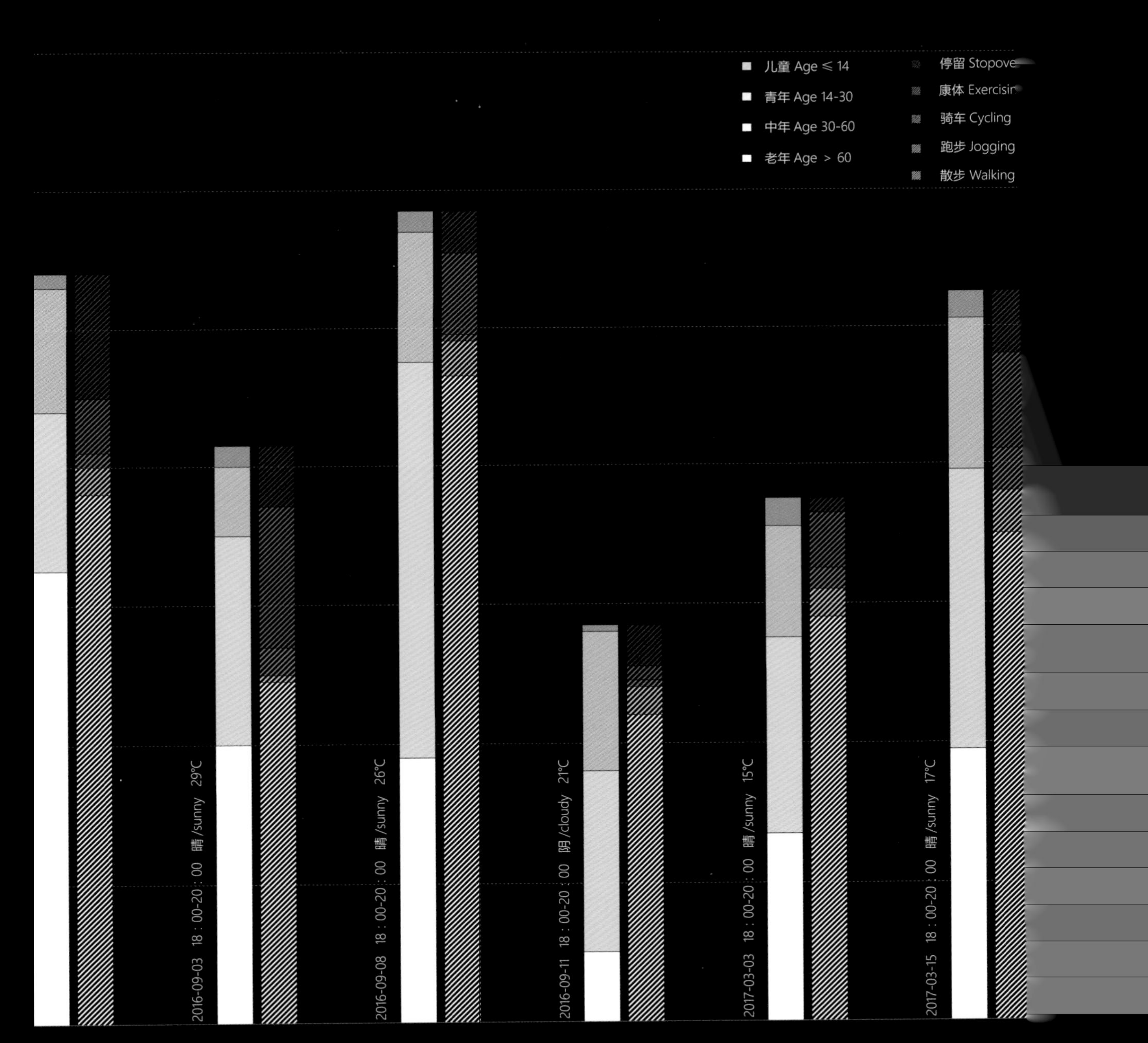

on

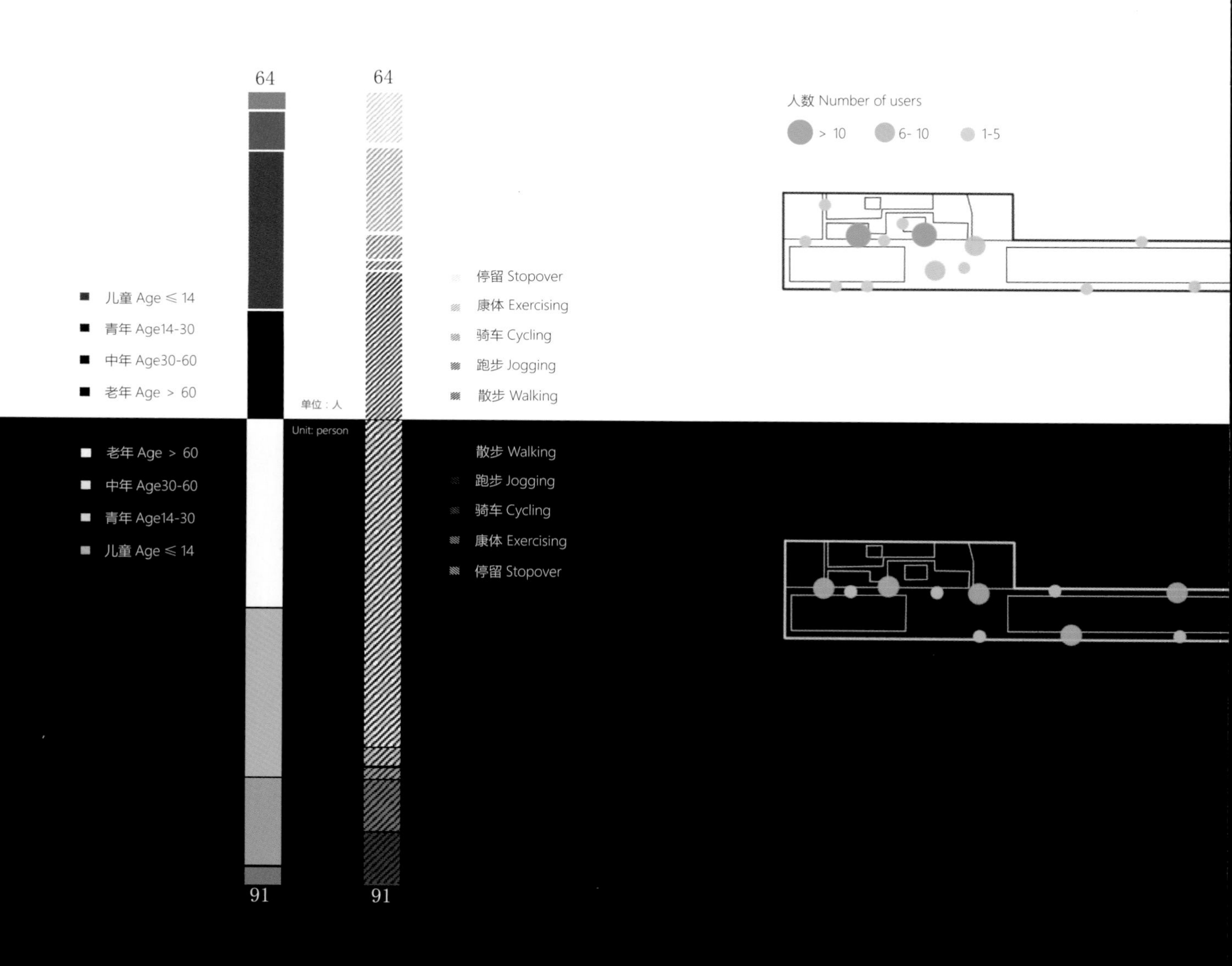
64
64
人数 Number of users
> 10
6- 10
1-5
停留 Stopover
康体 Exercising
骑车 Cycling
跑步 Jogging
散步 Walking
儿童 Age ≤ 14
青年 Age14-30
中年 Age30-60
老年 Age > 60
单位：人
Unit: person
老年 Age > 60
中年 Age30-60
青年 Age14-30
儿童 Age ≤ 14
散步 Walking
跑步 Jogging
骑车 Cycling
康体 Exercising
停留 Stopover
91
91

日常傍晚的总体使用人群规模要略高于周末下午，但两个时段使用人群年龄结构类似，均以中老年人为主。周末下午使用活动类型相对更多样，约有 45% 的人群开展康体与停留休息活动；而在日常傍晚，散步则成为绝对的主导活动类型。

Every day at nightfall, the number of users is slightly higher than that on weekend afternoons. However, the age structures during both time periods are similar in that most users are middle-aged and senior people. The types of activities are more diverse on weekend afternoons and about 45% of users exercise and stop over. Every day at nightfall, walking is absolutely the dominant activity.

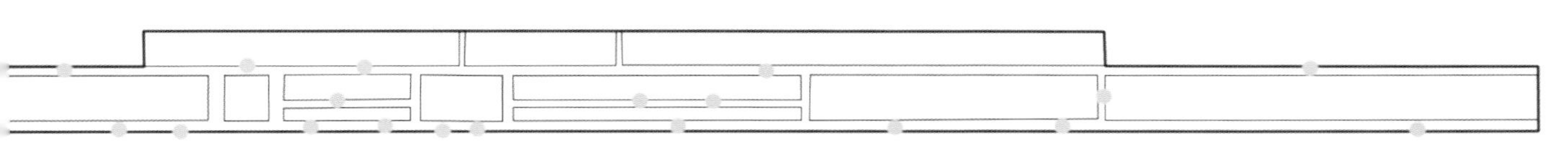

14:00-16:00 人群典型分布图
Typical distribution map of users

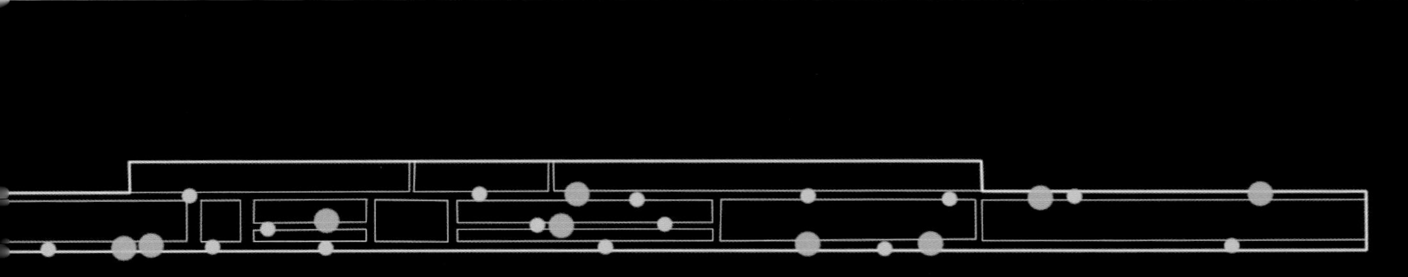

18:00-20:00 人群典型分布图
Typical distribution map of users

在人群空间分布上，周末下午的人群除了在绿道最宽处的场地上有所聚集外，基本沿该区段均匀分布，在滨河一侧会有使用人群停留钓鱼、健身等。在日常傍晚，使用人群活动类型则以散步为主，人群分布相对分散。

In terms of spatial distribution of users, on weekend afternoons, in addition to gathering around the widest place on the greenway, the crowd is basically evenly distributed along the segment. Some people stay on one side of the waterfront to fish and exercise. Every day at nightfall, those who use the greenway mainly walk, and user distribution in space is relatively dispersed.

区段 5（凤台路—中华门）实景图
Site view of S5 (FTR–ZHG)

区段 5：凤台路—中华门

调查结果显示，周末下午使用人群规模总体变化不大，在 170 人至 240 人之间变化，人群以中老年人为主，使用活动类型相对多样均衡，有一定规模的人群进行散步、康体和停留休息活动。

Segment 5: Fengtai Road-Zhonghua Gate

According to the result of the survey, the number of greenway users the segment on a weekend afternoon does not vary too much, with the data ranging from 170 to 240, and most are middle-aged and senior people. The types of activities are relatively diverse and balanced. A great many people walk, do exercise, or stop over.

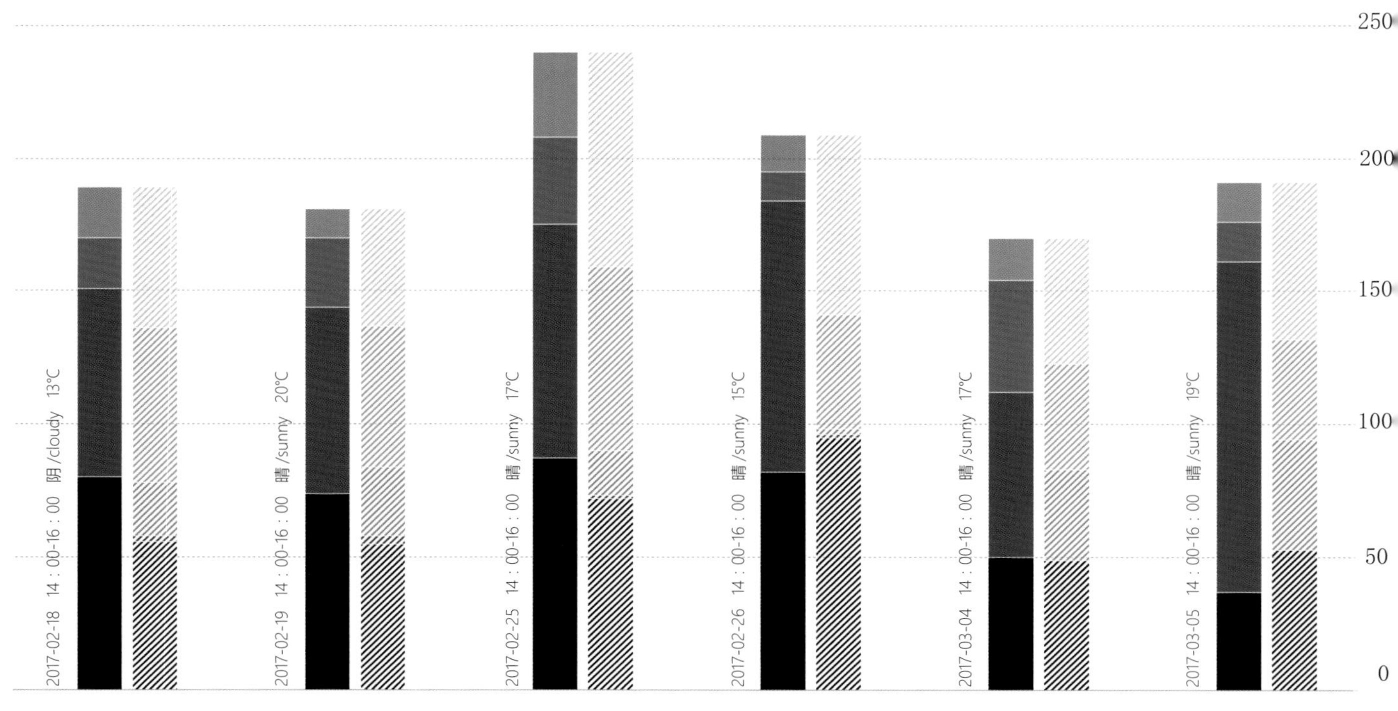

每日傍晚使用人群规模波动较大，在 185 人至 385 人之间变化，使用人群以中老年人为主，青年人也占据一定比重，总体使用活动类型以散步为主。

Every day at nightfall, the number of users fluctuates greatly, with the data ranging from 185 to 385. Most are middle-aged and senior people, but young people also hold a definite proportion. The dominant type of activity during this time period is walking.

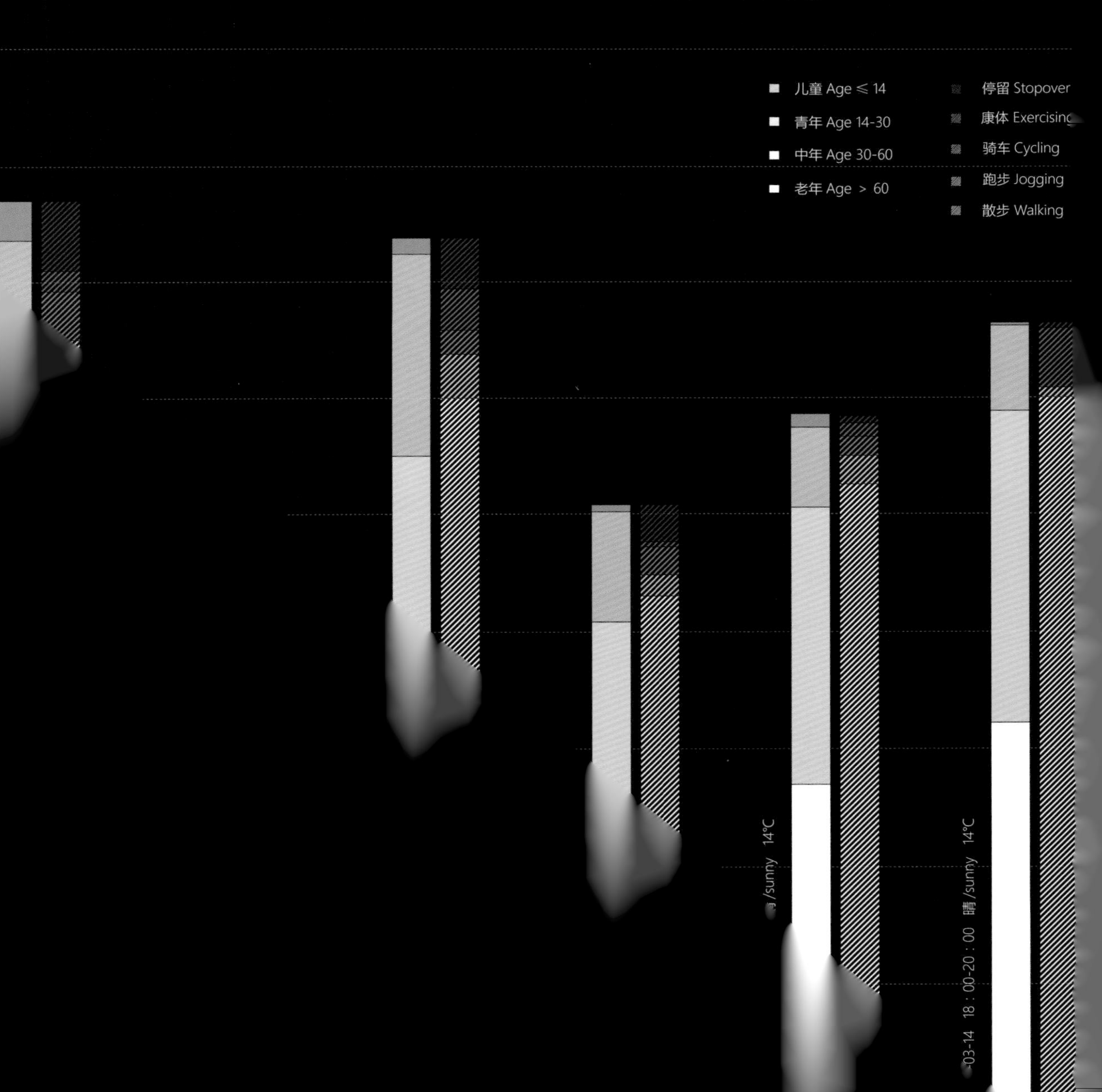

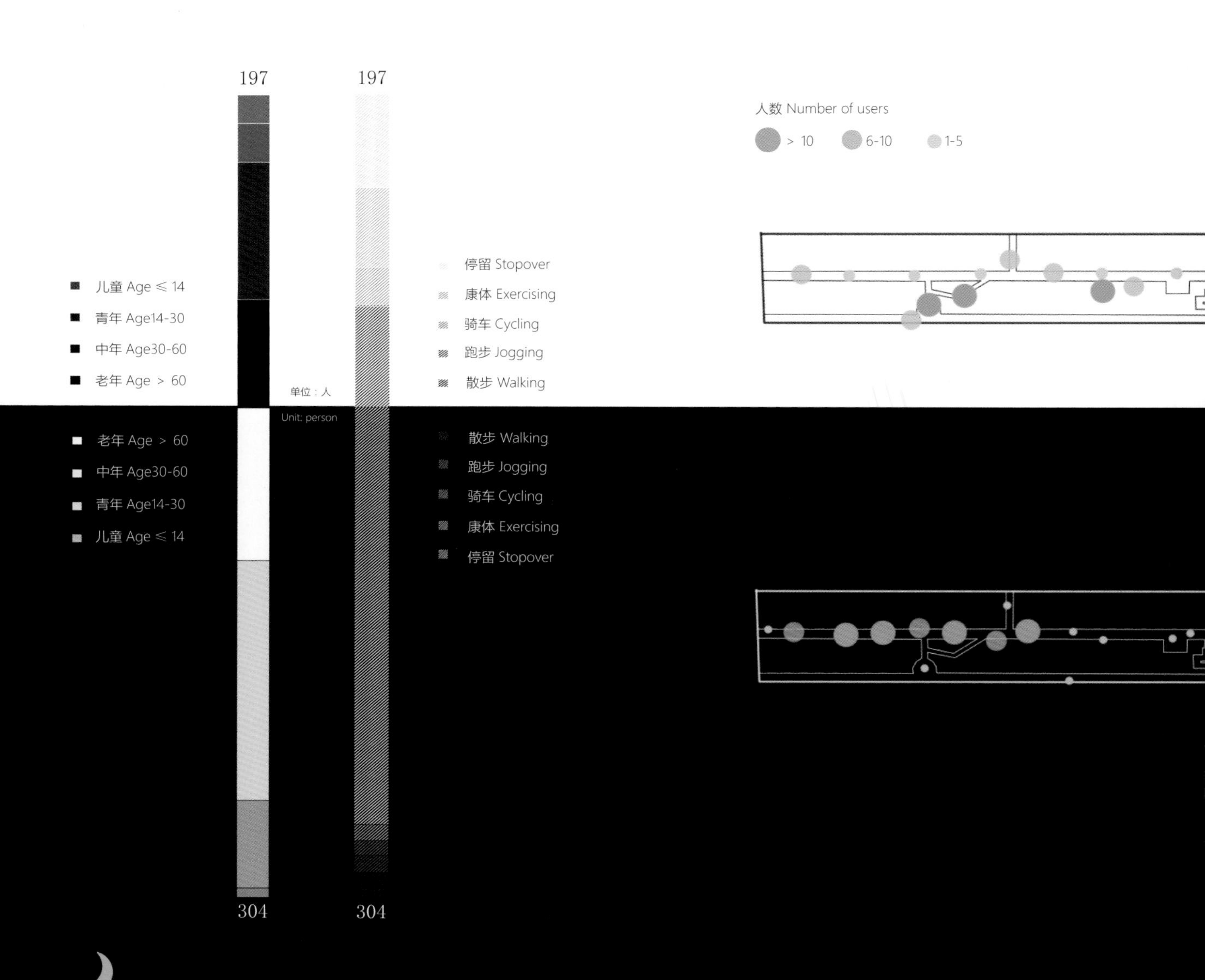
197
197
人数 Number of users
> 10
6-10
1-5
儿童 Age ≤ 14
青年 Age14-30
中年 Age30-60
老年 Age > 60
停留 Stopover
康体 Exercising
骑车 Cycling
跑步 Jogging
散步 Walking
单位：人
Unit: person
老年 Age > 60
中年 Age30-60
青年 Age14-30
儿童 Age ≤ 14
散步 Walking
跑步 Jogging
骑车 Cycling
康体 Exercising
停留 Stopover
304
304

从平均数据来看，日常傍晚的使用人群规模要略大于周末下午，人群年龄组成均以中老年人为主，日常傍晚人群的使用活动类型多为散步，而周末下午的使用活动类型更多样，进行散步、康体、停留活动的使用人群占比基本相当。

According to average data, the number of users every day at nightfall is slightly higher than that on weekend afternoons. Middle-aged and senior people dominate the age composition. Every day at nightfall, the main activity here is walking, while that on a weekend afternoon is rather diverse. The numbers of people who walk, exercise, and stop over are relatively the same in proportion.

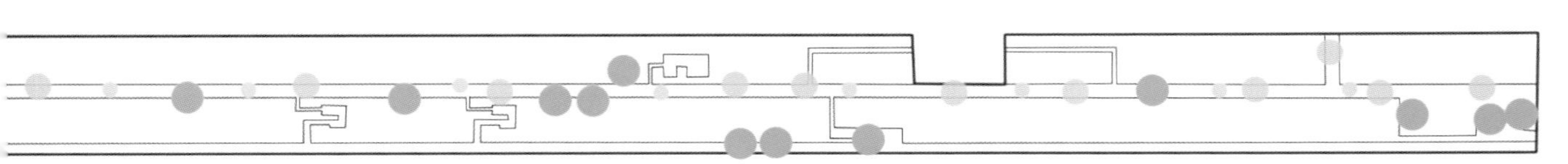

14:00-16:00 人群典型分布图
Typical distribution map of users

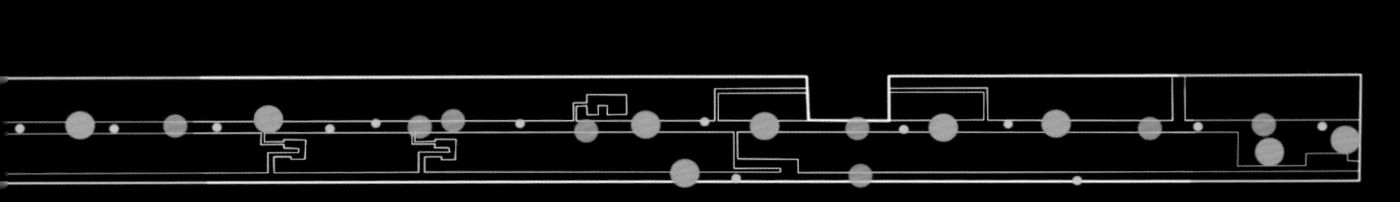

18:00-20:00 人群典型分布图
Typical distribution map of users

在人群空间分布上，该区段的人群分布在两个时段均较分散，人群主要集中在主游径及两侧集中硬质场地上，滨水一侧步行道上分布的人群要远低于主游径。

In terms of spatial distribution of population, user distribution on this segment is rather dispersed during both time periods. Most people gather around on the trails and open yards around. The number of people distributed along the footpaths at the waterfront is much smaller than that on the main trails.

区段 6（中华门—雨花门）实景图
Site view of S6（ZHG–YHG）

区段 6：中华门—雨花门

该区段周末下午使用人群规模在 230 人至 380 人之间，整体波动不大。使用人群以中老年人为主，开展散步和停留休息活动较多，开展康体活动的人群也有一定规模。

Segment 6: Zhonghua Gate-Yuhua Gate

On the afternoon of a weekend, the number of greenway users ranges from 230 to 380, without too much fluctuation. Most users are middle-aged and senior people who mostly walk or stop over. The number of people who exercise is also big.

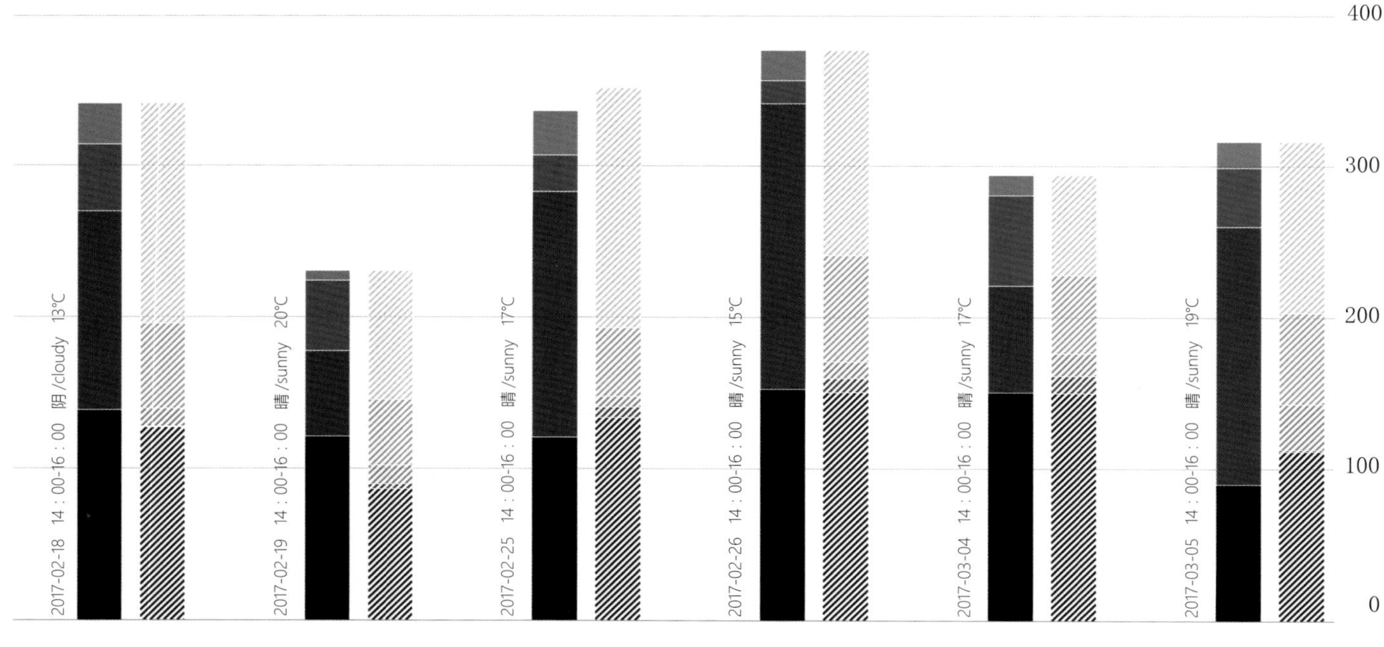

日常傍晚使用人群规模变化波动较大，除了一次达到约 750 人外，其余 5 次则在 195 人至 450 人之间变化。使用人群以中老年人为主，活动类型以散步为主，有时停留休息的人群也会达到一定规模。

Every day at nightfall, the number of users fluctuates greatly. Though that number reached about 750 once, the number ranges from 195 to 450 for the other five times. Most users are middle-aged and senior people who mainly walk. The number of people who stop over is also big.

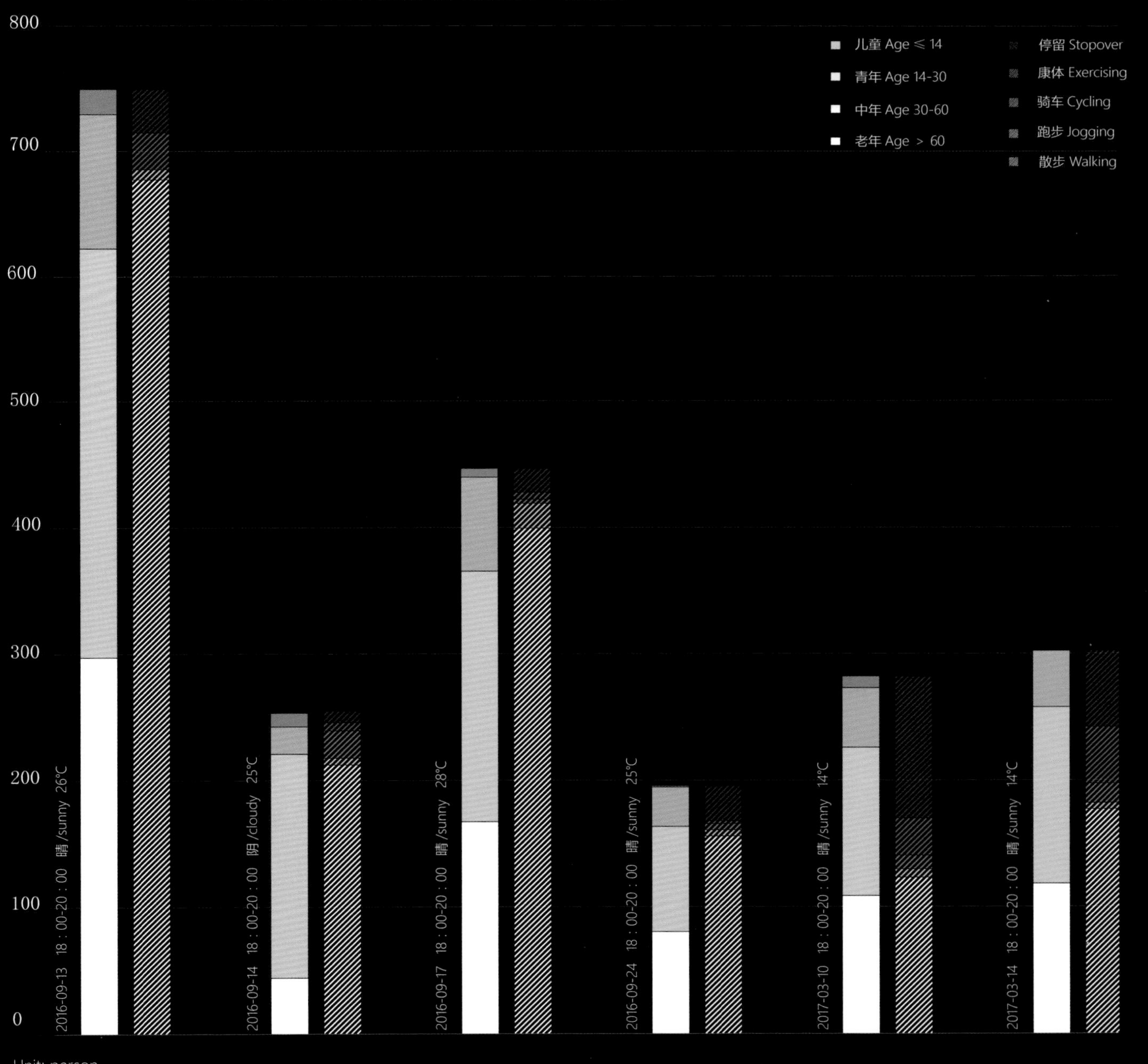

Unit: person

316
316
人数 Number of users
> 10
6-10
1-5
儿童 Age ≤ 14
青年 Age14-30
中年 Age30-60
老年 Age > 60
停留 Stopover
康体 Exercising
骑车 Cycling
跑步 Jogging
散步 Walking
单位：人
Unit: person
老年 Age > 60
中年 Age30-60
青年 Age14-30
儿童 Age ≤ 14
散步 Walking
跑步 Jogging
骑车 Cycling
康体 Exercising
停留 Stopover
372
372

平均数据显示，周末下午的使用人群规模与日常傍晚较接近，年龄结构也类似，但周末下午的使用人群活动类型更丰富，有较多中老年人停留休息、打牌下棋和围观；而日常傍晚主要以中老年人散步为主。

According to average data, the number of users on a weekend afternoon is close to that of every day at nightfall, with similar age structure. However, the types of activities on weekend afternoons are richer. More middle-aged and senior people like stopover, playing cards and Chinese chess or just watching, while most users every day at nightfall are middle-aged and senior people who mainly walk.

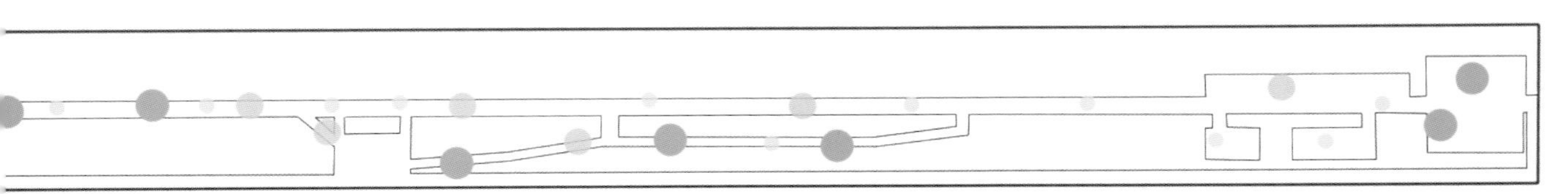

14:00-16:00 人群典型分布图
Typical distribution map of users

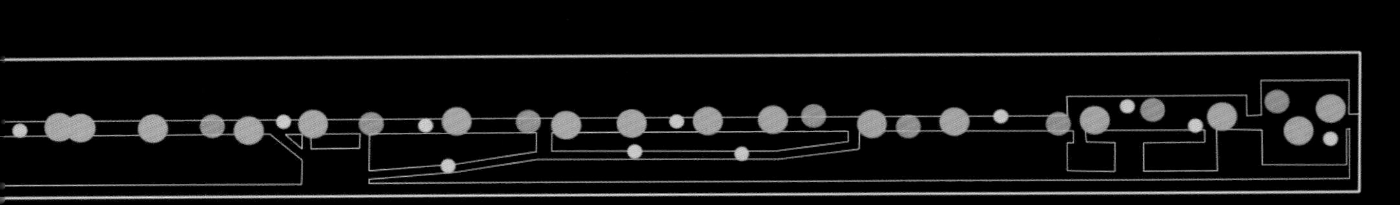

18:00-20:00 人群典型分布图
Typical distribution map of users

在空间分布上，周末下午有大量使用人群分布在该区段西侧的集中硬质场地上，以棋牌娱乐和围观行为为主；而日常傍晚人群分布则相对分散，主要集中在游径上。

In terms of spatial distribution, a large number of people are distributed among the big central sites on the west of the segment on the afternoon of a weekend. Most of them play chess or cards for fun, while some others prefer to watch. Every day at nightfall, user distribution in space is rather dispersed, with most people gathering around on the trails.

区段4：水西门—集庆门
S4:SXG-JQG

区段3：草场门—清凉门
S3:CCG-QLG

区段2：定淮门—草场门
S2:DHG-CCG

区段1：挹江门—华严岗门
S1:YJG-HYGG

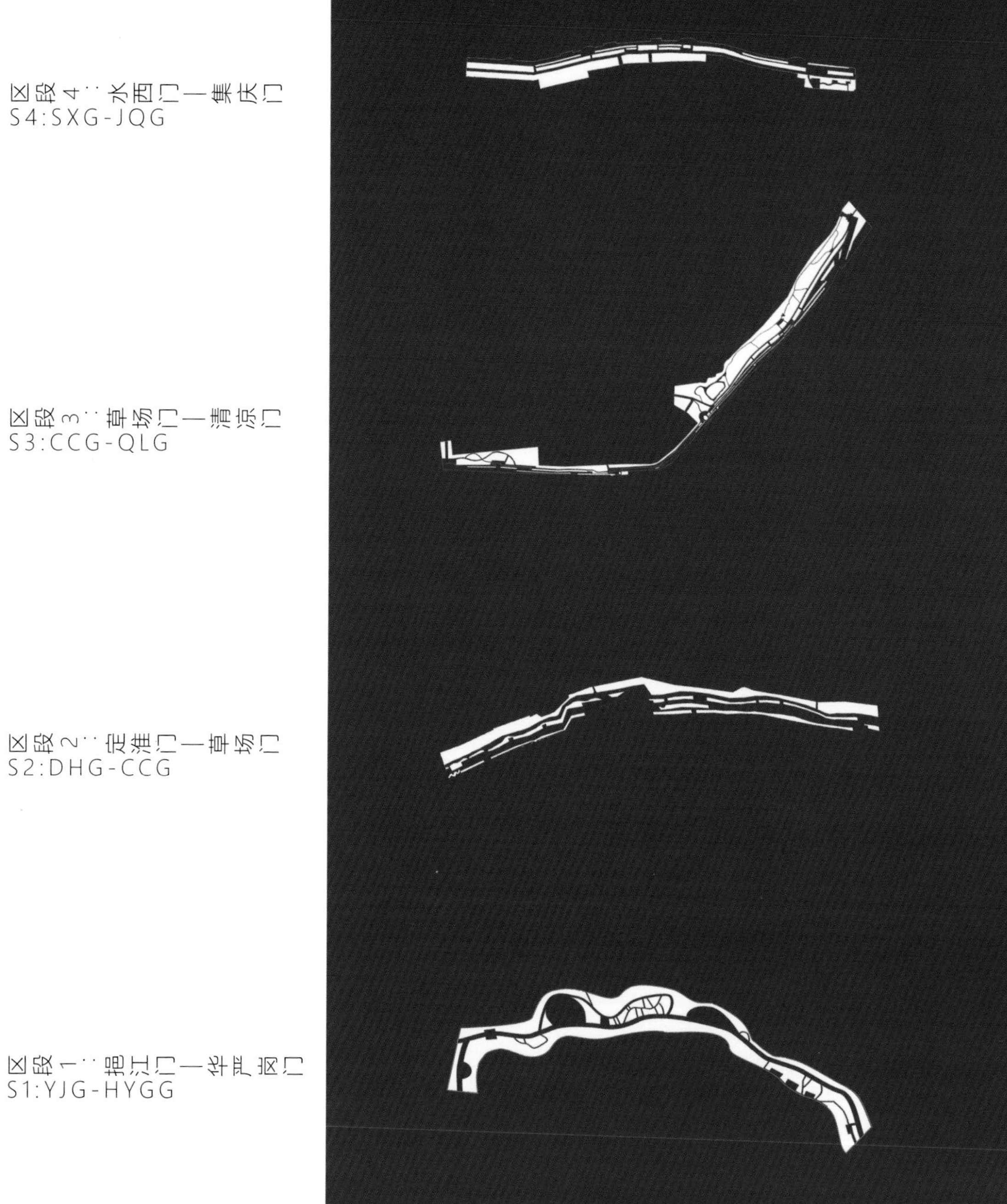

绿道内部空间特征分析

Characteristic Analysis of the Interior Space of the Greenway

区段5：凤台路—中华门
S5:FTR-ZHG

区段6：中华门—雨花门
S6:ZHG-YHG

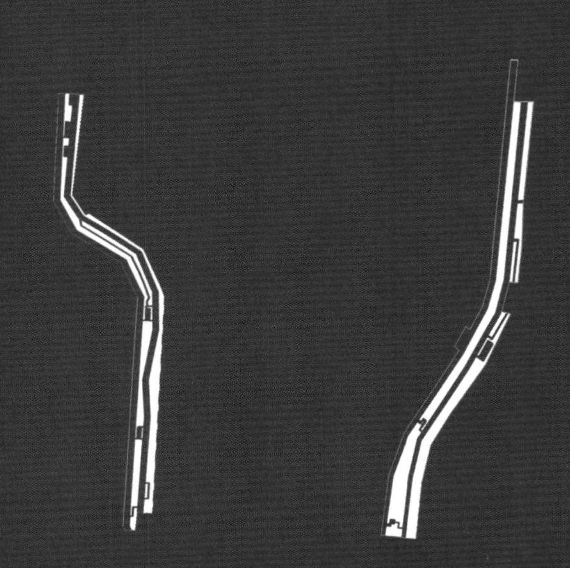

什么是绿道内部空间
What constitutes the interior space of a greenway

绿道内部空间主要由游径、绿化和设施三大部分组成。

绿道游径的空间组织形式主要有自行车道与步行道分开设置、合并设置两种形式。在操作实施空间比较宽裕的区段，通常会分开设置独立的自行车道与步行道，以降低两者的相互干扰。而在操作实施空间有限的区段或景区里则通常会将自行车道与步行道合并，组成综合慢行道。

绿道绿化是绿道景观环境品质的基本保障，同时也为部分活动提供了场所，如开敞草坪、滨水绿地等。由于绿道绿化总体上呈线形，部分绿道的绿化空间也是生物迁徙的重要廊道。

绿道的设施主要包含服务设施、标识设施和市政设施等类型。由于设施体系庞大且类型众多，本书仅对与人群使用直接关联，并在绿道内部空间扮演重要角色的服务设施空间特征进行分析和探讨。

The interior space of the greenway mainly consists of trails, greening, and facilities.

The spatial organization of the trails mainly includes the separate setting of bikeways and footpaths, as well as the merged setting of bikeways and footpaths. On the segments with enough space, independent bikeways and footpaths are usually separate, set up to minimize their mutual interference. On the segments with limited space or at scenic spots, bikeways and footpaths are usually merged to form comprehensive slow lanes.

In addition to providing a place for some activities, the greening of the greenway basically guarantees the quality of the environment, such as wide-open lawns, waterfront green spaces, etc. Since the pattern of most greening along greenway is linear, it can often be relied on as a migration corridor for wildlife in many instances.

Greenway facilities mainly include service facilities, identification facilities, and municipal facilities. Since the content of the facility system is gigantic and covers too many types, we only analyze and discuss in our study the characteristics of the service facilities that are directly linked to use and play key roles in the service performance of the greenway.

综合慢行道 Comprehensive slow lanes
设施 Facilities
绿化 Greening

自行车道 Bikeways
绿化 Greening
步行道 Footpaths
设施 Facilities

绿道结构图

Structure drawing of the greenway

硬地面积（单位：m²）

Area of hard sites (Unit : m²)

绿地面积（单位：m²）

Area of green space (Unit : m²)

座位数（单位：个）

Number of bench seats (Unit : piece)

年龄结构分布图（单位：人）

Age structure (Unit : person)

活动类型分布图（单位：人）

Structure of activity type (Unit : person)

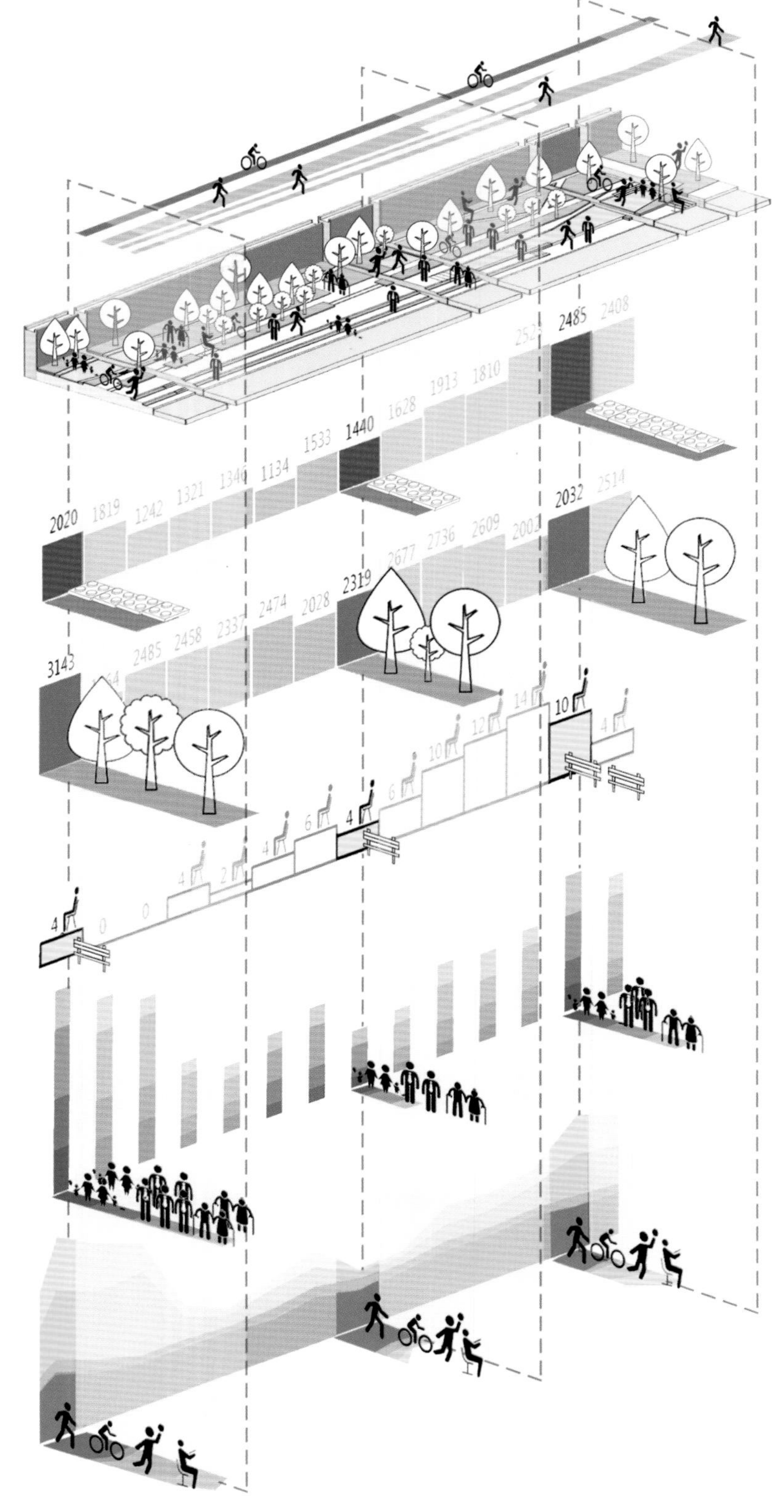

怎样关联分析绿道内部空间特征与服务绩效

How to correlate and analyze the characteristics of greenway interior space and service performance

由于绿道内部空间要素众多，且对人群使用的作用方式各异，如何有效梳理和分析各因素对于绿道服务绩效的影响成为一大难题。

为便于对绿道内部各空间要素进行独立和组合的空间分析，我们效仿医疗检查中进行切片分析的方式，将每个绿道区段以50m为单位划分成20-30段切片，并将每个切片单元上的人群使用信息和空间要素信息数据量化，然后逐项分解和对应排列。从而能直观地把握绿道区段不同位置使用人群的状态，并能对应绿道内部空间各要素的变化情况来分析和推导两者的内在联系，实际上为相对客观地对绿道内部空间要素与人群使用特征的关联分析创造了条件。

绿道的服务绩效主要通过人群使用强度（等同于人群使用密度）指标来衡量，即绿道单位长度的使用人数。由于本书中将切片长度统一设置为50m，因而绿道区段在切片上的使用人群规模指标的变化情况实际上反映了使用人群密度的变化。鉴于此，在进行切片分析过程中主要以其使用人群规模指标作为衡量该切片单元服务绩效的主要参照指标。

Since the interior space of the greenway is constituted by diverse elements with different modes of action toward utilization, how to effectively put in order and analyze the influence of all the elements on the service performance of the greenway has become a huge difficulty.

To facilitate independent and combined spatial analysis of the elements inside the greenway, we follow the example of a slice in medical examination, divide each greenway segment into twenty to thirty 50m-slices, quantify the status of utilization and spatial elements of the greenway on each slice, and break up and arrange them item by item in graphs, so as to directly get hold of utilization and element statuses on different parts of the greenway segments and analyze and deduce the inner relationship between the two according to the correlative changing of the data, which in fact provides a pathway to correlate and analyze the characteristics of greenway interior space and greenway service performance.

One of the important indicators to measure service performance of the greenway is use intensity (equal to user density), namely, the number of users in unit length of the greenway. Since we uniformly set the length of the slices as 50m, the changing situation of user number on the slice in fact reflects the change of user density. In light of this, during the process of slice analysis, we mainly consider the user number as the main reference index to measure the service performance of the unit slice.

绿道内部空间特征与服务绩效分析

Analysis of characteristics of the greenway interior space and service performance

切片分析法是本次研究绿道内部空间特征与服务绩效分析的主要方法，也是将绿道内部空间分析过程及结果可视化、直观化表述的一次探索和尝试。从应用特征来看，该分析方法应用在绿道空间内部分析中能将其丰富的空间要素进行分解量化，并与使用人群状态的量化数据进行关联，使绿道各个部位的使用特征和问题一目了然，具有较高的适宜性。

但在切片分析法的应用过程中，首先碰到的问题是绿道区段大多并非直线，而如何将蜿蜒的绿道进行切片划分并整合关联，为切片分析法的应用带来了些许挑战。为解决该问题，我们将航拍照片按照切片尺寸进行分解和线性重组，同时辅以线性带状的区段空间结构图与切片单元对应进行关联分析。

此外，受研究条件所限，本次切片分析所能量化的空间要素数据仅限于绿道的绿地面积、硬地面积、游径和休息服务设施规模。而绿道内部空间的布局特征，实际上很难通过量化的方式来进行全方位描述，因此主要通过抽象的空间结构图对绿道内部空间要素的布局特质进行提取，以便于与空间要素量化数据进行直观对照。

从切片分析法的应用过程来看，它能非常直观清晰地暴露和锁定绿道空间使用中的问题。通过切片分析图，可以直观发现人群活动聚集与空间要素聚集地段，并能发现其中的错位关系，从而引导我们去探索和分析造成该问题的内在原因。但切片分析法虽能直观暴露问题，却很难直接揭示产生问题的原因。这主要是因为影响人在绿道上行为活动的空间因素过于复杂，在目前的技术条件下，还很难完全通过定量途径对切片空间特征以及人在空间中的感知状态等方面进行客观描述。这也是为何在应用切片分析法研究绿道内部空间时，还需同时依托绿道结构图和实景照片进行空间辅助分析的主要原因。

Slice analysis is a main tool this study employ to analyze the greenway interior space characteristics and greenway service performance. It is also an exploration and attempt, through which this study presents the analytical process and results of the interior space of the greenway in a visualized manner. Judging from the effect of the application, we can see that the application of this tool to the analysis of the interior space of the greenway can help decompose and quantify the rich spatial elements and link them to the quantitative data of the users with a high suitability, so that the characteristics and problems of the use of all parts of the greenway are clear at a glance.

During the application of slice analysis, the first problem that we come across is that most greenway segments are non-linear. How to divide the zigzagging greenway into slices and integrate and correlate the slices has brought some challenges to the application of slice analysis. To solve this problem, we decompose the aerial photos according to the sizes of slices, slightly deform and linearly recombine them in a linear framework, while making again an abstractive drawing showing the space structure and the layout of the key elements as a supplement for correlation analysis of slice units.

Besides, limited by study conditions, the quantifiable spatial element data in the analysis of slices are confined to the area of green space, the area of hard sites, the size of trails, and the number of service facilities. In fact, we find it very difficult to describe all-roundly the characteristics of the layout of the interior space of the greenway in a quantifiable way. Therefore, we extract the characteristics of the layout of the internal spatial elements of the greenway mainly through the abstractive drawings of space structure, so as to facilitate a visualized contrast to the quantitative data of spatial elements.

Judging from the application of slice analysis, we find that it can visually and clearly expose and lock in problems arising out of the use of the greenway space. According to the slice analysis diagram, we can visually find out the areas that aggregate user groups and spatial elements and find out the dislocation relationship therein, so as to use the findings to guide us to explore and analyze the internal causes of the problem. Although slice analysis can bring a problem to light visually, it cannot easily reveal the cause of the problem directly. This is mainly because the spatial elements that influence human behaviors and activities on the greenway are too complex to objectively describe the characteristics of slice spaces, human perception in the spaces, and other such aspects completely by a quantitative approach, under the current technical conditions. This is why we also need to rely on the structure drawing of the greenway and site photos to conduct spatial and auxiliary analyses when applying slice analysis .

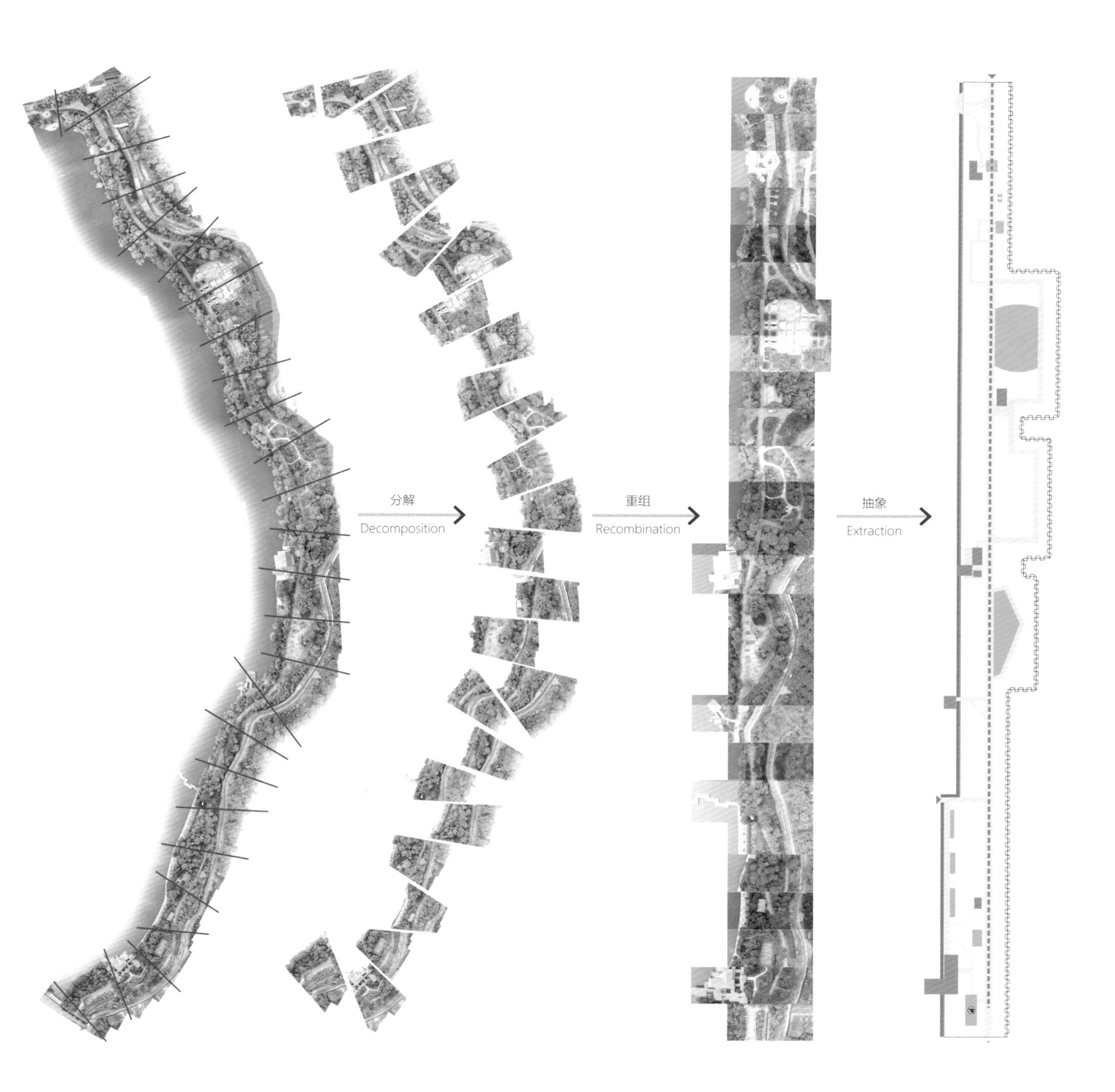
分解
Decomposition
重组
Recombination
抽象
Extraction

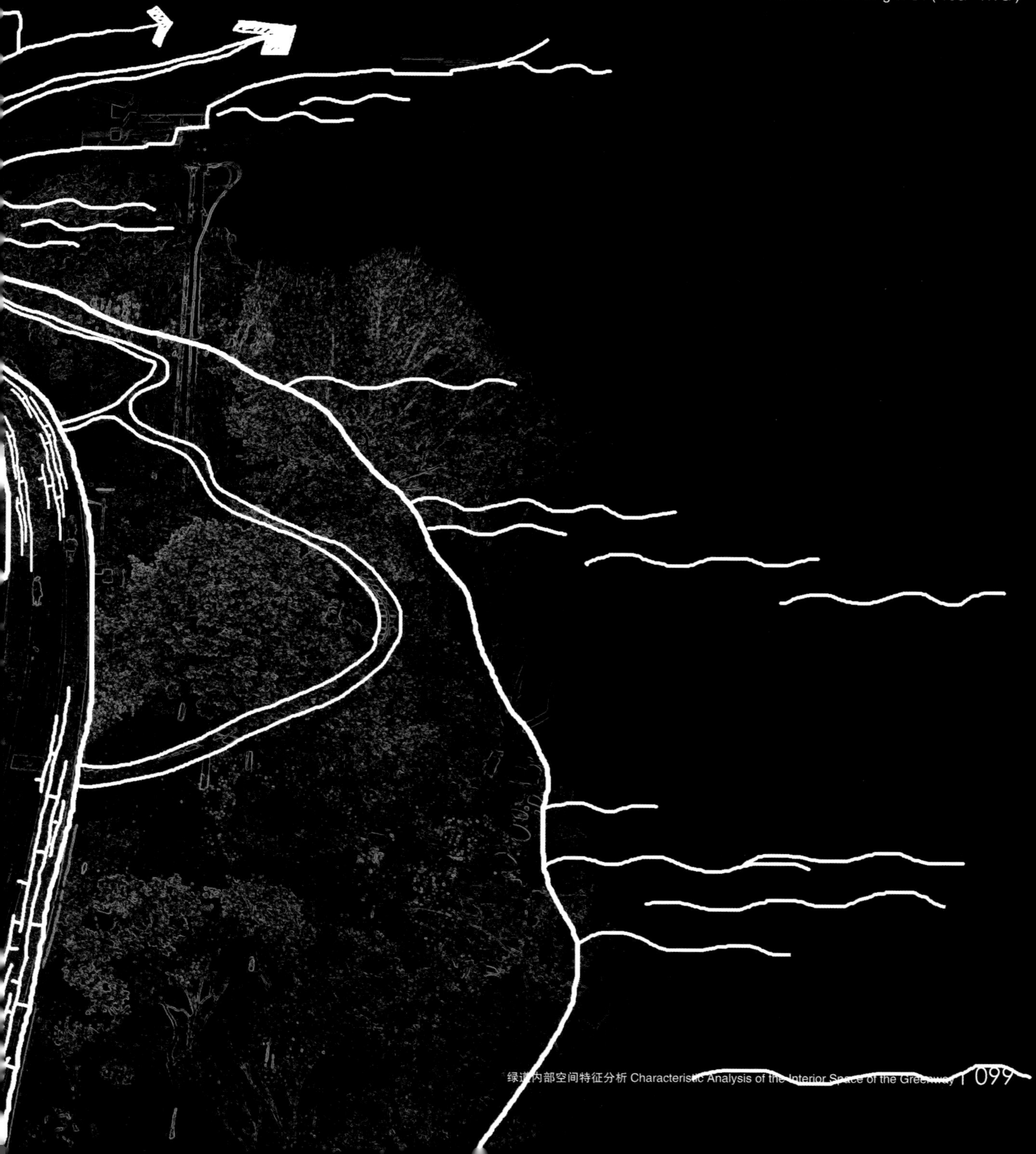

区段 1（挹江门—华严岗门）抽象图
Abstractive drawing of S1 (YJG - HYG)

区段 1：挹江门—华严岗门
Segment 1: Yijiang Gate-Huayangang Gate

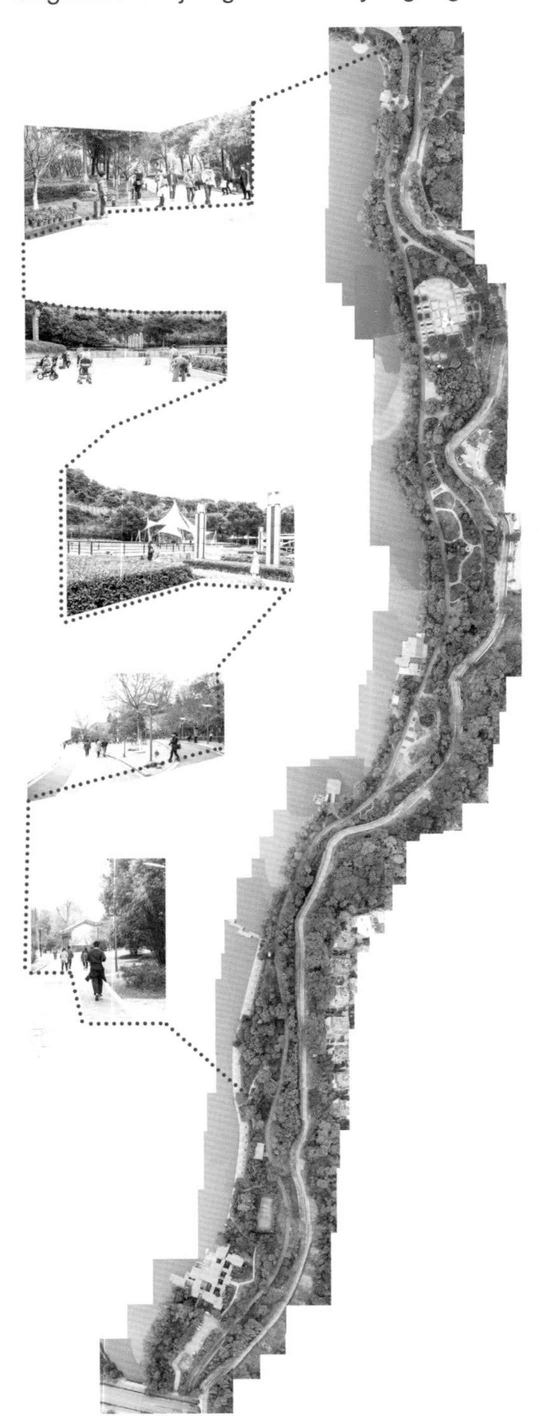

内部空间特征概述

区段 1（挹江门—华严岗门）是一条充满活力的区段，东倚明城墙，西临护城河，并与边上的带状公园相结合。该区段要素内容和空间变化较丰富，集中活动场地和停留空间充足，能够吸引大量人群使用。

Overview of the characteristics of the interior space

S1 YJM-HYGG is full of vitality. It leans on the Ming Dynasty City Wall to the east, faces the city moat to the west, and combines with the belt-shaped park on the side. This segment is rich in elements and patterns of space, with enough open yards for recreational activities and stopover so that it has both good condition and capacity to hold a large number of users.

设施类别 Type of facilities	具体设施 Specific facilities	数据 Data
游径及交通设施 Trails and transportation facilities	步行道 Footpaths	4.5m
	自行车道 Bikeways	4.5m
	停车场 Parking lot	260m^2
	无障碍道 Barrier-free roads	√
游憩服务设施 Recreational service facilities	游乐器具 Amusement equipment	√
	休息设施 Rest facilities	206 个
	文体活动场地 Cultural and sports field	4386m^2
	无线网 Wi-Fi	X
管理服务设施 Management service facilities	管理中心 Management Center	X
	游客服务中心 Tourist Service Center	X
	信息咨询点 Information consultant	X
商业服务设施 Commercial service facilities	售卖点 Selling	√
	自行车租赁点 Bike rentals	X
	自动售货机 Vendingmachines	X
科普教育设施 Popular science education facilities	科普宣传、展览设施 Popularization and propaganda of science, Exhibition facilities	X
	解说设施 Commentary facilities	X
环境卫生设施 Environmental hygiene facilities	公厕 Restrooms	√
	垃圾桶 Trash cans	√
安全保障设施 Safety & security facilities	照明 Lighting	√
	治安消防 Public security & fire control	√
标识信息设施 Signs	信息、指路、规章标识等 Information, directions, signs, etc.	√

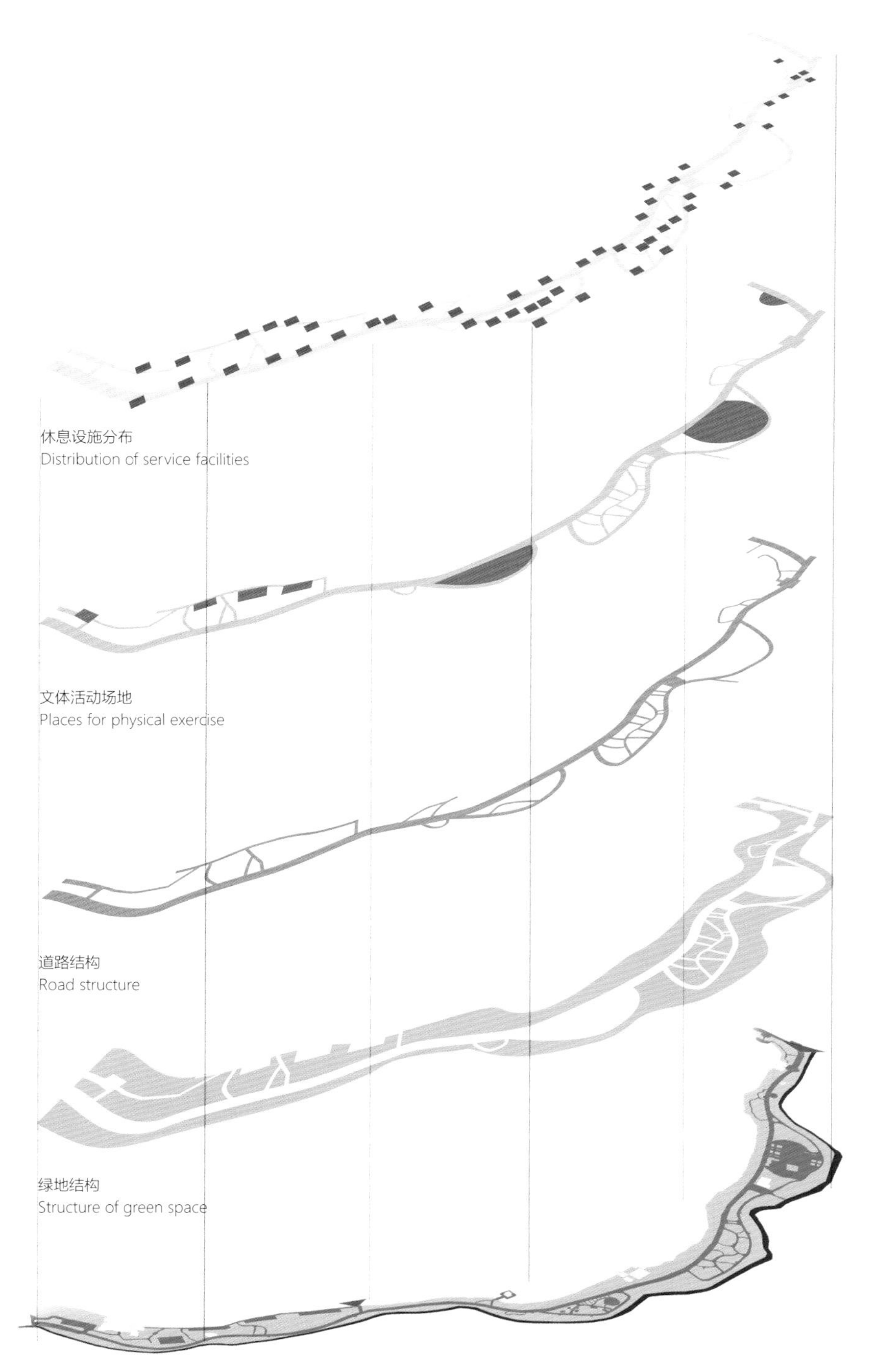

区段1（挹江门—华严岗门）主游径是一条4.5m宽的综合慢行道，并沿岸线展开，在部分大型活动场地会有次级游步道连接。该区段城墙走向蜿蜒曲折，在区段中营造出了空间开闭的丰富效果。该区段内的两处大面积硬质场地均是结合城墙蜿蜒外凸空间设置。滨水地段以小型亲水平台场地为主，主游径沿线和硬质场地上的座椅和游憩服务设施较多。

The main trail of S1 YJM–HYGG is a comprehensive slow lane that is 4.5m wide. It spreads out along the waterfront and is connected to some large open sites via secondary walking trails. The city wall of this segment zigzags so as to generate a rich effect of open and closed space. Two large open sites on this segment are both combined with the zigzagging outwardly convex space of the city wall. The waterfront areas are mainly small hydrophilic platforms. The benches and other recreational service facilities of this segment are mainly distributed along the main trail and in the open sites.

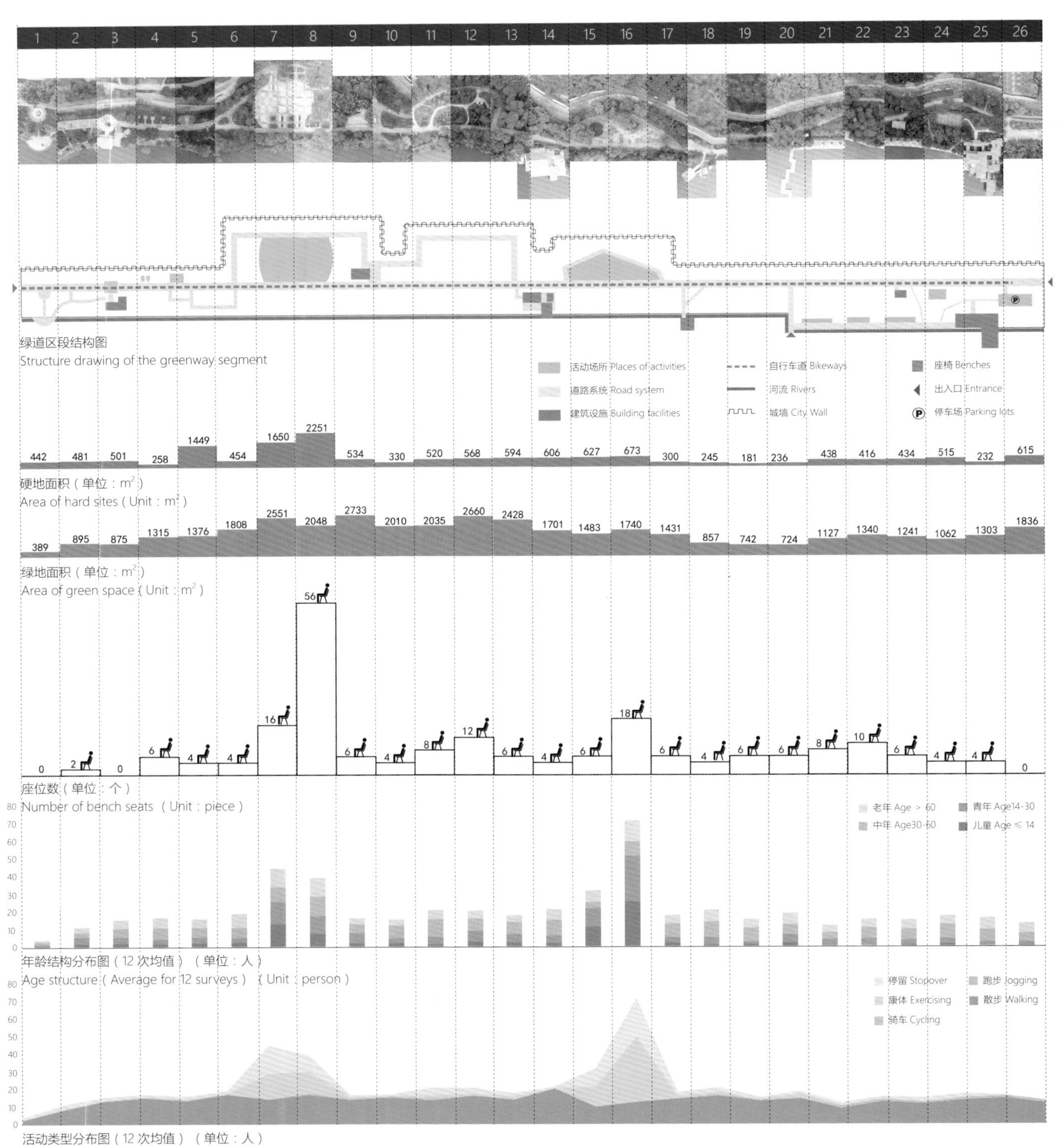

1 2 3 4 5 6 7 8 9 10 11 12 13 14 15 16 17 18 19 20 21 22 23 24 25 26
绿道区段结构图
Structure drawing of the greenway segment
活动场所 Places of activities
道路系统 Road system
建筑设施 Building facilities
自行车道 Bikeways
河流 Rivers
城墙 City Wall
座椅 Benches
出入口 Entrance
停车场 Parking lots
442 481 501 258 1449 454 1650 2251 534 330 520 568 594 606 627 673 300 245 181 236 438 416 434 515 232 615
硬地面积（单位：m²）
Area of hard sites (Unit : m²)
389 895 875 1315 1376 1808 2551 2048 2733 2010 2035 2660 2428 1701 1483 1740 1431 857 742 724 1127 1340 1241 1062 1303 1836
绿地面积（单位：m²）
Area of green space (Unit : m²)
0 2 0 6 4 4 16 56 6 4 8 12 6 4 6 18 6 4 6 6 8 10 6 4 4 0
座位数（单位：个）
Number of bench seats (Unit : piece)
老年 Age＞60
中年 Age30-60
青年 Age14-30
儿童 Age≤14
0 10 20 30 40 50 60 70 80
年龄结构分布图（12 次均值）（单位：人）
Age structure (Average for 12 surveys) (Unit : person)
停留 Stopover
康体 Exercising
骑车 Cycling
跑步 Jogging
散步 Walking
0 10 20 30 40 50 60 70 80
活动类型分布图（12 次均值）（单位：人）
Structure of activity type (Average for 12 surveys) (Unit : person)

内部空间特征总体分析

区段 1（挹江门—华严岗门）长约 1.3km，共等分为 26 段切片。该区段内部空间较为丰富，切片 6-17 由于城墙外凸增加了绿道的宽度，形成了三个各具特色的节点空间。

从使用人群数据的变化可以看出，切片 6-8 以及切片 15-17 是该区段主要的两处活动场地。其中使用人数最多的是切片 15，其次为切片 7 和切片 8。除切片 1 人数明显较少外，其他切片使用人数变化不大。两处集中活动场地（切片 6-8 以及切片 15-17）能够吸引各个年龄层次的人群停留下来开展一些康体休闲以及停留休息活动。其他切片则以中老年人散步活动为主。

在硬地分布上，切片 5、切片 7 和切片 8 拥有较大的硬地面积，并且比其他切片硬地面积高出两倍以上。而绿地面积在各个切片中的波动则较小，其中切片 7-13 中的绿地面积均超过 2000m^2，仅有切片 1-3 以及切片 18-20 两部分切片单元的绿地面积在 1000m^2 以下。

在座椅分布上，切片 8 的座椅座位数目达到 56 个，是一个集中停留休息区域，其规模远超排位第二和第三的切片。除了切片 1、切片 3 和切片 26 中无座椅座椅座位分布外，其他切片单元中的座椅座椅座位分布较均衡。

Analysis of the characteristics of the interior space

S1 YJM-HYGG is about 1.3km long, equally divided into 26 slices. The patterns of the interior space of this segment are very diverse. The space from slice 6 to slice 17 generate three distinctive nodes as a result of the increased greenway width due to the outset of the city wall.

Judging from change of user data, we've found out that slices 6-8 and slices 15-17 are the two main active places of this segment. Most people gather in slice 15, with the second being slices 7 and 8. Except slice 1 where the number of people is noticeably small, the number of people using the greenway on the rest of the slices does not change too much. The two central active places (slices 6-8 and slices 15-17) can attract different age groups to stop and exercise. On the other slices are middle-aged and senior people who mainly walk.

In terms of hard site distribution, slices 5, 7, and 8 have large areas of hard sites that are more than twice those of other slices. The green spaces on all the slices do not fluctuate too much. The green spaces from slice 7 to slice 13 all exceed 2,000m^2. Only the green spaces of slices 1- 3 and slices 17-19 are below 1,000m^2.

In terms of bench distribution, the number of bench seats on slice 8 reaches 56, thus making it a central area for stopover whose size outruns those of slices ranking the 2nd and the 3rd. Except for slices 1, 3, and 26, where there are no bench seats, the distribution of bench seats on the other slices is relatively balanced.

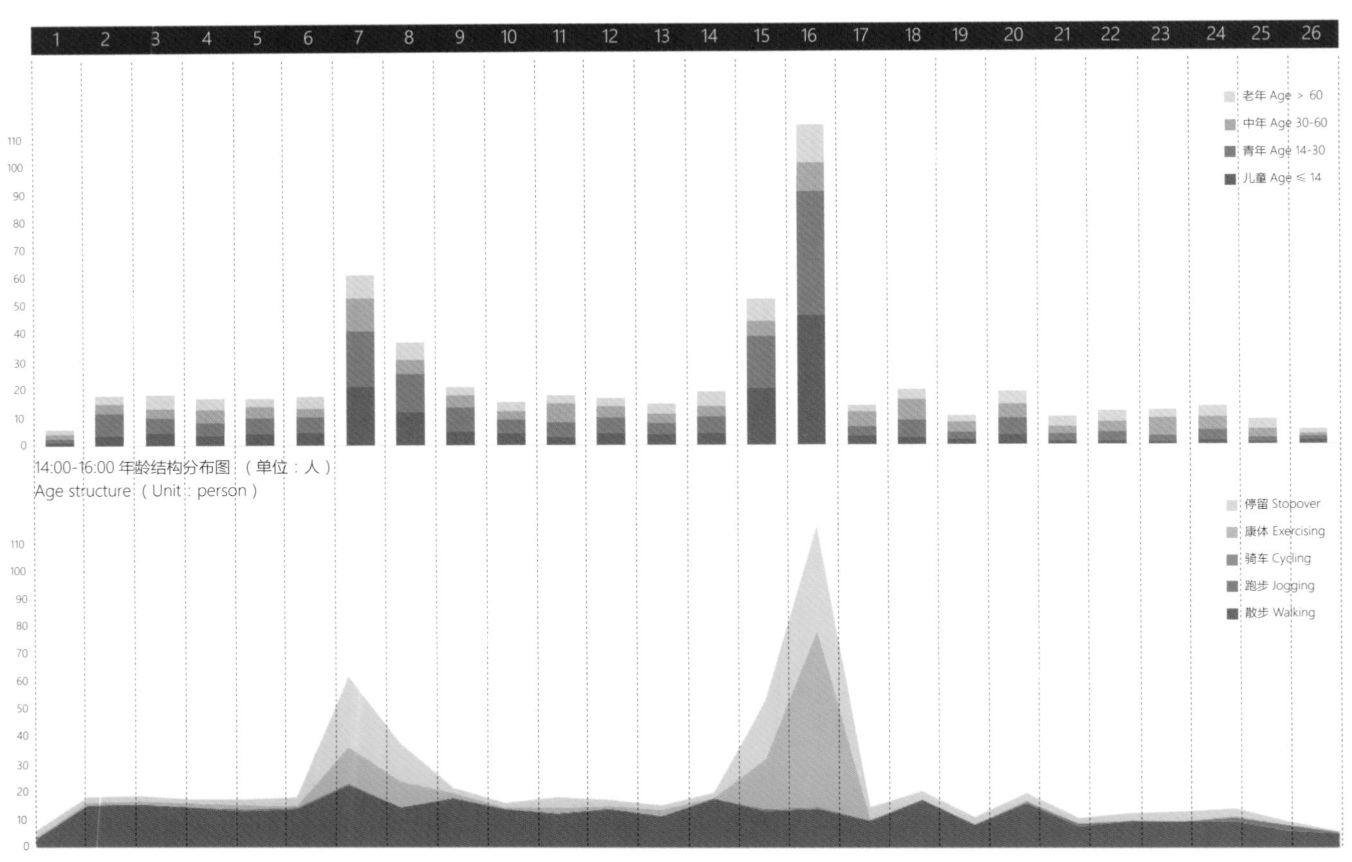

14:00-16:00 年龄结构分布图（单位：人）
Age structure (Unit : person)

14:00-16:00 活动类型分布图（单位：人）
Structure of activity type (Unit : person)

两个时段的服务绩效分析

周末下午（14:00-16:00）使用人群分布主要在切片 7-8 以及切片 15-16 出现两个波峰，在其他切片中则较均衡。其中切片 16 使用人群规模最大，人群以青年和儿童为主，使用活动类型则以康体和停留休息为主。而切片 7 和切片 8 使用人群年龄分布则较均衡。

Service performance analysis of two time periods

On a weekend afternoon (14:00-16:00), two peaks appear on slices 7-8 and slices 15-16 in terms of user distribution in space, but user distribution is relatively balanced on the other slices. The biggest number of users appears on slices 16. They are mainly young people and children, who mainly play for fun, exercise and stop over. The age distribution on slices 7 and 8 is relatively balanced.

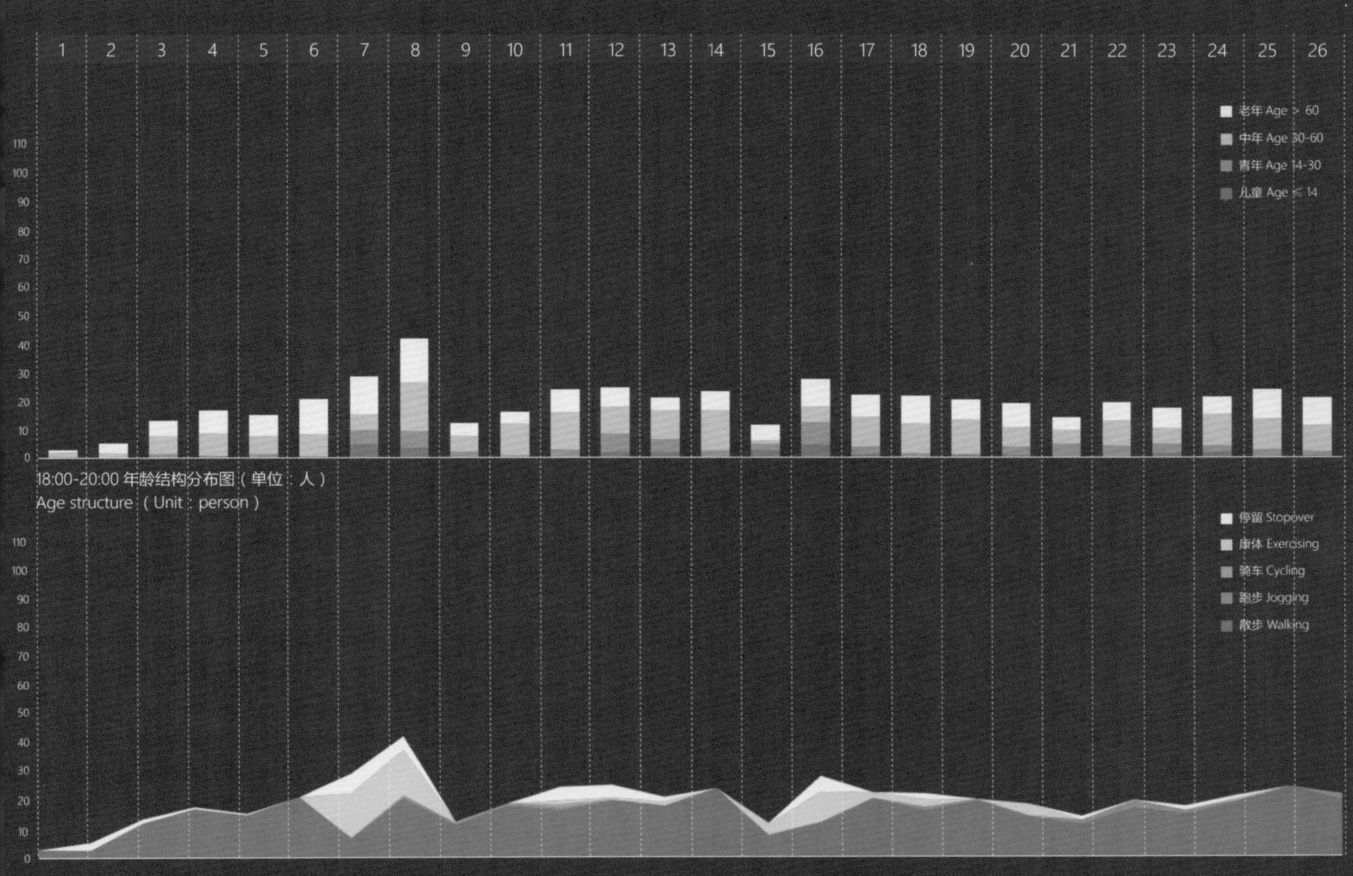

18:00-20:00 年龄结构分布图（单位：人）
Age structure（Unit：person）

18:00-20:00 活动类型分布图（单位：人）
Structure activity type（Unit：person）

日常傍晚（18:00-20:00）各切片单元使用人群规模差异相对较小，以中老年人群散步为主。切片 8 人数最多，其中进行康体活动（以广场舞为主）的人数较其他切片单元明显增加。

Every day at nightfall (18:00-20:00), the difference in the number of people using the greenway is relatively small on all slice units. Most users are middle-aged and senior people who mainly walk. Slices 8 have the biggest number of users, with a noticeable increase of the number of people who exercise (mainly square dance), compared to the other slice units.

切片 8　Slice 8
切片 7　Slice 7

典型切片分析 1

切片 7 切片 8 是整条绿道中最宽、硬地面积最大的地段，也是座椅座位数目最多的地段，但该地段总体使用人数和周末下午使用人数（即使用强度）却并非最高。较之使用人数总体水平较高的切片 15-16，可以发现该地段空间尺度过大，虽高差较多但空间缺乏变化，且内部要素单一，这均在一定程度上减弱了其吸引力。但在日常傍晚，由于拥有较集中的硬地，其为居民开展集体康体活动（如广场舞等）创造了条件，因此在该时段使用人数最高。反观该地段内部，切片 7 与切片 8 虽空间要素数据接近，但人群活动状态却也存在差异。对比周末下午和日常傍晚使用人群规模与活动特征，切片 7 在周末下午的使用人群规模比区段 8 高出约 50%。从左图中也能清晰看出使用人群在空间分布上的差异。

解析对比两者空间特征可发现，切片 8 结合树池设置了大量座椅座位，以外向型独立座椅为主。切片 7 的“座椅”仅为一些可坐空间，例如可坐下的平台、台阶等，这些可坐空间分布于下沉广场周围，形成了一个内向型的空间。这为使用人群围绕下沉广场开展康体活动、停留休息和观看创造了机会，使得下沉广场及周边空间对家庭、团体等人群的吸引力陡增。

切片 7
slice 7

切片 8
slice 8

切片 7
slice 7

切片 8
slice 8

Analysis 1 of typical slices

Slices 7 -8 have the largest areas of hard sites and are the widest parts of the entire greenway, and they also have the largest number of bench seats. However, the total number of users and the number of users on weekend afternoons is not the largest. Compared to slices 15-16, which are generally at a higher level in terms of the number of users, the space of this area is too spacious. With a big difference in altitude, this slice does not change too much in terms of space pattern, and its internal elements are organized in a single way, all of which weakens its attraction to potential users. However, every day at nightfall, since it has large and centralized hard sites, this slice is able to provide residents with enough space and condition to exercise like square dancing. Therefore, the number of users has a chance to reach to the largest during this time period.

Reflecting on the inner area of this place, we've found out that the data of spatial elements of slices 7 and 8 are similar, but they differ in activity status of users. By comparison of user number and activity characteristics on weekend afternoons and every day at nightfall, we've found out that the number of people using slice 7 is about 50% larger than that using slice 8 on a weekend afternoon. The difference in spatial distribution of users can be clearly seen from the picture on the left.

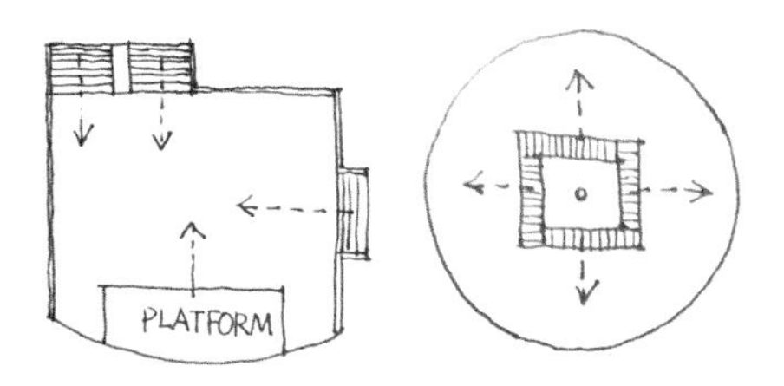

Through analysis and comparison of the spatial characteristics of two slices, we can find that slice 8 is set up with a lot of bench seats in combination with a tree pool, and the bench seats are mainly outward independently designed and installed. Slice 7 only possesses some spaces for sitting, such as platforms and steps. These sit-able spaces are distributed around the sunken plaza and form an inward space. In this way, it provides users with opportunities to gather around the plaza to exercise, have a rest, and watch around, which would be of significant attraction to families and groups instantly.

切片 15 Slice 15
切片 16 Slice 16

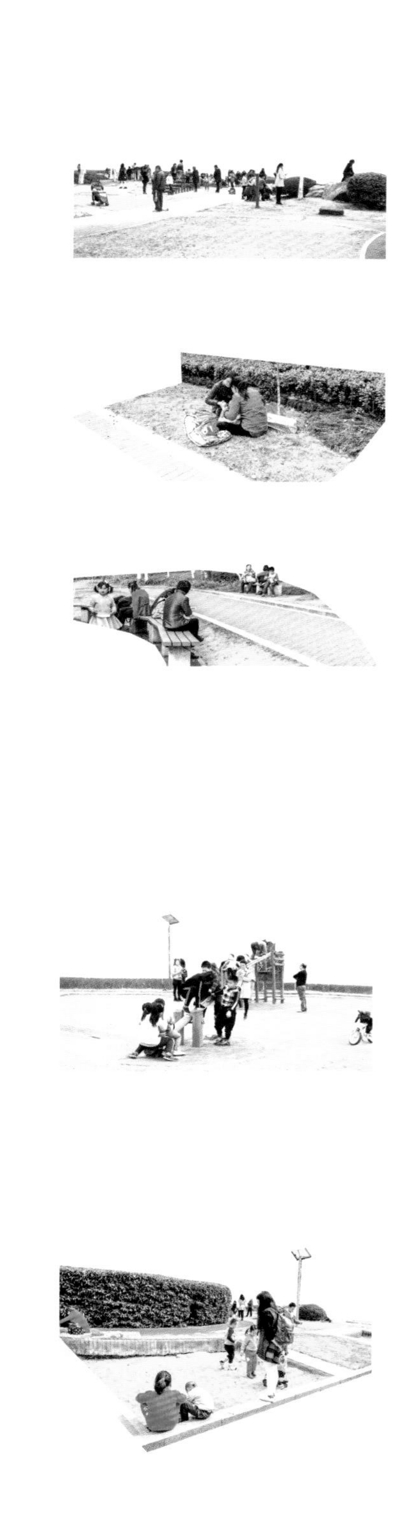

典型切片分析 2

切片 15-16 是一个梭形场地，周末午后时段是该区段中最活跃的地方，但日常傍晚的使用人数相对较少。较之切片 7-8，该地段在各类要素指标上并无明显优势，但其空间变化较丰富，内部小径众多，整个场地被划分成了若干个空间特质各异的子空间，并且为儿童配置了一系列游乐设施，从而为儿童聚集玩耍和亲子类活动的开展创造了条件。

场地上最具吸引力的是紧贴道路的四个沙池，沙池与草地间隔设置，当孩子们在沙池中玩耍时，家长们可以坐在一旁的草地中休息。因此，在周末午后使用人数较多的时段内，以青年和儿童人群为主。

由于傍晚整个绿道区段的使用人群以中老年人为主，该区段的空间特质与老人活动需求明显不符，这也造成了其在周末午后与日常傍晚使用人数的巨大反差。

Analysis 2 of typical slices

Slices 15-16 form a fusiform site, which is the most popular place on the entire segment on a weekend afternoon. However, every day at nightfall, the number of users is rather small. Compared to slices 7-8, this area does not have any advantages in terms of element indicators, but it boasts of rich spatial variations and inside pathways. It divides the entire site into several sub-spaces that have different spatial characteristics and provides children with a series of entertainment facilities, so as to create good conditions for family outings and play.

The most attractive on the site is four sand pits close to the trail. The sand pits and lawns are set up at intervals. When children are playing in the sand pits, their parents can sit on the nearby lawns for a rest. Therefore, young people and children are the majority of people occupying the space and using the facilities on this slice on weekend afternoons when it is most active in the segment.

Since users of the entire segment are middle-aged and senior people at nightfall, the spatial characteristics of this slice evidently do not match the activity demand of the senior, resulting in the great contrast between a weekend afternoon and nightfall every day in the number of users.

区段 2（定淮门—草场门）抽象图
Abstractive drawing of S2 (DHG - CCG)

区段 2：定淮门—草场门
Segment 2: Dinghuai Gate-Caochang Gate

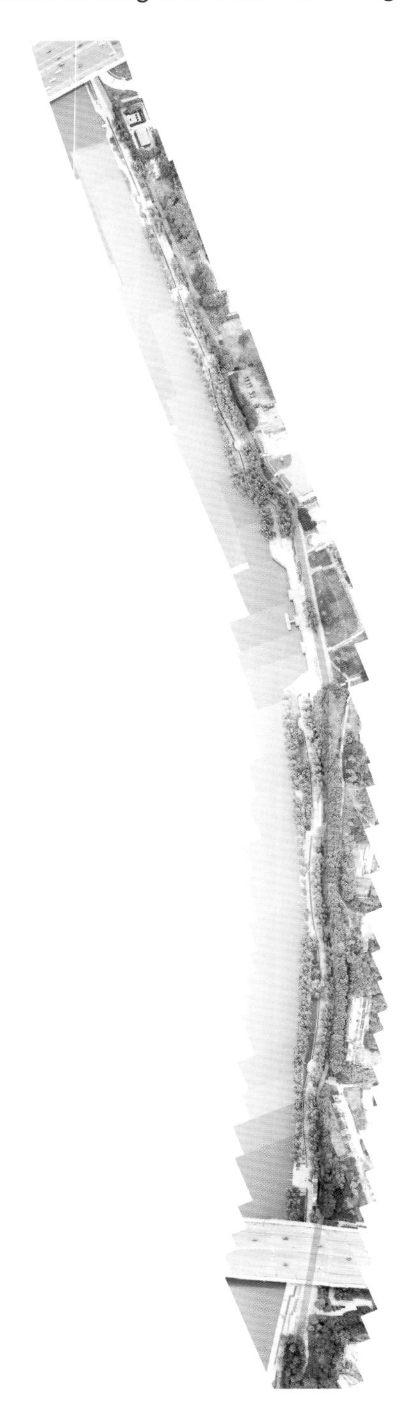

内部空间特征概述

区段 2（定淮门—草场门）北起定淮门，南至草场门大桥。明城墙在该区段缺失中断，因此绿道空间一侧的空间界定明显减弱。该区段绿道结合河岸的三级亲水平台展开，区段中心有一个游船码头，并散布一定数目的座椅。

Overview of the characteristics of the interior space

S2 DHG-CCG starts at Dinghuai Gate on the north and spreads to Caochang Gate Bridge on the south. The Ming Dynasty City Wall is missing and cut short on this segment, so spatial demarcation of one side of the greenway space has been weakened noticeably. The greenway of this segment spreads out in combination with a tertiary hydrophilic platform at the riverbank. There is a cruise terminal at the middle of the segment where a certain amount of bench seats are dispersed.

设施类别 Type of facilities	具体设施 Specific facilities	数据 Data
游径及交通设施 Trails and transportation facilities	步行道 Footpaths	5.0m
	自行车道 Bikeways	4.5m
	停车场 Parking lot	520m^2
	无障碍道 Barrier-free roads	√
游憩服务设施 Recreational service facilities	游乐器具 Amusement equipment	X
	休息设施 Rest facilities	22 个
	文体活动场地 Cultural and sports field	1760m^2
	无线网 Wi-Fi	X
管理服务设施 Management service facilities	管理中心 Management Center	X
	游客服务中心 Tourist Service Center	X
	信息咨询点 Information consultant	X
商业服务设施 Commercial service facilities	售卖点 Selling	X
	自行车租赁点 Bike rentals	X
	自动售货机 Vendingmachines	X
科普教育设施 Popular science education facilities	科普宣传、展览设施 Popularization and propaganda of science, Exhibition facilities	X
	解说设施 Commentary facilities	X
环境卫生设施 Environmental hygiene facilities	公厕 Restrooms	√
	垃圾桶 Trash cans	√
安全保障设施 Safety & security facilities	照明 Lighting	√
	治安消防 Public security & fire control	X
标识信息设施 Signs	信息、指路、规 章标识等 Information, directions, signs, etc.	√

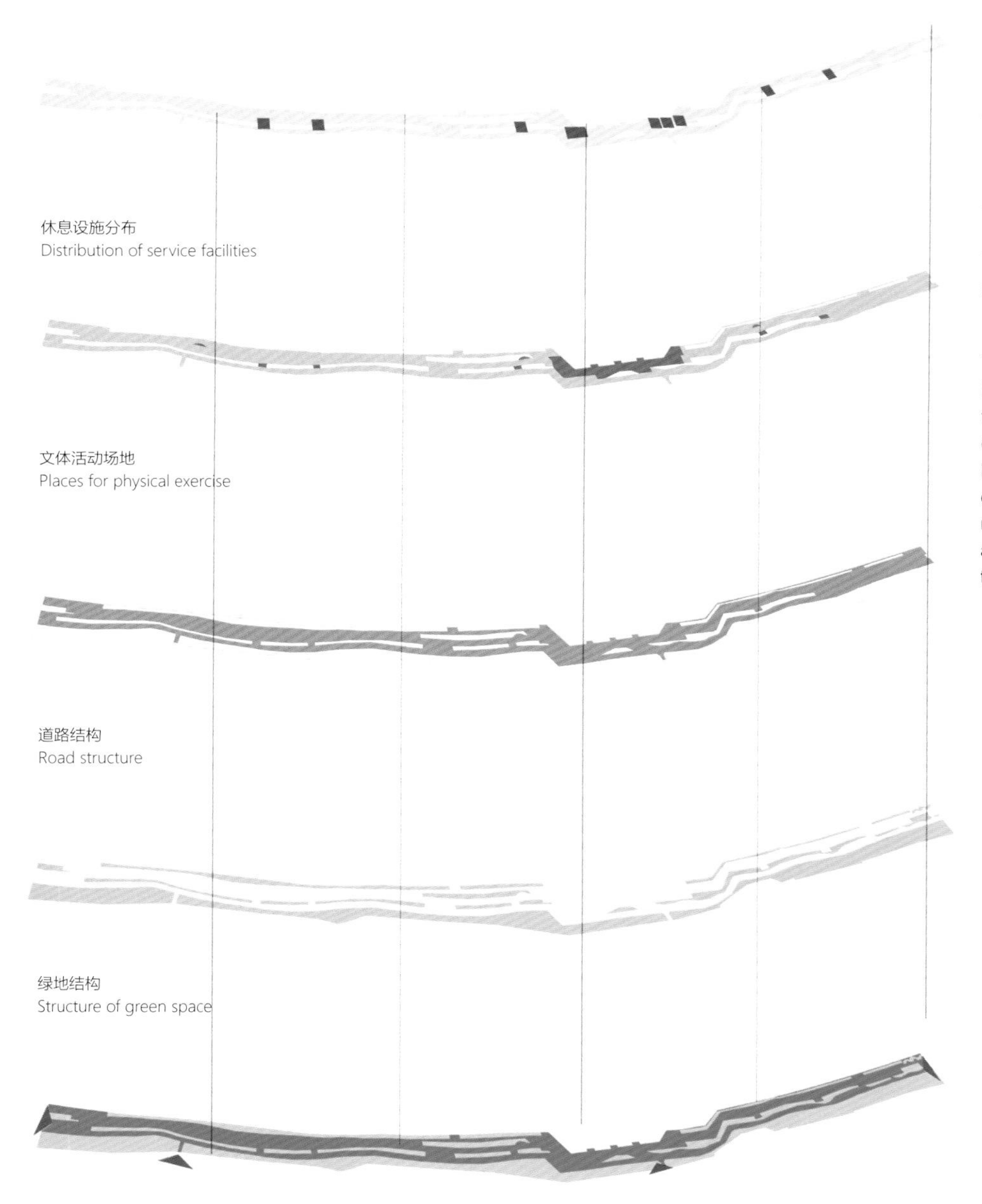

区段 2（定淮门—草场门）内部空间整体较均质，除了中部游船码头岸线局部内凹外，并没有巨大的变化。绿道的自行车道、步行道和堤岸游步道分别沿河岸三级台地展开。该区段的硬质场地集中在游船码头周边。

The interior space of S2 DHG-CCG is generally homogeneous. Except for the partial shoreline of the cruise terminal at the middle being inwardly concave, there are not many changes. The bikeways, footpaths, and waterfront trails spread out along the tertiary hydrophilic platform at the riverbank. The large open sites of this segment are all located at the surroundings of the cruise terminal.

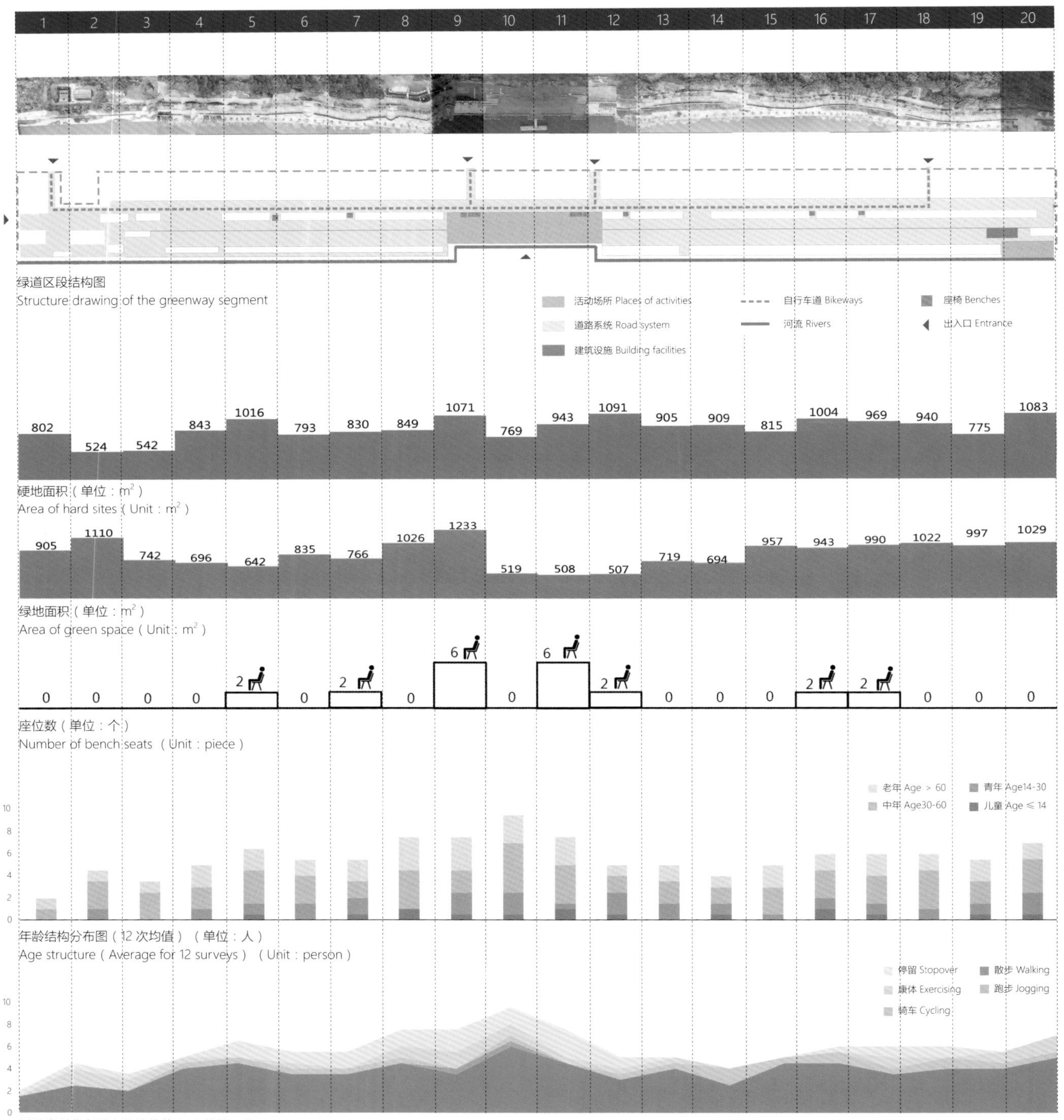

1
2
3
4
5
6
7
8
9
10
11
12
13
14
15
16
17
18
19
20
绿道区段结构图
Structure drawing of the greenway segment
活动场所 Places of activities
道路系统 Road system
建筑设施 Building facilities
自行车道 Bikeways
河流 Rivers
座椅 Benches
出入口 Entrance
802
524
542
843
1016
793
830
849
1071
769
943
1091
905
909
815
1004
969
940
775
1083
硬地面积（单位：m²）
Area of hard sites（Unit：m²）
905
1110
742
696
642
835
766
1026
1233
519
508
507
719
694
957
943
990
1022
997
1029
绿地面积（单位：m²）
Area of green space（Unit：m²）
0
0
0
0
2
0
2
0
6
0
6
2
0
0
0
2
2
0
0
0
座位数（单位：个）
Number of bench seats （Unit：piece）
老年 Age＞60
中年 Age30-60
青年 Age14-30
儿童 Age ≤ 14
年龄结构分布图（12 次均值）（单位：人）
Age structure（Average for 12 surveys）（Unit：person）
停留 Stopover
康体 Exercising
骑车 Cycling
散步 Walking
跑步 Jogging
活动类型分布图（12 次均值）（单位：人）
Struture of activity type（Average for 12 surveys）（Unit：person）

内部空间特征总体分析

区段2（定淮门—草场门）长约1.0km，共等分为20段切片。该区段内部空间较相对均质，沿河岸三级台地展开。该区段在切片9-12向内凹进，形成一个开阔的广场平台。

该区段使用人群在空间上的变化差异相对较小，人群数量较集中的地段为切片10及两侧地段。切片1-3以及切片14周边使用人群较少。该区段使用人群以中老年人为主，散步仍是主要活动类型，而在切片5-12以及切片14-15会有一定的停留休息人群。

在硬地和绿地分布上，整个区段绿道由于空间收放变化较小，且没有太多的集中活动场地，因此没有太大的波动。

在座椅设置上，该区段座椅数量座椅座位数目较少，大部分地段没有座椅，在切片9和切片11座椅数目座椅座位数目相对较多，但每个地段也仅有6个。

Analysis of the characteristics of the interior space

S2 DHG-CCG is about 1.0km long, equally divided into 20 slices. The interior space of this segment is relatively homogeneous and spreads out along the tertiary platform at the riverbank. This segment is inwardly concave from slice 9 to slice 12, so as to form a wide-open plaza.

The number of people using this segment does not change too much. Slice 10 and the area on both sides are where the crowd is gathered. There are not many people from slice 1 to slice 3, as well as the surrounding areas of slice 14. Most people using this segment are mainly middle-aged and senior pedestrians. However, a certain amount of people choose to stop over from slice 5 to slice 12, as well as the area of slices 14 and 15.

In terms of distribution of hard sites and green space, since the space of this greenway segment does not change much, nor does it have much large space for activities, the data of these two elements do not fluctuate too much along the segment.

In addition, not too many bench seats are placed along this segment. Most of the slices don't even have any bench seats. On slices 9 and 11, the number of seats is relatively bigger, but no more than six bench seats are in each.

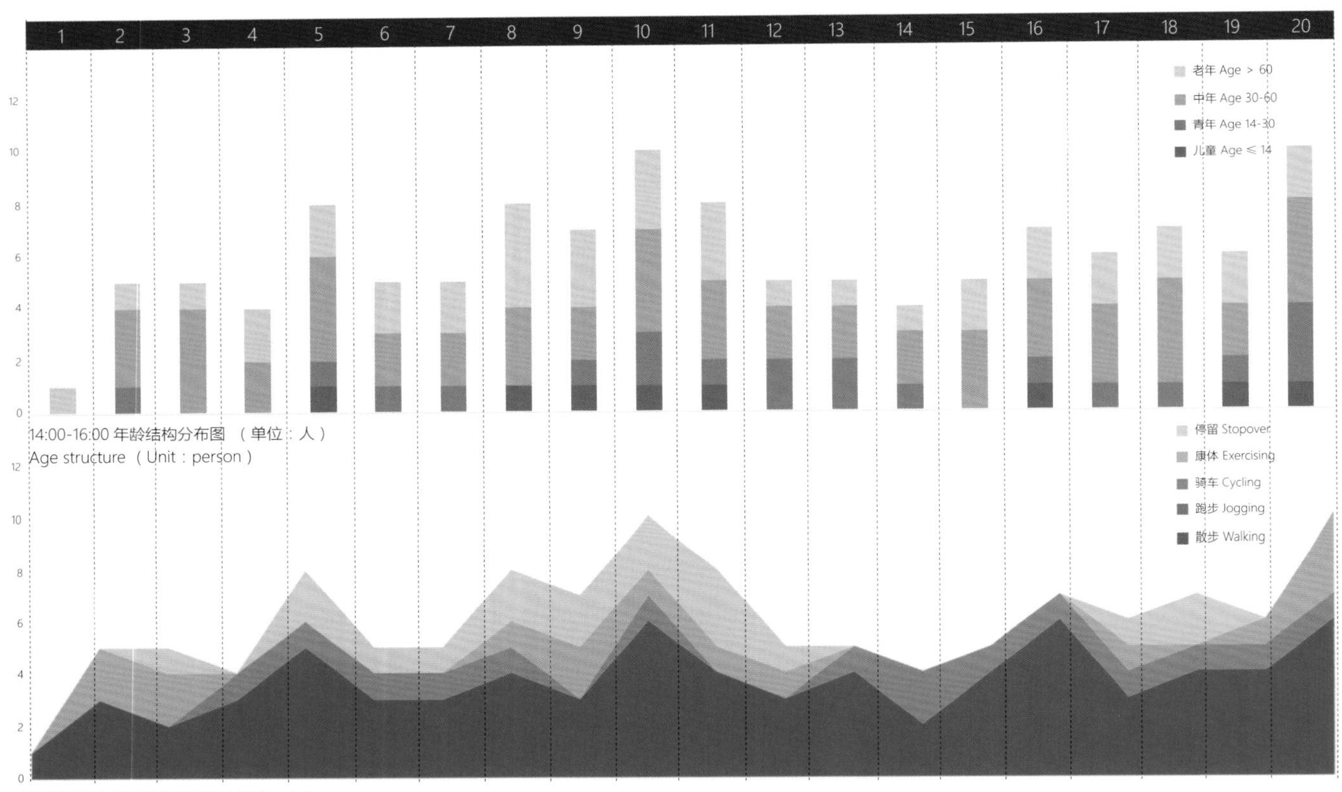

14:00-16:00 年龄结构分布图 （单位：人）
Age structure （Unit：person）

14:00-16:00 活动类型分布图（单位：人）
Structure of activity type（Unit：person）

两个时段的服务绩效分析

周末下午（14:00-16:00）使用人群主要分布在切片 5、切片 8-11 和切片 20 三个地段，在其他切片中则较均衡。其中切片 10、切片 20 使用人群规模最大，人群以中老年人为主，活动类型以散步和停留休息为主。

在人群年龄结构上，中老年人为该区段的主要使用人群。在绿道使用活动类型上，散步是该区段周末午后主要的活动类型，停留人群主要集中于切片 4-5、切片 8-11 及切片 18-19 三个地段，进行康体活动的人群则主要集中在切片 1-3、切片 8-10 以及切片 20。

The analysis of the service performance of two time periods

On a weekend afternoon (14:00-16:00), users are mainly distributed on slice 5, slices 8-11, and slice 20, with user distribution being relatively balanced on the other slices. Slices 10 and 20 have the largest numbers of users, and most of them are middle-aged and senior people who mainly walk or stop over.

In terms of age structure, middle-aged and senior people are the majority of users on this segment. Walking is the main activity type on this segment on a weekend afternoon. Those who stop over are gathered around on slices 4-5, slices 8-11, and slices 18-19. Those who exercise are mainly gathered around on slices 1-3, slices 8-10, and slice 20.

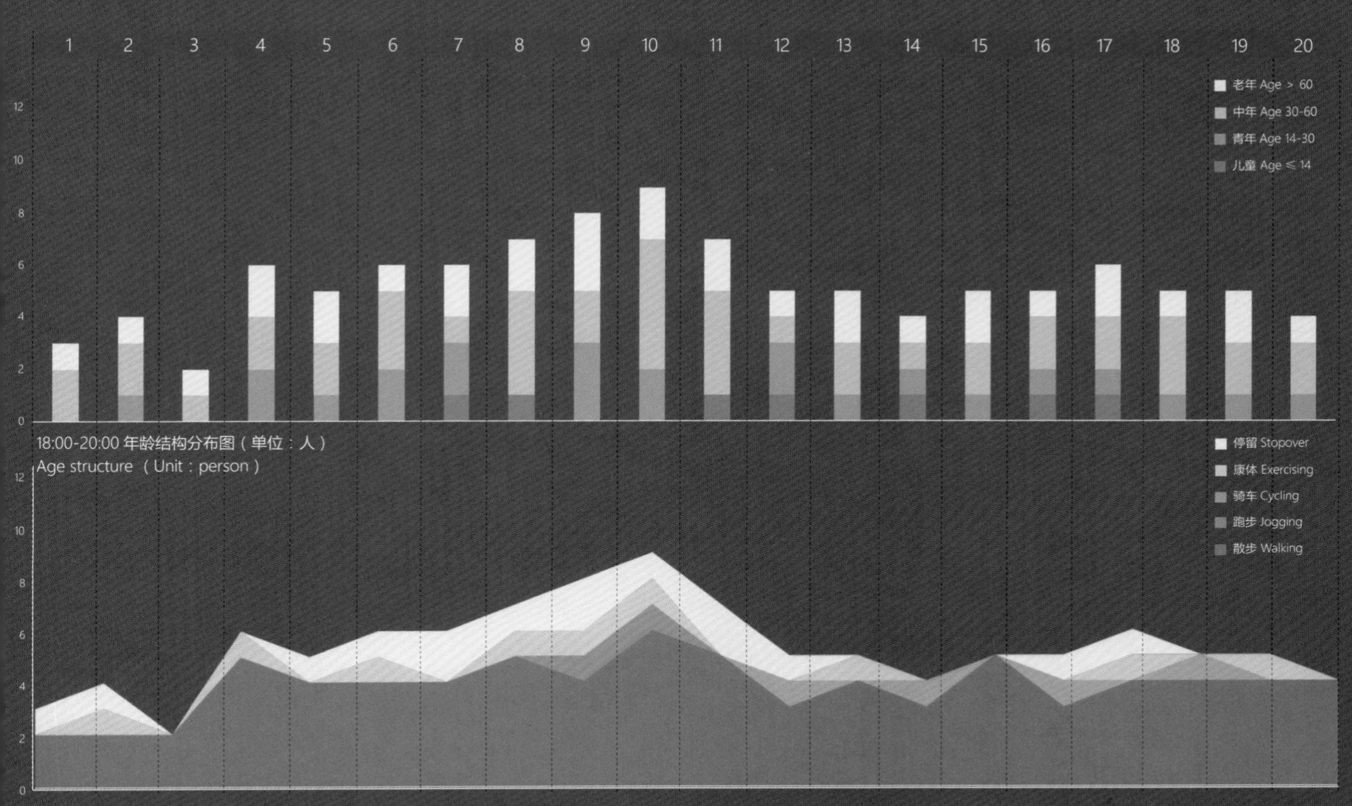

日常傍晚（18:00-20:00）各切片单元使用人群规模波动相对较小，以中老年人群散步为主，进行停留休息和康体活动的人群主要在切片 1-2、切片 5-10 以及切片 17-18 三个地段，各种使用活动类型集中地段与周末下午基本一致。

Every day at nightfall (18:00-20:00), the number of users on all the slices does not fluctuate too much. Most users are middle-aged and senior people who mainly walk. Those who choose to stop over and exercise are mainly on slices 1-2, slices 5-10, and slices 17-18. The place of the most active areas for recreational activities is basically consistent with those on a weekend afternoon.

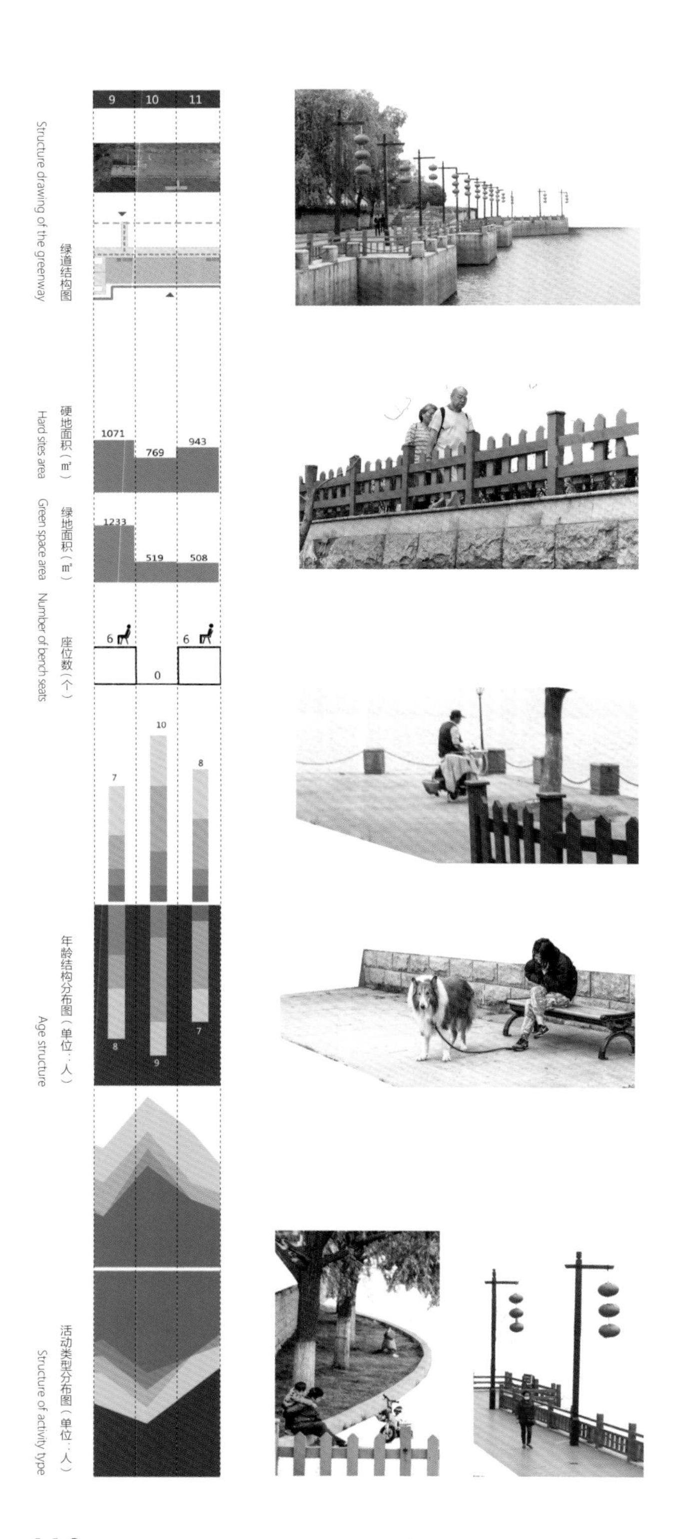

典型切片分析 1

切片 10-12 是该区段中使用人群相对集中的地段。从整体上看，这三个切片单元是该区段中硬质活动场地最集中的地段，也是该区段中使用人群停留休息、开展康体活动的重点地段。该地段的硬地、座椅等设施配置也是该区段当中最高的。

该地段在整个区段中相对均质的空间流线中形成突然开敞的效果，是该区段内最重要的硬质活动场地。该地段在周末下午和日常傍晚均能吸引较多人群停留休息和康体活动，是该区段内人气和活力最高的地段。

Analysis 1 of typical slices

Slices 10-12 cover an area where users are mostly gathered. Overall, these three slice units constitute the most centralized area for activities on hard sites and also occupy a very important area for users to stay and exercise on this segment. According to the survey, the hard site area, bench seats, and other facilities in this area are greatest on this segment.

This area gives an effect of sudden openness in the relatively homogeneous spatial streamline, thus it is the most important hard site for recreational activities on this segment. It can attract a lot of people to stop over or exercise both a weekend afternoon and every day at nightfall, so this area remains highly popular and full of vitality all the time.

切片 10　Slice 10
切片 11　Slice 11
切片 12　Slice 12

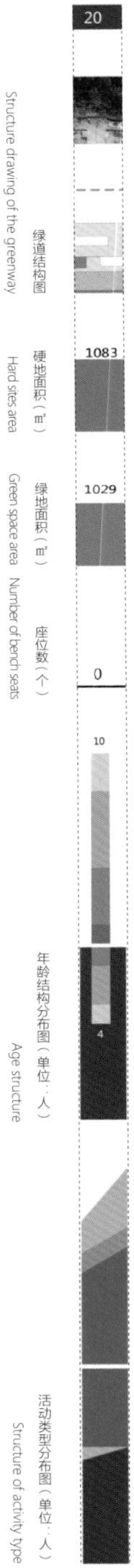

典型切片分析 2

切片 20 是该区段和城市空间联系较紧密的地段，使用人数明显高于周边切片，并成为周末下午使用人数最多的地段，人群中以年轻人居多，使用类型也相对多样。使用人群主要通过草场门大桥和南京艺术学院后街之间的入口广场进入绿道。

该切片硬地面积较大，但绿地和其他设施配置条件与其他切片相比并无特殊之处。日常傍晚的人群使用情况也与周边切片相差不大，唯独周末下午会有较多人群聚集，这在很大程度上与其处于绿道与城市接口的位置相关。该区段的核心要素为一个方形亲水平台，视野开阔，能够在周末下午吸引人群聚集停留和康体活动。

Analysis 2 of typical slices

Slice 20 is where this segment is closely linked to outside surroundings. The number of its users is evidently larger than other slices nearby and has become the area where the number of users is the largest on a weekend afternoon. Most of the people are middle-aged and young, doing diverse activities. Users enter the greenway mainly through the entrance plaza between Caochang Gate Bridge and the back street of Nanjing University of The Arts.

This slice has a large area of hard sites, but the quantity of green space and other facilities is not very distinctive, compared to that of the other slices. Every day at nightfall, its service condition is like that of the other slices. A weekend afternoon is the only time period when more people are gathered here, which, to a large extent, has something to do with the fact that it is located at the connecting point of the greenway and the city. The core element of this segment is a square hydrophilic platform with a wide view. On a weekend afternoon, it can attract crowds of people to gather, stay here, and exercise.

20

绿道结构图 Structure drawing of the greenway

硬地面积（㎡） Hard sites area 1083

绿地面积（㎡） Green space area 1029

座位数（个） Number of bench seats 0

年龄结构分布图（单位：人） Age structure 10 4

活动类型分布图（单位：人） Structure of activity type

区段 3（草场门—清凉门）抽象图
Abstractive drawing of S3（CCG-QLG）

区段 3: 草场门—清凉门
Segment 3: Caochang Gate-Qingliang Gate

内部空间特征概述

“鬼脸照镜”是该区段最重要的景点。该区段绿道穿越了石头城公园，来往游客络绎不绝。该区段在石头城公园以北部分狭长且单调，延续了区段 2（定淮门—草场门）均匀单调的线性空间特质，使用人数也较少。但随着秦淮河转向，在秦淮河拐点以南地段空间豁然开朗，绿道与石头城公园相互交融，使用人数也迅速增加。

Overview of the characteristics of the interior space

The most important scenic spot of this segment is “Historical Stone City, Funny Face Mirrors.” The greenway of S3 CCG-QLG goes through Stone City Park, where visitors come and go in an endless stream. The north of the segment is at the north of Stone City Park, narrow and dull. It continues the linear spatial characteristics of S2 DHG-CCG, so the number of users is small. However, as the Qinhuai River turns, the area to the south of the inflection point of the Qinhuai River is open and clear. The greenway and Stone City Park are intermingled, which results in a drastic increase in the number of users.

设施类别 Type of facilities	具体设施 Specific facilities	数据 Data
游径及交通设施 Trails and transportation facilities	步行道 Footpaths	5.0m
	自行车道 Bikeways	4.5m
	停车场 Parking lot	X
	无障碍道 Barrier-free roads	√
游憩服务设施 Recreational service facilities	游乐器具 Amusement equipment	√
	休息设施 Rest facilities	136 个
	文体活动场地 Cultural and sports field	6278m^2
	无线网 Wi-Fi	X
管理服务设施 Management service facilities	管理中心 Management Center	X
	游客服务中心 Tourist Service Center	√
	信息咨询点 Information consultant	X
商业服务设施 Commercial service facilities	售卖点 Selling	X
	自行车租赁点 Bike rentals	X
	自动售货机 Vendingmachines	√
科普教育设施 Popular science education facilities	科普宣传、展览设施 Popularization and propaganda of science, Exhibition facilities	√
	解说设施 Commentary facilities	X
环境卫生设施 Environmental hygiene facilities	公厕 Restrooms	X
	垃圾桶 Trash cans	√
安全保障设施 Safety & security facilities	照明 Lighting	√
	治安消防 Public security & fire control	√
标识信息设施 Signs	信息、指路、规章标识等 Information, directions, signs, etc.	√

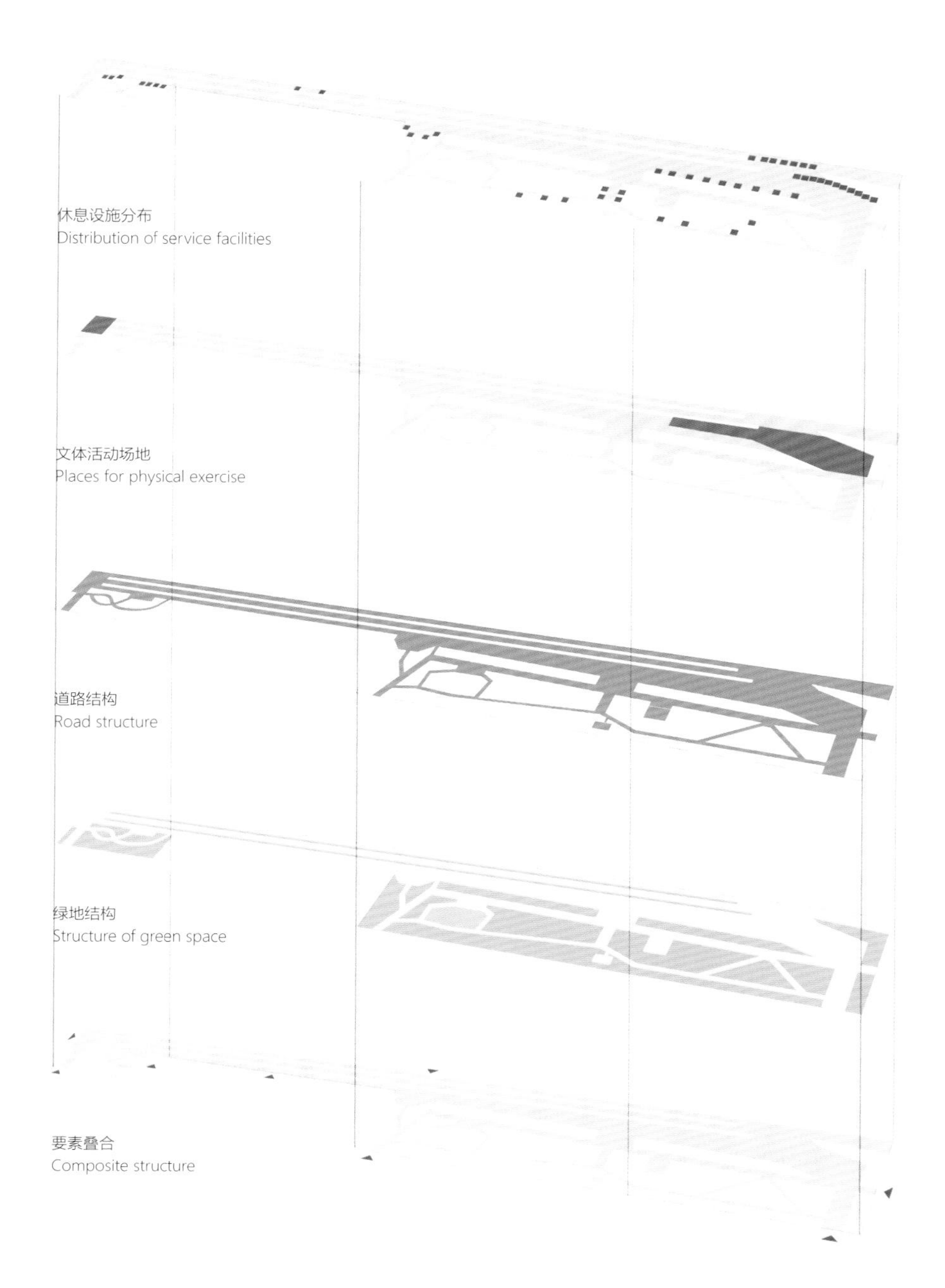

区段 3 根据其结构特征可分为南北两部分。北部与区段 2（定淮门—草场门）类似，是沿河岸形成的三级台地，座椅间距较大；南部则较开敞，与石头城公园相接，活动场地大，并且分布密集，游径与城墙之间有大片的开敞草坪和游步道，座椅等设施分布也较密集。

S3 CCG-QLG is divided into southern and northern parts according to its structural characteristics. Similar to S2 DHG-CCG, the northern part is a tertiary platform formed along the riverbank, with great intervals between bench seats. The southern part is more open and connected to Stone City Park. There are plenty of large sites for activities, which are densely distributed. Between the trails and the city wall are large areas of open lawns and footpaths. There is a high density of bench seats and other such facilities.

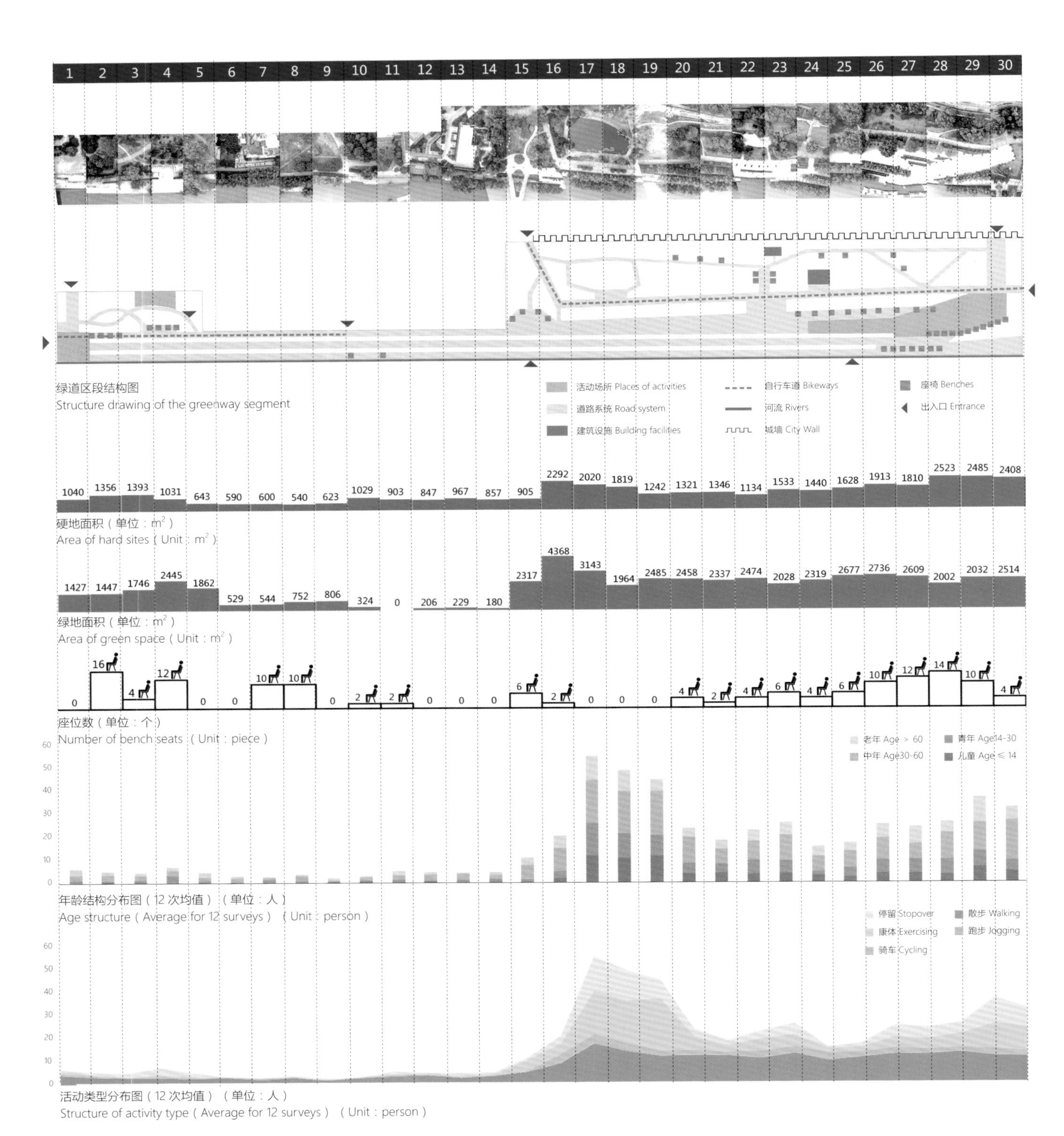

1 2 3 4 5 6 7 8 9 10 11 12 13 14 15 16 17 18 19 20 21 22 23 24 25 26 27 28 29 30
绿道区段结构图
Structure drawing of the greenway segment
活动场所 Places of activities
道路系统 Road system
建筑设施 Building facilities
自行车道 Bikeways
河流 Rivers
城墙 City Wall
座椅 Benches
出入口 Entrance
1040 1356 1393 1031 643 590 600 540 623 1029 903 847 967 857 905 2292 2020 1819 1242 1321 1346 1134 1533 1440 1628 1913 1810 2523 2485 2408
硬地面积（单位：m²）
Area of hard sites (Unit: m²)
1427 1447 1746 2445 1862 529 544 752 806 324 0 206 229 180 2317 4368 3143 1964 2485 2458 2337 2474 2028 2319 2677 2736 2609 2002 2032 2514
绿地面积（单位：m²）
Area of green space (Unit: m²)
0 16 4 12 0 0 10 10 0 2 2 0 0 0 6 2 0 0 0 4 2 4 6 4 6 10 12 14 10 4
座位数（单位：个）
Number of bench seats (Unit: piece)
老年 Age > 60
青年 Age14-30
中年 Age30-60
儿童 Age ≤ 14
年龄结构分布图（12 次均值）（单位：人）
Age structure (Average for 12 surveys) (Unit: person)
停留 Stopover
散步 Walking
康体 Exercising
跑步 Jogging
骑车 Cycling
活动类型分布图（12 次均值）（单位：人）
Structure of activity type (Average for 12 surveys) (Unit: person)

内部空间特征总体分析

区段 3（草场门—清凉门）长约 1.5km，共等分为 30 段切片。该区段呈南北走向，以中部人行天桥为界（切片 15），南北空间特征差异明显。南侧靠近石头城公园，空间宽阔开敞，节点空间丰富；北侧靠近居民区，以狭窄线性空间为主。

该区段空间的收放变化，直接反映在使用人数的变化趋势上。切片 1-14 的使用人数一直处在低位，使用类型以散步为主。从切片 15 开始使用人数陡然上升，一直到切片 30 都保持在相对高位，使用类型也开始多样化，并且部分切片（如切片 17-19、切片 30）停留休息和康体活动的人数都接近和超过了散步人数。切片 17-19 以及切片 28-30 是整个区段使用人数的最集中地段，亦是该区段主要的两处主要活动场地，能吸引不同年龄层次的人群停留休息或开展康体活动，其中切片 17-19 更是吸引大量儿童和青年驻足。

该区段硬地和绿地指标数据也集中体现了整个区段空间的收放变化，呈现南多北少的特点。除靠近草场门大桥地段切片（切片 1-5）外，在切片 15 北面几乎所有切片的硬地和绿地面积均不超过 1000m²。而在切片 15 南面切片中，除了切片 22 外，所有切片的硬地面积均在 1200m² 以上，并且除切片 18 外，绿地面积均在 2000m² 以上，特征差异十分明显。

在座椅分布上，切片 20-30 座椅分布较连续，座椅座位数目较大；切片 1-19 座椅分布不连续，主要在节点空间放置较多，但其利用率相对南部要低。

Analysis of the characteristics of the interior space

S3 CCG-QLG is about 1.5km long, equally divided into 30 slices. This segment spreads from the north to the south, with a sharp difference in spatial characteristics between the north and the south. The south side is close to Stone City Park, with wide-open space, and is rich in node sites, while the north side is close to a residential area, mainly with narrow and linear space.

The spatial change of this segment directly reflects the change in number of users. The number of users on slices 1-14 remains small, and users mainly walk. From slice 15, the number of people using the greenway increases drastically. Until slice 30, the numbers of users on the slices in between remain large, and the types of activities are diverse. For some slices (e.g. slices 17-19, slice 30), the number of people who stop over and exercise exceeds the number of people who walk. In terms of the number of people using this segment, peaks appear on the area of slices 17-19 and slices 28-30, making both areas major places for recreational activities on this segment. They attract different age groups to stop over or exercise. Slices 17-19 attract a lot of children and young people.

The index data of the hard sites and green space on this segment show spatial change of the entire segment. That is to say, more changes take place on the south side than on the north side. Except for in the slices close to the slice of Caochang Gate Bridge, the areas of hard sites and green space of all the slices to the north of slice 15 are no more than 1,000m². In the rest of the slices to the south of slice 15—with the exception of slice 22—the areas of all the hard sites exceed 1,200m²; and, except for in slice 18, the areas of green space exceed 2,000m², with a noticeable difference in characteristics.

The distribution of bench seats on slices 20-30 is relatively continual, and the number of bench seats is large. That on slices 1-19 is not continual, and the number of bench seats is large around node sites. Overall, the utilization ratio of the bench seats on slices 1-19 is much lower than that on slices 20-30.

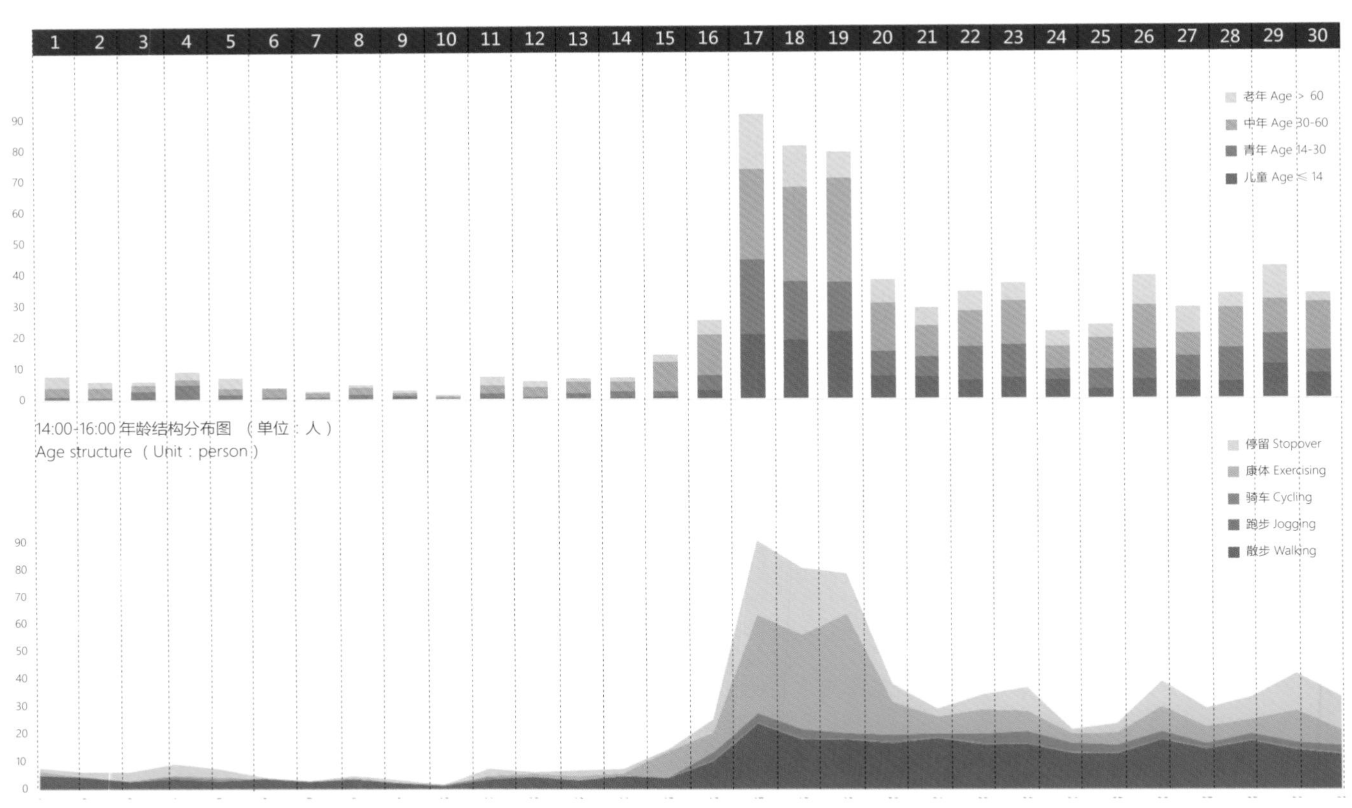

14:00-16:00 年龄结构分布图 （单位：人）
Age structure （Unit：person）

14:00-16:00 活动类型分布图（单位：人）
Structure of activity type（Unit：person）

两个时段的服务绩效分析

周末下午（14:00-16:00）使用人群在场地上的分布出现了较大波动，北部（切片 1-15）活动人数极少，且主要以散步为主，而南部（切片 16-30）使用人数较多，除了散步外，出现了大量进行康体活动与停留休息的人群，并在切片 17-19 达到最高点。该地段切片单元内各个年龄阶段的人群比例趋向于均衡，使用活动类型多样。

Analysis of service performance of two time periods

On a weekend afternoon (14:00-16:00), the distribution of users on this area fluctuates greatly. There are not many people on the north side (slices 1-15), and those who prefer to visit this area mainly walk. However, the number of users on the south side (slices 16-30) is larger. Apart from walking, a lot of people prefer to exercise and stop over, with the peak appearing on slices 17-19. Among the slice units of this area, the ratio of different age groups tends to be balanced, and the types of activities are diverse.

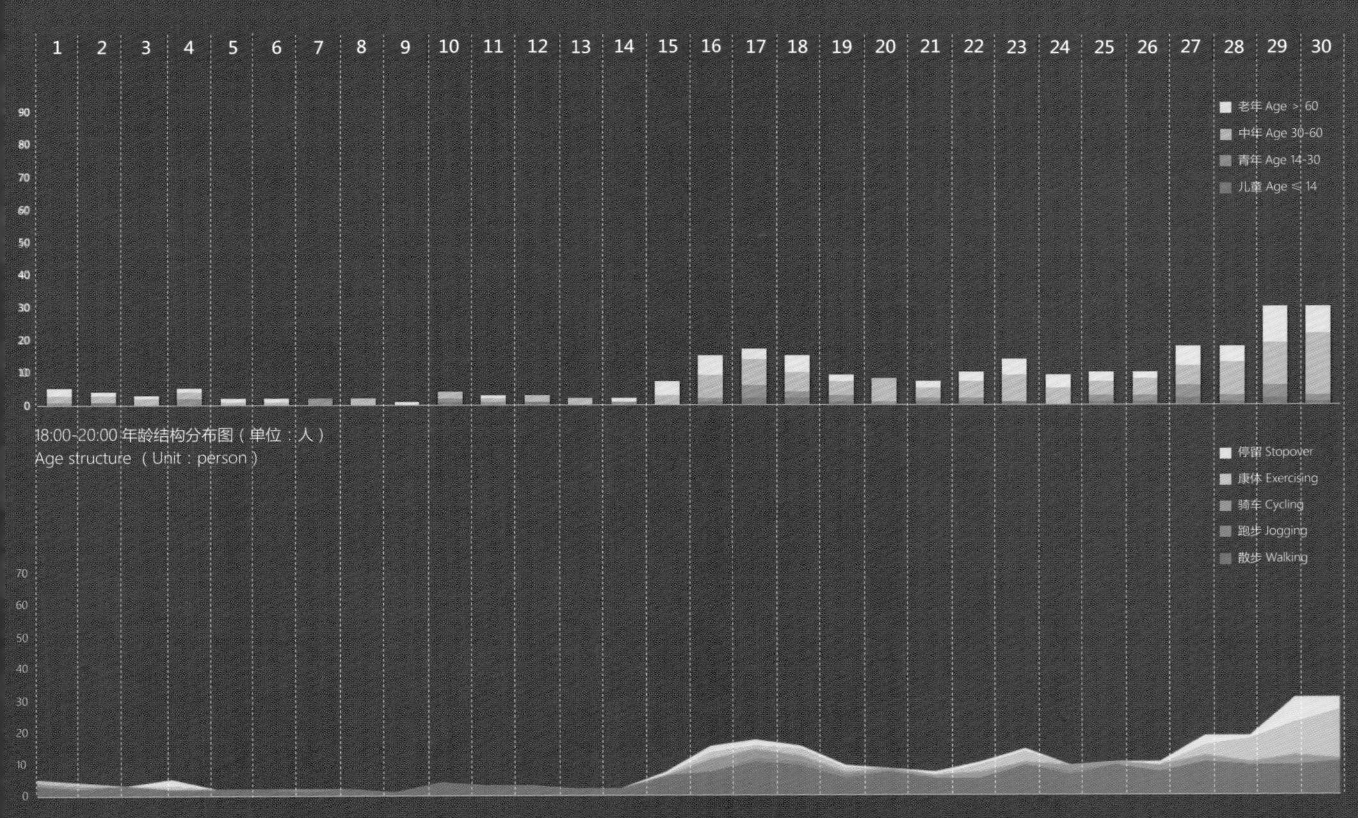

18:00-20:00 年龄结构分布图（单位：人）
Age structure （Unit：person）

18:00-20:00 活动类型分布图（单位：人）
Structure of activity type（Unit：person）

日常傍晚（18:00-20:00） 使用人群在该区段上的分布依然是南多北少，但波动幅度较之周末下午要小得多。使用人数最高区段也转移到了切片 29-30，整个区段以中老年人散步为主，从切片 26 开始停留休息和康体活动的人数开始明显增加。

场地上使用人数在切片 16-18，切片 29-30 形成两个高点。虽然这两个高点地段人群都以中老年人为主，但停留休息和康体活动的人数在切片 29-30 明显较高。

Every day at nightfall (18□00-20□00), there are still more users on the south side than on the north side, but the fluctuation range is much smaller than that on a weekend afternoon. The area with the biggest number of users also moves to the area of slice 29 and slice 30, marked by middle-aged and senior people who walk. From slice 26, the number of people who stop over and exercise increases drastically.

On the site, the number of users on slices 16-18 and on slices 29-30 form two peaks. Although most people on these two peaks are middle-aged and senior, the number of people who stop over and exercise on slices 29-30 is noticeably larger than that in slices 16-18.

切片 17 Slice 17
切片 18 Slice 18
切片 19 Slice 19

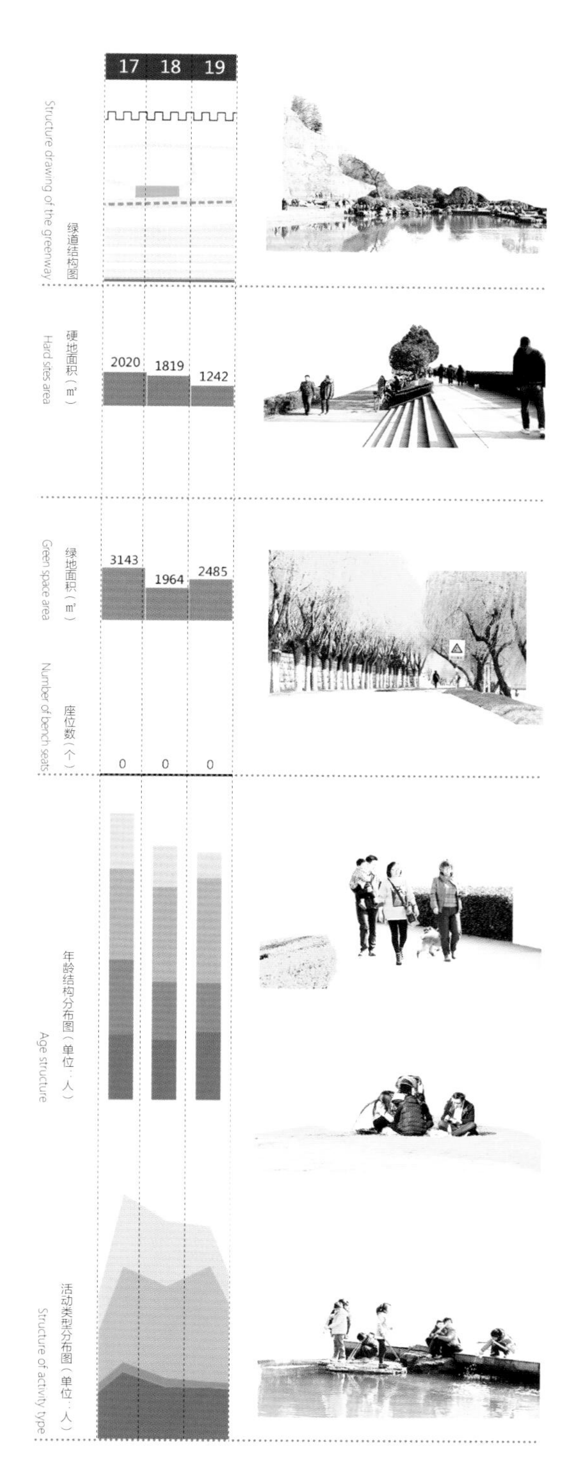

典型切片分析 1

切片 17-19 是该区段中周末下午使用人数最多的地段。该区段最具特色的景点“鬼脸照镜”就在此处，经常能吸引大量游客和儿童来此观光游玩，因此使用人群的年龄结构趋向年轻化，使用人群活动类型则以康体活动和停留休息为主。

与其他切片单元对比，该地段座椅寥寥，也没有集中宽敞的活动场地，却能吸引大量使用人群在此聚集。解析该地段的空间和人群分布特征我们发现，使用人群主要集中在镜子湖周围、鬼脸石壁前以及龟背巨石边。下午阳光灿烂的时候，孩子们围着镜子湖戏水打闹，大人们就坐在一边的石头上看护，几乎能把整个湖岸围得水泄不通。鬼脸石壁前也吸引了不少晒太阳、休息、拍照的人群，地上的石头、台阶成为首选的休息座椅。南侧草地上的龟背巨石成为孩子们的游乐园，以至于周边草地都能看出明显的践踏痕迹。因此，除了绿道整体空间在此骤然放开外，该地段成为人群集中逗留地的主要因素还有多元化的景观要素及其背后的文化积淀和内涵，这些均对外来游客和儿童产生了巨大的吸引力。

Analysis 1 of typical slices

The area of slices 17-19 is where the number of users is the largest on weekend afternoons. The most characteristic scenic spot “Funny Face Mirrors” of this segment is located in this area. It often attracts a lot of tourists and children to come, so the age structure of users of this area tends to be young. People go to this area mainly for the purpose of sightseeing, playing, and stopover.

Compared to the other slice units, we can find that this area does not have many bench seats, nor does it have any centralized spacious places for activities, but it can attract a lot of users to gather around. According to the analysis of characteristics of space and user distribution in this area, we can find that users mainly gather around Mirror Lake, the front of Funny Face on Stone City Wall, and the side of the Giant Turtleback Rock. On a sunny afternoon, small children gather around the Mirror Lake to play, and their parents sit on nearby rocks to watch out for them. The entire lakeshore is crowded with people. The front of Funny Face on Stone City Wall also attracts a lot of people who come here to bask in the sunshine, have a rest and converse, and take photos. The rocks on the ground and the steps become their primary bench seats for taking a break. The Giant Turtleback Rock on the grassy area of the south side becomes an amusement park for children, so we can see that trampling marks are all over the surrounding grassy area. Therefore, apart from sudden opening of the entire greenway space, the probable reasons why people are gathered around this area also refers to the big attraction, especially to the tourists and children, generated by the diversified landscape elements and the culture heritage and connotation behind them.

切片 28 Slice 28
切片 11 Slice 11
切片 29 Slice 29
切片 30 Slice 30

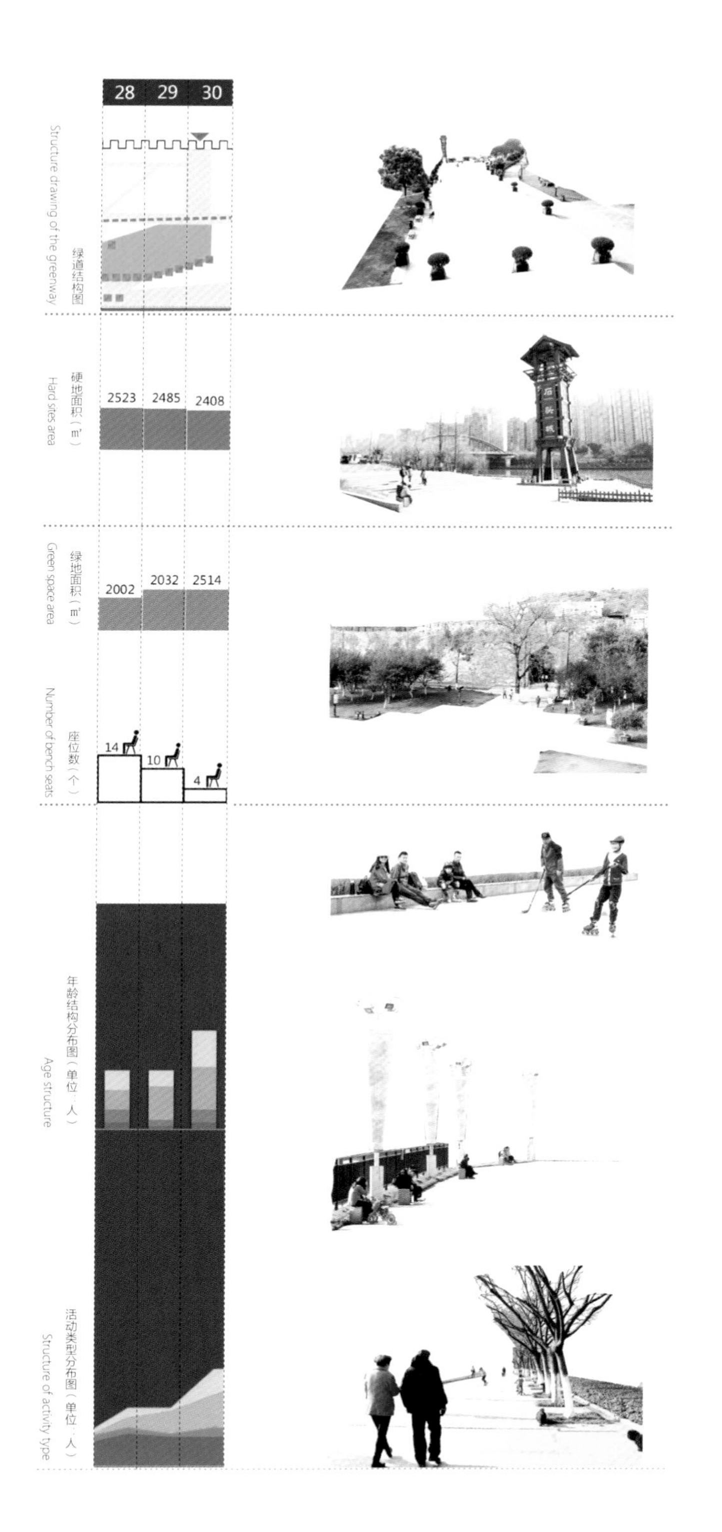

典型切片分析 2

切片 28-30 是该区段日常傍晚使用人数最多的地段。该切片位于绿道入口处，外秦淮河游船码头也设在这里。从指标上来看，该地段是整个区段硬地最集中的区域，这是由于结合游船码头设置了近 3000m² 面积开阔的广场，适合各种康体活动的开展。

通过空间和人群使用特征分析可以发现，在日常傍晚该广场是健身人群的聚集地，良好的灯光照明条件有力促进了傍晚活动的开展，以妇女广场舞为主，抖空竹的老人、溜旱冰的孩子、做健身运动的中年人等为该地段增添了大量的城市生活气息，同时广场一侧设置的整排座椅，在此时也会座无虚席。

在周末下午由于该地段的广场过于开阔而没有遮蔽，除了一些健身人群，整体使用人数相对较少。

Analysis 2 of typical slices

The area on slices 28-30 is where the number of users is the largest every day at nightfall. It is located at the entrance to the greenway, and the cruise terminal of the outer ring of Qinhuai River is also located here. Judging from indicators, we can find that this area is the most centralized for hard sites for the reason that a wide-open plaza of about 3,000m² is set up here in combination with the cruise terminal, making it ideal for doing all kinds of physical exercises.

According to the analysis of characteristics of space and users, we can find that this area is a gathering place for those who exercise every day at nightfall. Good lighting strongly promotes the development of evening activities which mainly include square dancing led by ladies, senior people playing diabolo, children going roller-skating, and middle-aged people exercising, thus adding a great deal of vitality to the city. Besides, one side of the plaza is equipped with an entire row of bench seats, which are all occupied during this time of the day.

On weekend afternoons, since the plaza of this area is too wide-open to have any shading, the number of users is relatively small, except for some who prefer to exercise.

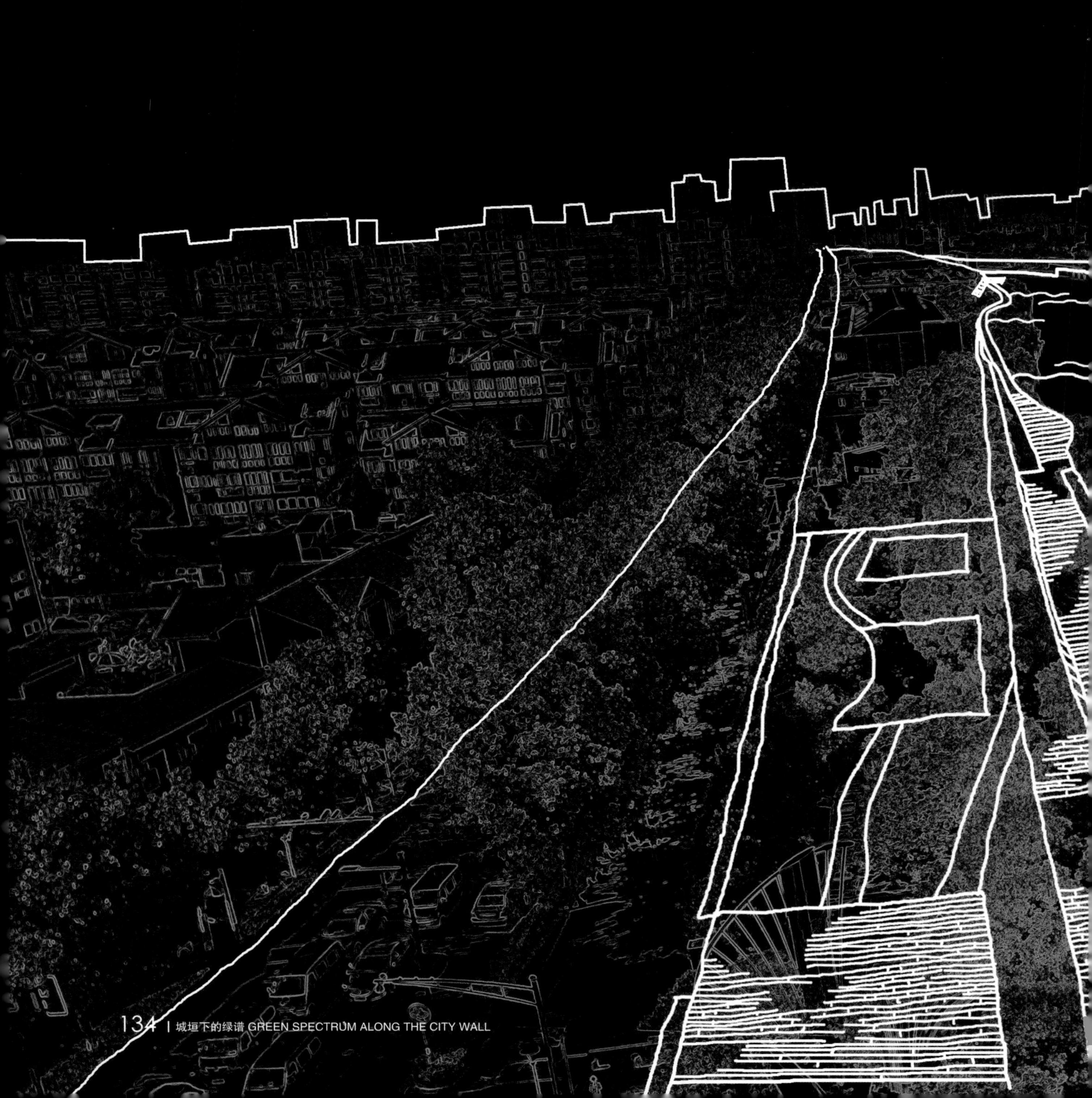

区段 4（水西门一集庆门）抽象图
Abstractive drawing of S4 (SSG – JQG)

区段 4：水西门—集庆门

Segment 4: Shuixi Gate-Jiqing Gate

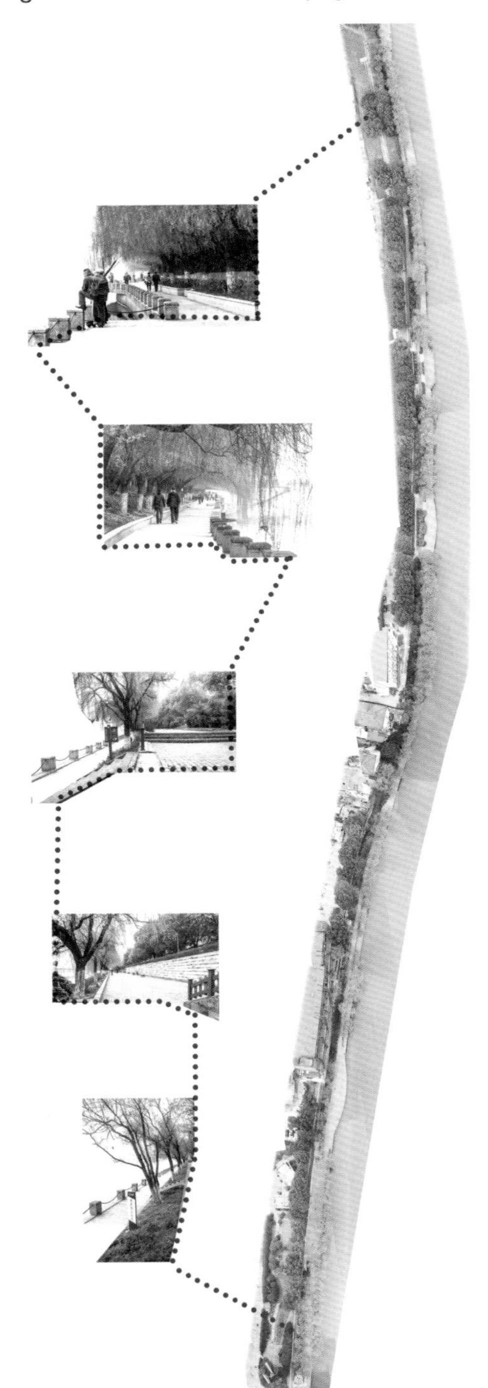

内部空间特征概述

区段 4（水西门—集庆门）是明城墙绿道当中空间位置比较特别的区段，受秦淮河东岸明城墙一侧空间所限，该区段绿道被设置在秦淮河西岸，与明城墙隔河相望。该区段自身空间较窄，中间停留场所也相对较少，除较少地段能吸引人群停留和活动外，其他空间利用率相对较低。

Overview of the characteristics of the interior space

Along the Nanjing Ming Dynasty City Wall Greenway, S4 SXG-JQG is very special in terms of spatial location. Limited by the space on one side of the Ming Dynasty City Wall on the east riverbank of the Qinhuai River, the greenway of this segment is set up on the west riverbank of the Qinhuai River, right across the river from the Ming Dynasty City Wall. Except for a few points in the middle, this segment has narrow space that is too limited for people to stop over, which also leads to a pretty low utilization ratio.

设施类别 Type of facilities	具体设施 Specific facilities	数据 Data
游径及交通设施 Trails and transportation facilities	步行道 Footpaths	2.0m
	自行车道 Bikeways	2.6m
	停车场 Parking lot	X
	无障碍道 Barrier-free roads	√
游憩服务设施 Recreational service facilities	游乐器具 Amusement equipment	X
	休息设施 Rest facilities	122 个
	文体活动场地 Cultural and sports field	1398m^2
	无线网 Wi-Fi	X
管理服务设施 Management service facilities	管理中心 Management Center	X
	游客服务中心 Tourist Service Center	X
	信息咨询点 Information consultant	X
商业服务设施 Commercial service facilities	售卖点 Selling	X
	自行车租赁点 Bike rentals	X
	自动售货机 Vendingmachines	X
科普教育设施 Popular science education facilities	科普宣传、展览设施 Popularization and propaganda of science，Exhibition facilities	X
	解说设施 Commentary facilities	X
环境卫生设施 Environmental hygiene facilities	公厕 Restrooms	X
	垃圾桶 Trash cans	√
安全保障设施 Safety & security facilities	照明 Lighting	√
	治安消防 Public security & fire control	√
标识信息设施 Signs	信息、指路、规章标识等 Information, directions, signs, etc.	√

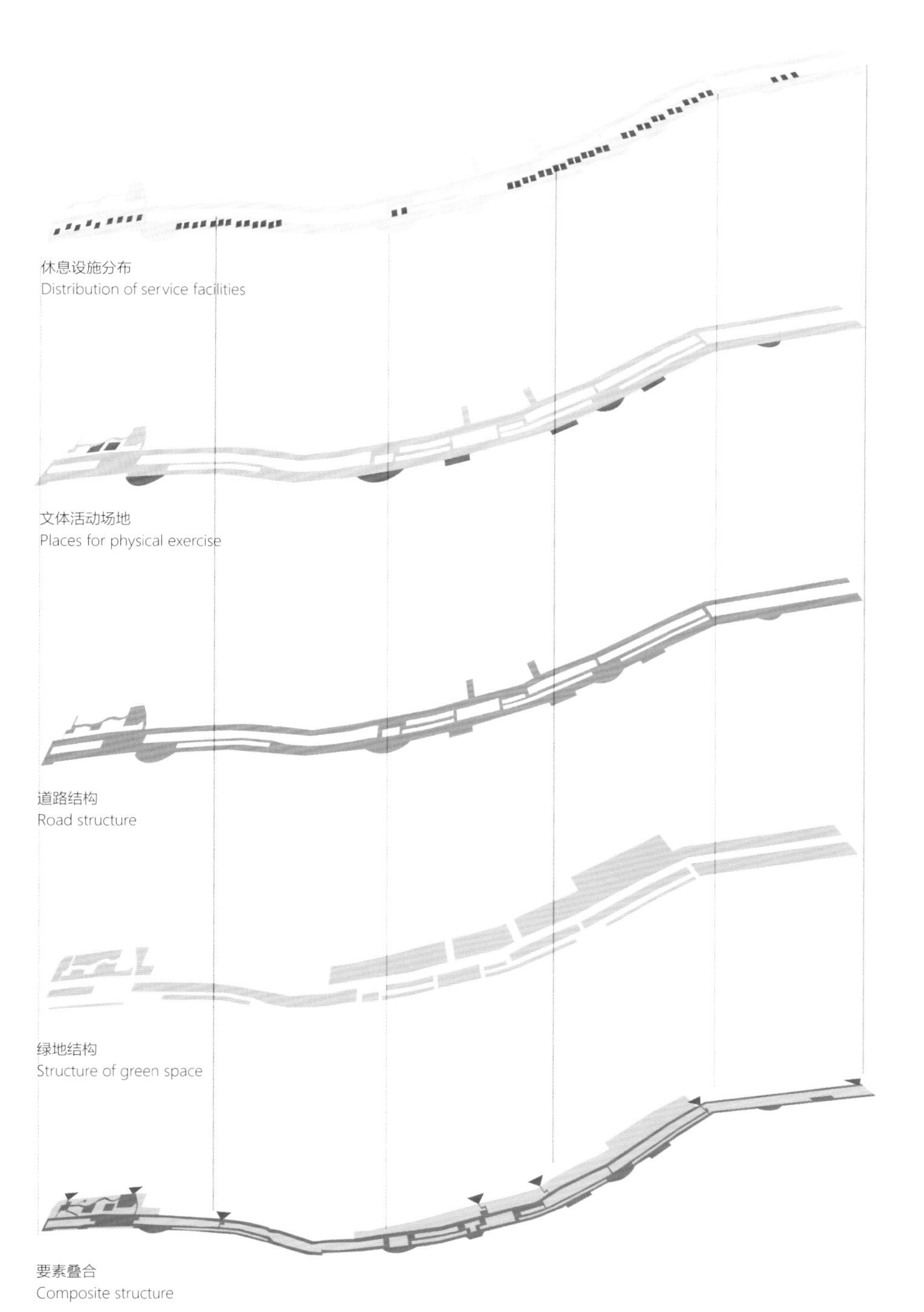

区段 4（水西门—集庆门）空间结构较狭长，沿河的主游径与一条紧贴西侧居民区的次游径构成绿道交通系统。沿河的主游径以综合慢行道为主，中间局部地段步行道和自行车道分离。该区段座椅座位密度较高，达到了 97 个 /km。该区段沿河一侧局部外凸形成一系列空间节点，南侧亲水平台是该区段中最大的硬质活动场地。

The spatial pattern of S4 SXG-JQG is dull and narrow. The main trail along the river and a sub trail right next to a residential area on the west form a greenway transportation system. The main trail along the river is mainly a comprehensive slow lane with the exception of the middle section, at which the footpath is separated from the bikeway. The bench seats of this segment are densely distributed at 97 per km. Part of the segment along the river is outwardly convex to form a series of spatial nodes. The hydrophilic platform on the south side is the biggest hard site in the entire segment.

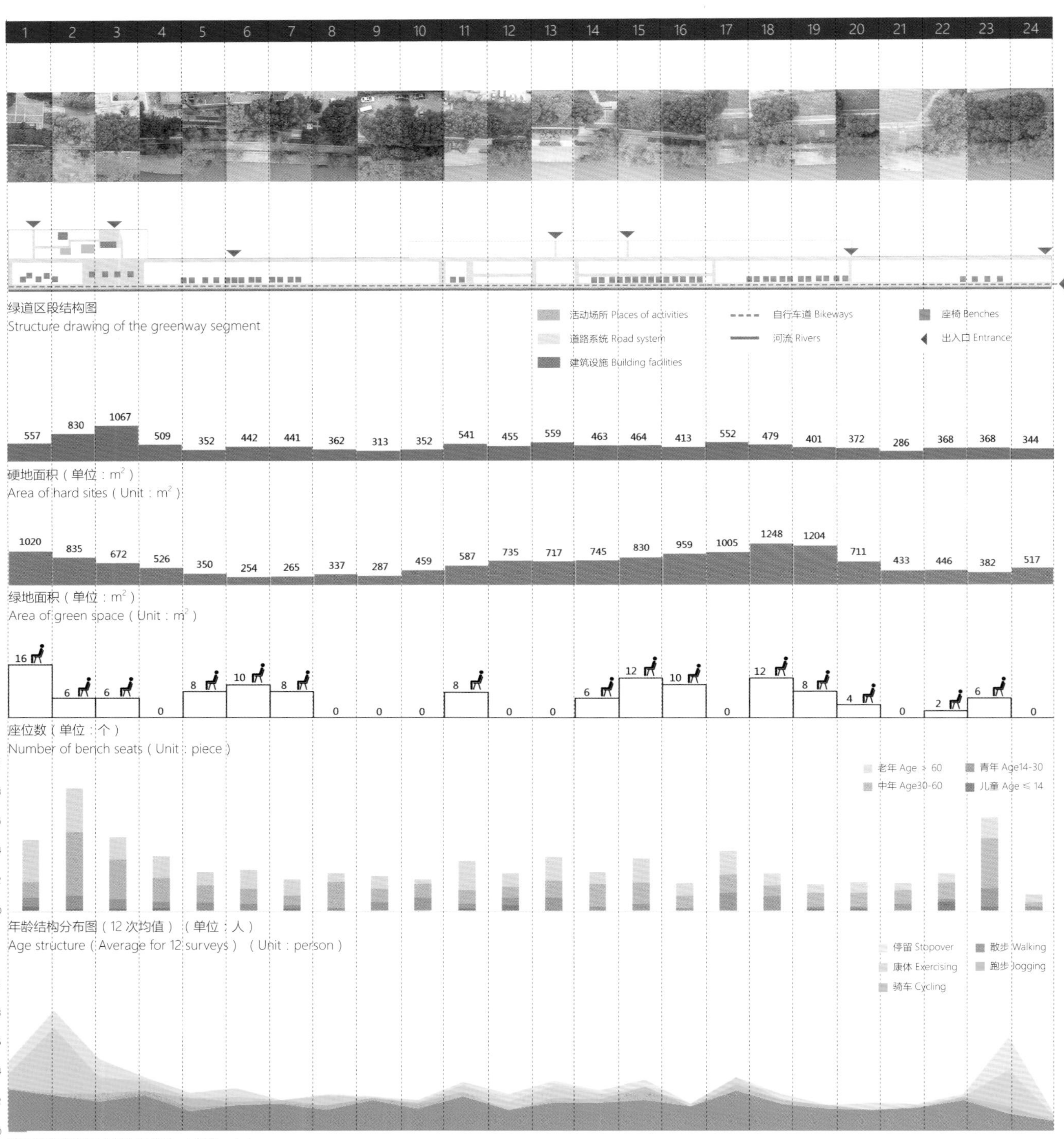

1 2 3 4 5 6 7 8 9 10 11 12 13 14 15 16 17 18 19 20 21 22 23 24
绿道区段结构图
Structure drawing of the greenway segment
活动场所 Places of activities
道路系统 Road system
建筑设施 Building facilities
自行车道 Bikeways
河流 Rivers
座椅 Benches
出入口 Entrance
557 830 1067 509 352 442 441 362 313 352 541 455 559 463 464 413 552 479 401 372 286 368 368 344
硬地面积（单位：m²）
Area of hard sites（Unit：m²）
1020 835 672 526 350 254 265 337 287 459 587 735 717 745 830 959 1005 1248 1204 711 433 446 382 517
绿地面积（单位：m²）
Area of green space（Unit：m²）
16 6 6 0 8 10 8 0 0 0 8 0 0 6 12 10 0 12 8 4 0 2 6 0
座位数（单位：个）
Number of bench seats（Unit：piece）
10 8 6 4 2 0
老年 Age ＞ 60
中年 Age30-60
青年 Age14-30
儿童 Age ≤ 14
年龄结构分布图（12 次均值）（单位：人）
Age structure（Average for 12 surveys）（Unit：person）
停留 Stopover
康体 Exercising
骑车 Cycling
散步 Walking
跑步 Jogging
10 8 6 4 2 0
活动类型分布图（12 次均值）（单位：人）
Structure of activity type（Average for 12 surveys）（Unit：person）

内部空间特征总体分析

区段 4（水西门—集庆门）长约 1.2 km，共等分为 24 段切片。该区段在六个区段中最为狭窄，绿道总宽度通常在 10 m 出头，局部空间会有小幅度放大。整个区段空间较为均质，在沿河一侧有少量出挑的亲水平台，形成驻留节点空间。整个区段仅在切片 1-3 有较集中的活动场地。

从指标分布情况可以看出，切片 1-3 以及切片 23 是该区段最主要的两处活动场地。除此之外，其他切片使用人群规模变化较小。该区段也是本书所调查的 6 个区段中使用人群总体规模最小的区段。

该区段绿地面积与硬地面积接近，由于空间较均质，数据上并未有明显波动。该区段中部有多个较宽的亲水平台节点，如切片 4、切片 11、切片 13、切片 16-18，但使用率却高低不一，如切片 4 有较多人群使用，切片 11 却较少。虽然整体空间较为狭窄，但是座椅座位密度相对较高，整个区段座位总数达到 116 个，且座椅座位密度较高节点分布均匀，如切片 1-3、切片 5-7 等地段。

Analysis of the characteristics of the interior space

S4 SXG-JQG is about 1.2km long, equally divided into 24 slices. This segment is the narrowest one of the six segments. The total width of the greenway is usually just more than 10m, with partial space being slightly enlarged. The spatial pattern of the entire segment is relatively homogeneous. Some hydrophilic platforms at the riverside form the key node spaces for people to stop over. Only slices 1-3 have relatively centralized sites offering chances for different activities.

Judging from data distribution, we can see that slices 1-3 and slice 23 are the two areas that are most attractive to recreational activities on this segment. Besides, the other slices do not change too much in terms of the number of users. The number of users of this segment is the smallest of the six segments.

The area of green space is close to that of the hard sites without too much fluctuation because of the homogeneous space. The middle of the segment is embedded with several wide hydrophilic platform nodes, such as slices 4, 11, 13, 16, 17, and 18, but utilization ratios of the platforms fluctuate significantly. For example, slice 4 is used by a lot of people, while slice 11 is usually empty. Although most space is relatively narrow, bench seats are densely distributed. The number of bench seats in the entire segment is up to 116, and the node spaces where bench seats are located are evenly distributed, such as in slices 1-3, slices 5-7, etc.

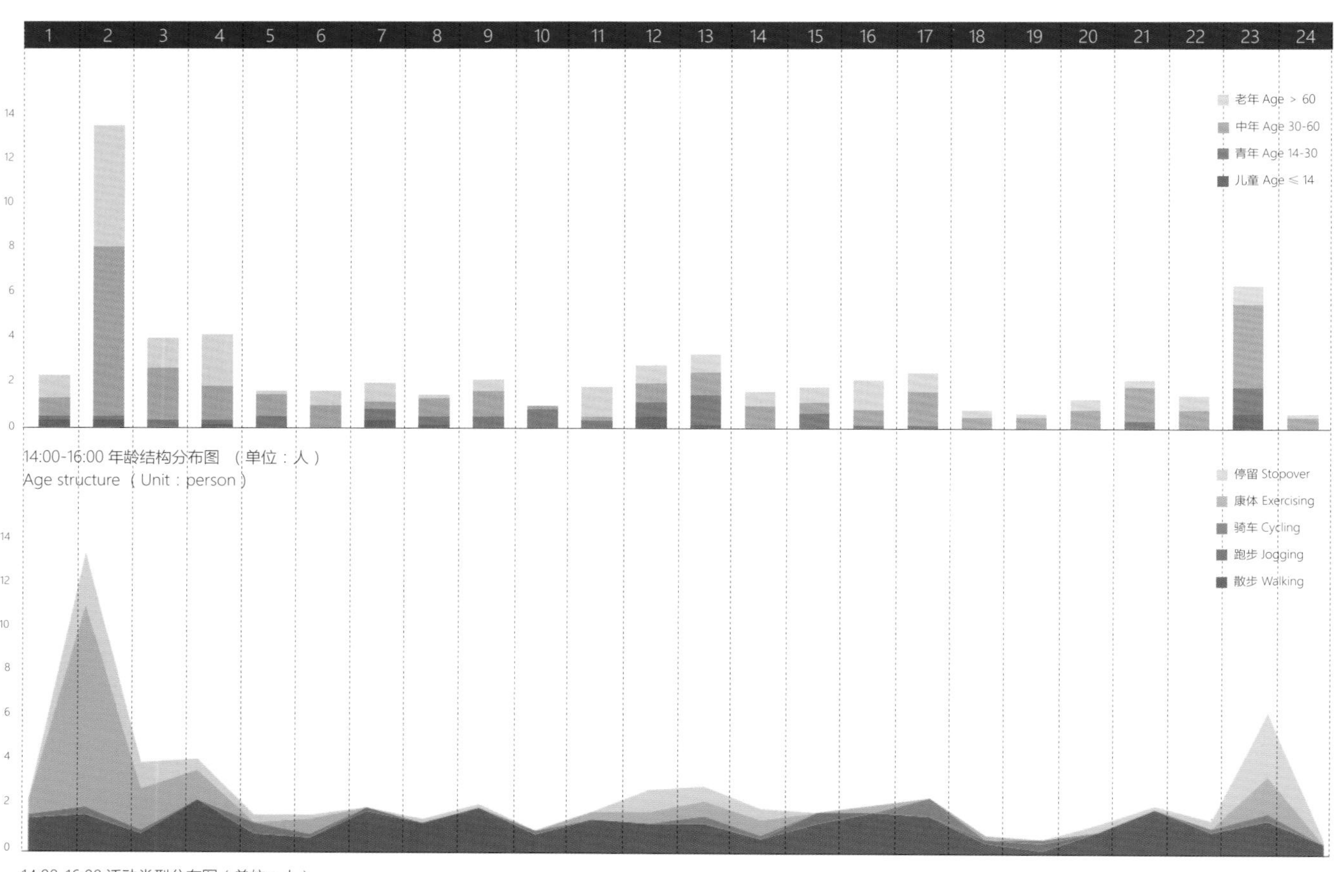

14:00-16:00 年龄结构分布图（单位：人）
Age structure（Unit：person）

14:00-16:00 活动类型分布图（单位：人）
Structure of activity type（Unit：person）

两个时段的服务绩效分析

周末下午（14:00-16:00）使用人群在场地上的分布呈现两端高、中间低的趋势，切片 2 与切片 23 是使用人群最集中的地段。整体而言，该区段的使用以中老年人散步为主，进行康体活动与停留休息的人群主要集中出现在切片 2-3、切片 12-14、切片 23 地段。

Analysis of service performance of two time periods

On a weekend afternoon (14:00-16:00), users are distributed in such a way that the number of people on both ends is large while that in the middle is small. Slices 2 and 23 are where most people gather. Overall, most people using this segment are seniors who mainly walk, while those who exercise or stop over mostly gather around slices 2-3, slices 12-14, and slice 23.

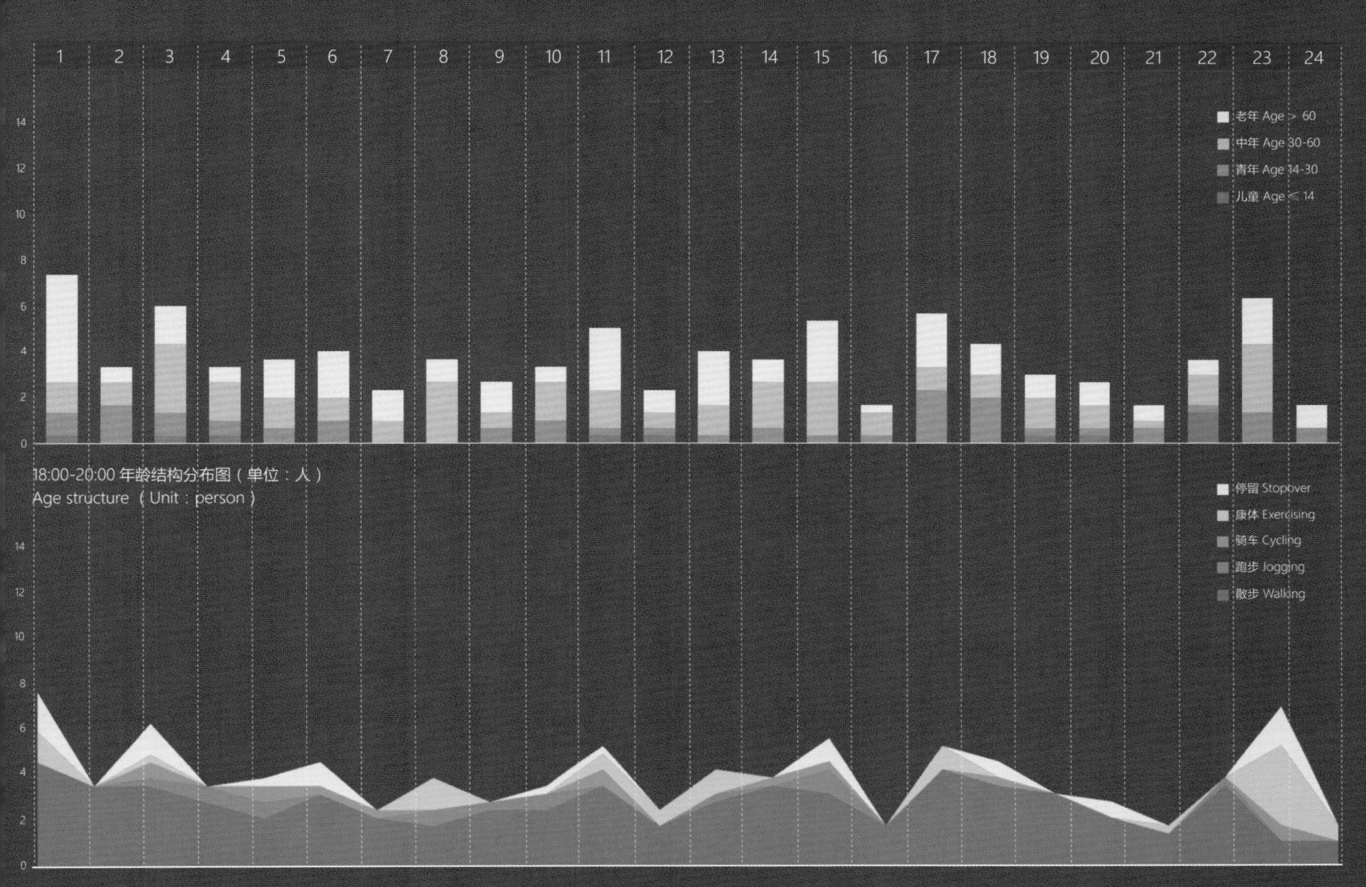

18:00-20:00 年龄结构分布图（单位：人）
Age structure （Unit：person）

18:00-20:00 活动类型分布图（单位：人）
Structure of activity type（Unit：person）

日常傍晚（18:00-19:00）该区段总体使用人数高于周末下午，人群依然以中老年人为主，但使用活动多样性降低，散步成为主导使用活动类型，也导致使用人数在该区段中均匀流动，波动较周末下午大幅度减小。与周末下午相比，切片 2 的使用人数明显降低，但切片 23 却对人群维持了较高吸引力，并聚集了该区段最多的停留休息和康体活动的人群。

Every day at nightfall (18:00-19:00), the total number of people using this segment is larger than that on a weekend afternoon. The people who use this segment are mainly senior people, but the types of activities are simple. Walking becomes the dominant activity, resulting in a balanced flow of users on this segment and a significant reduction of fluctuation in comparison to that on a weekend afternoon. Compared to weekend afternoons, the number of users on slice 2 decreases drastically. However, slice 23 remains very popular among users and attracts the most users who are fond of staying and exercising there.

切片 11
Slice 11

切片 17
Slice 17

切片 23
Slice 23

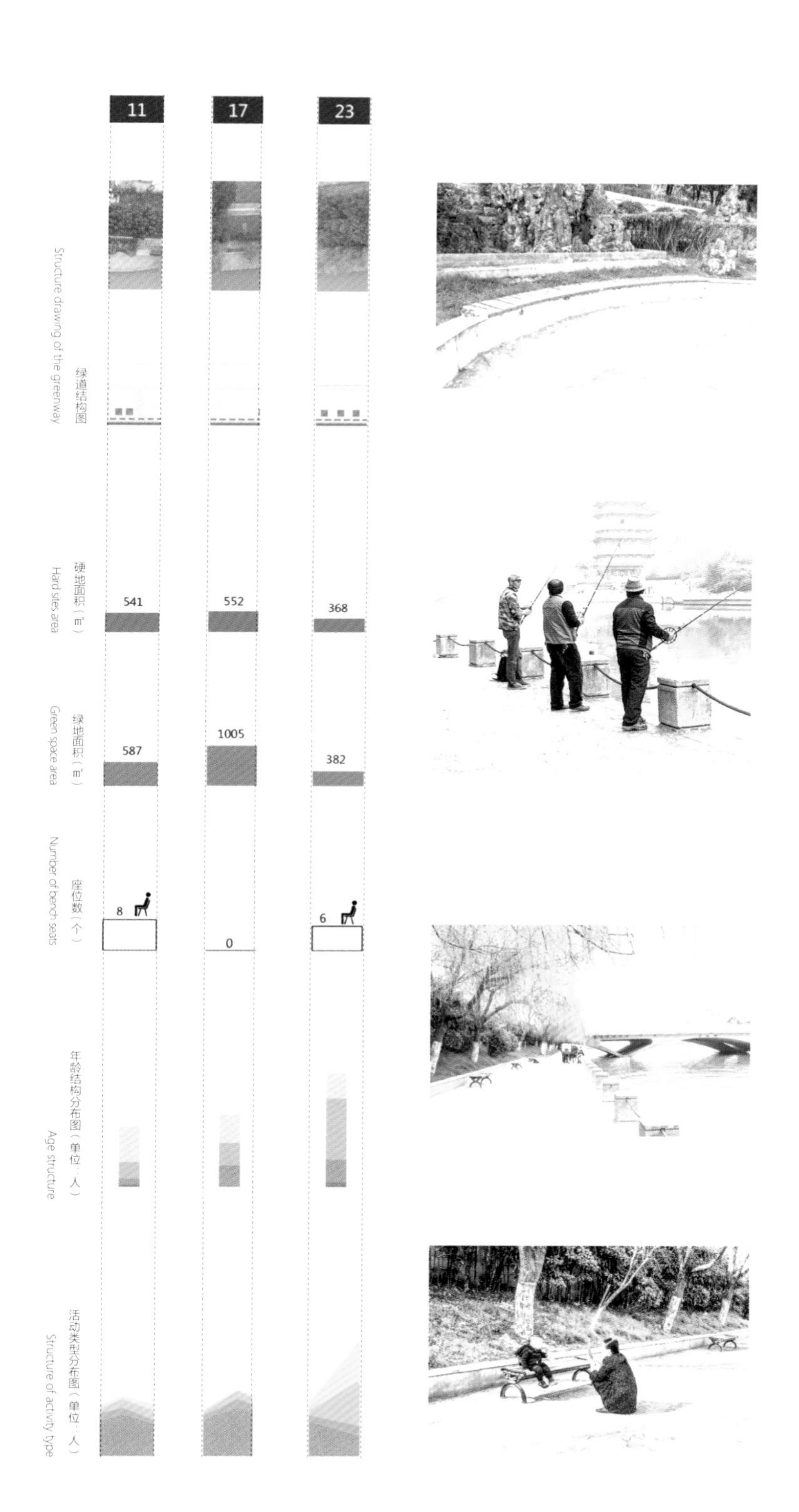

典型切片分析 1

从硬地和绿地要素指标上来看，切片 11 和切片 17 均优于切片 23，但在使用人数上切片 23 则最多，并且使用活动类型也最丰富，形成了强烈反差。

其中，这三个切片主要活动场地均在亲水地段，其中切片 11 和切片 17 略微出挑，整个空间形式和给人的感知仍是交通型为主的滨水步道，但切片 23 硬地则向内侧凹进，并明确界定出了停留空间和滨水步道空间，同时切片 23 种植坛边缘设计成可坐型边缘，并能在座椅座位被占用时继续吸引人群坐下停留休息。这些空间界定和设施设计的细节差异，在很大程度上造成看似相近的两类空间对使用人群吸引力的不同。

Analysis 1 of typical slices

Reviewing the indicators of the area of hard sites and green space, we find that slices 11 and 17 are better than slice 23. However, the number of people using slice 23 is the largest, and the types of activities are the richest, so as to show a sharp contrast in between.

The sites of activity of all three slices are located in waterfront areas. Slices 11 and 17 are slightly and outwardly convex over the water. In terms of the entire spatial pattern and human perception of visitors, both slices mainly consist of waterfront footpaths that are used for traffic. However, the sites of slice 23 are concave and clearly define the space for stopover and waterfront footpath. Meanwhile, the edge of the planting beds on slice 23 is designed in a way for people to sit on so that there is more seating when all the bench seats are occupied. These detailed differences in spatial definition and facilities design bring about a huge difference in the attraction of these two seemingly similar types of spaces to people.

2

绿道结构图 Structure drawing of the greenway

硬地面积（m²）Hard sites area
830

绿地面积（m²）Green space area
835

座位数（个）Number of bench seats
6

年龄结构分布图（单位：人）Age structure

活动类型分布图（单位：人）Structure of activity type

典型切片分析 2

切片 2 设置有两块较集中的活动场地，场地周边绿荫密布，是周末下午使用人群最集中的场所。此处是整个区段中硬质场地面积最大的地段，因此常见的场景是在天气适宜的午后，大量老人在此抖空竹、打羽毛球等。

解析周末下午和日常傍晚使用人数差异的原因发现，日常傍晚该地段的照明设施不足，导致傍晚时较为阴暗，空间特质无法满足作为该时段使用主体的中老年人聚集开展活动的需求，从而很大程度上使得该地段人气骤减。

Analysis 2 of typical slices

Slice 2 is equipped with two centralized places for recreational activities, with the surroundings dotted by green shades. This area is where most people go on a weekend afternoon. With the biggest area of hard sites on the segment, this slice is able to offer plenty of different recreational opportunities. Among all the activities, diabolo and badminton played by senior people are the most popular ones, especially on a fine day.

According to the analysis of the reason for the difference in the number of users between the two time periods in the survey, we can find that this slice lacks enough lighting every day at nightfall, so that it is very dim in the evening which can hardly meet the needs of the group activities organized by middle-aged and senior people who are the majority of users during this time period.

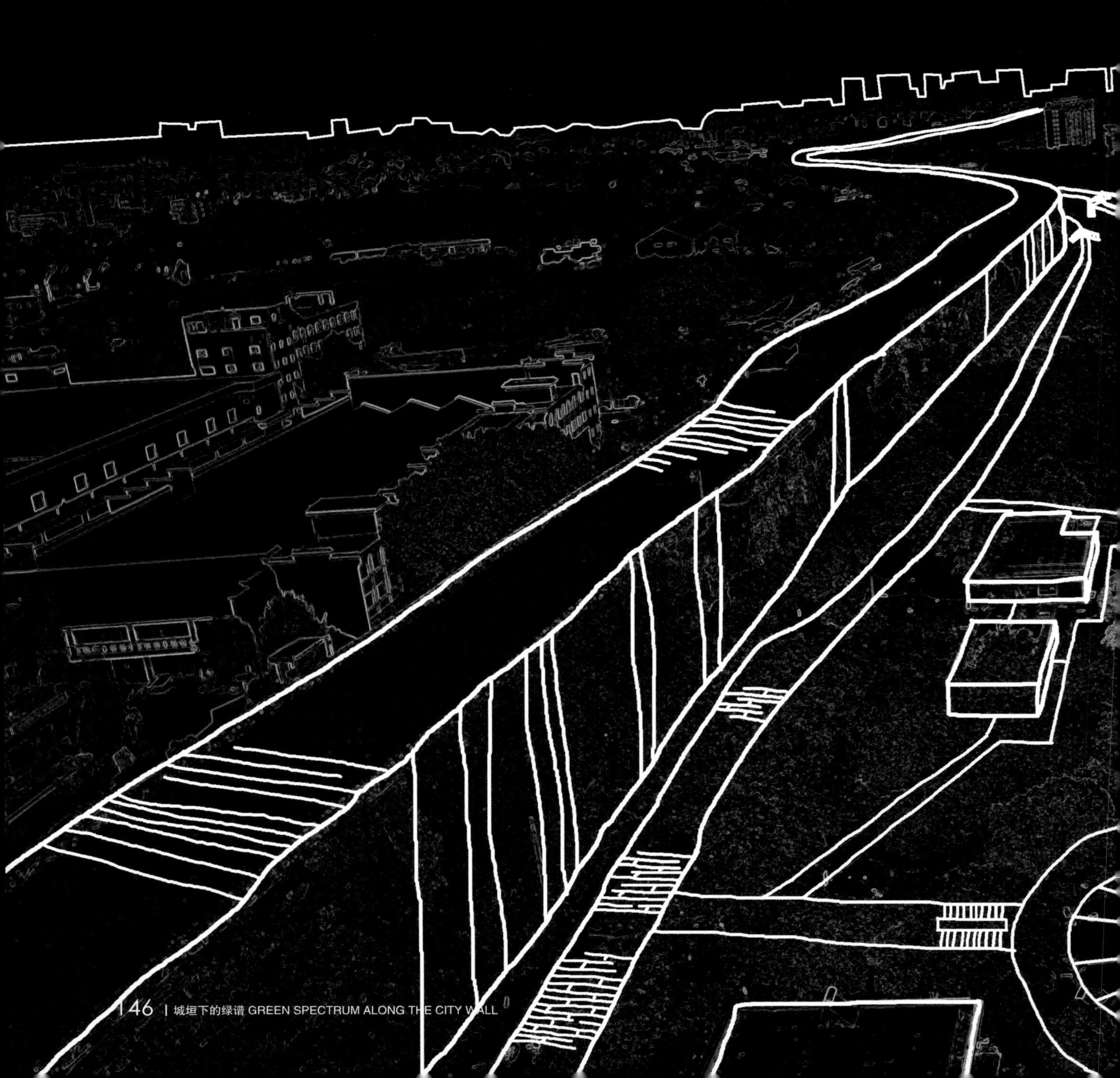

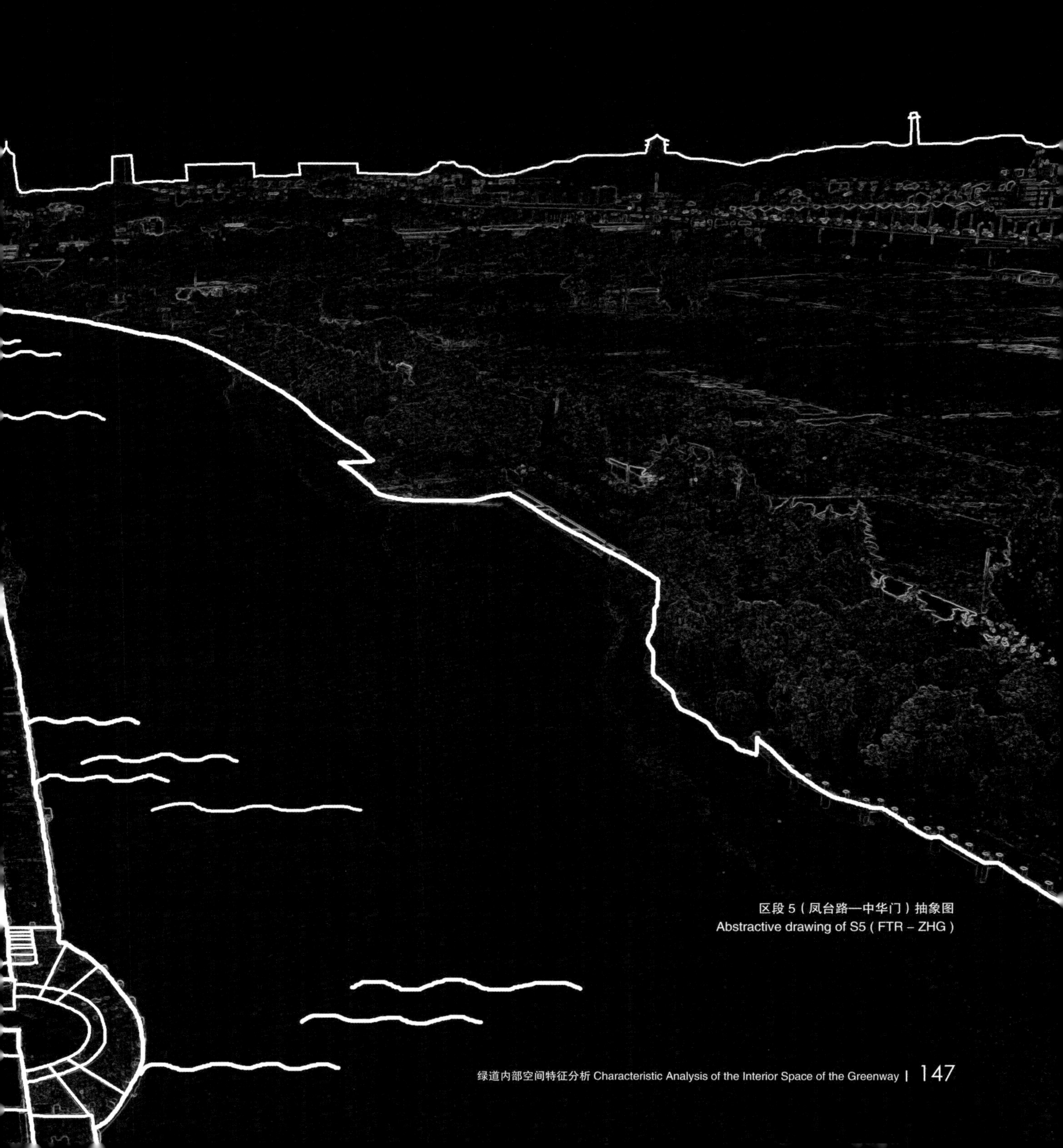

区段 5（凤台路—中华门）抽象图
Abstractive drawing of S5 (FTR – ZHG)

区段 5：凤台路—中华门
Segment 5: Fengtai Road-Zhonghua Gate

设施类别 Type of facilities	具体设施 Specific facilities	数据 Data
游径及交通设施 Trails and transportation facilities	步行道 Footpaths	4.2m
	自行车道 Bikeways	4.2m
	停车场 Parking lot	176m^2
	无障碍道 Barrier-free roads	√
游憩服务设施 Recreational service facilities	游乐器具 Amusement equipment	X
	休息设施 Rest facilities	44 个
	文体活动场地 Cultural and sports field	1836m^2
	无线网 Wi-Fi	X
管理服务设施 Management service facilities	管理中心 Management Center	√
	游客服务中心 Tourist Service Center	X
	信息咨询点 Information consultant	√
商业服务设施 Commercial service facilities	售卖点 Selling	X
	自行车租赁点 Bike rentals	X
	自动售货机 Vendingmachines	X
科普教育设施 Popular science education facilities	科普宣传、展览设施 Popularization and propaganda of science、 Exhibition facilities	X
	解说设施 Commentary facilities	X
环境卫生设施 Environmental hygiene facilities	公厕 Restrooms	√
	垃圾桶 Trash cans	√
安全保障设施 Safety & security facilities	照明 Lighting	√
	治安消防 Public security & fire control	√
标识信息设施 Signs	信息、指路、规章标识等 Information, directions, signs, etc.	√

内部空间特征概述

秦淮之北，城墙之南，区段 5（凤台路—中华门）位于中华门瓮城以西、凤台路以东。该区段贴近南京老城中心区，周边绿地面积较少，绿道成为居民日常游憩的主要场地。在这里能看到充满本地特色的休闲活动场景，例如，大量老年人自带小板凳，围成若干组团打牌谈天，饶有意趣。

Overview of the characteristics of the interior space

At the north of the Qinhuai River and at the south of the City Wall, S5 FTR-ZHG starts from the west of the barbican entrance to Zhonghua Gate and ends at the east of Fengtai Road. This segment is close to old downtown Nanjing, with a serious lack of green space in the surroundings. Therefore, the greenway becomes the main daily recreational destination for nearby residents. On this segment, people can experience a local leisure atmosphere and culture with a lot of activities sponsored by local people. For instance, a lot of senior people bring their own stools to form several enclosed spaces with groups and play cards or chat, which can usually go on for the whole afternoon.

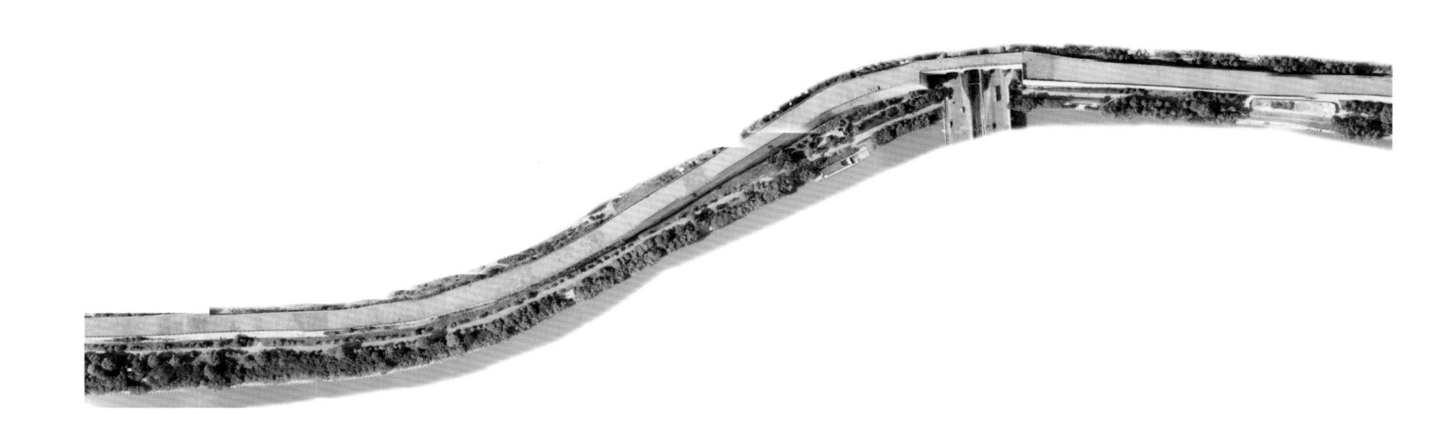

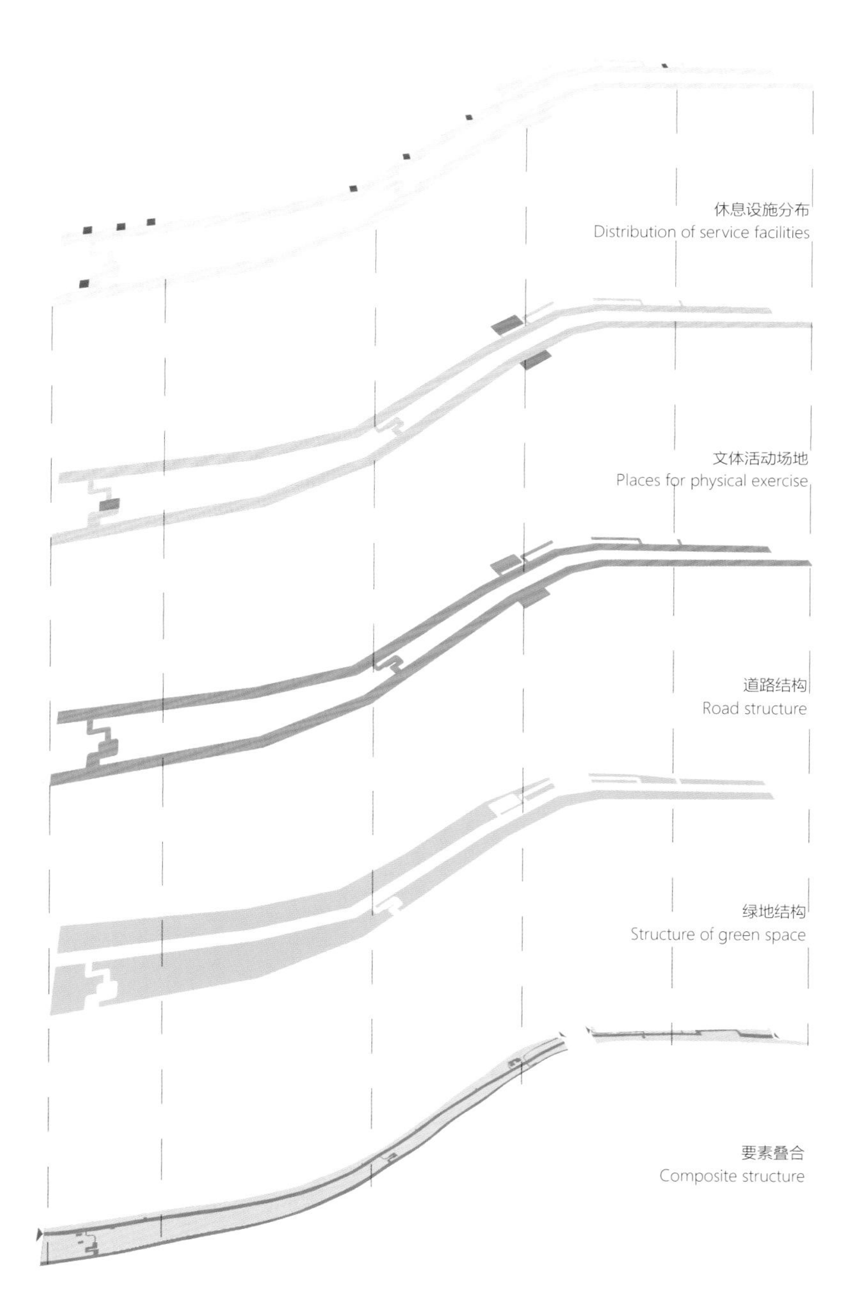

区段 5（凤台路—中华门）结构简单清晰，一条较宽的综合慢行道沿城墙布置，两侧是较宽的绿地，靠近秦淮河有一条亲水游步道。绿道每 200-300m 布有联系上下层的台阶。绿道的东西两端出入口以及与饮马桥交接处有面积不大的集中硬质活动场地和服务建筑。绿道区段上的座椅散布，相互间距离较大。

S5 FTR-ZHG owns a simple and clear structure. It is equipped with a wide comprehensive slow lane along the City Wall, with a wide green space on both sides. There is one more hydrophilic footpath close to the Qinhuai River. There are steps to a vertically connected terrace space at the riverbank every 200-300 m. Although not very large, several centralized sites for recreational activities and service buildings can be found at the greenway entrances of the east and west ends and the intersection of this greenway segment and Yinma Bridge. Bench seats are dispersed at large intervals.

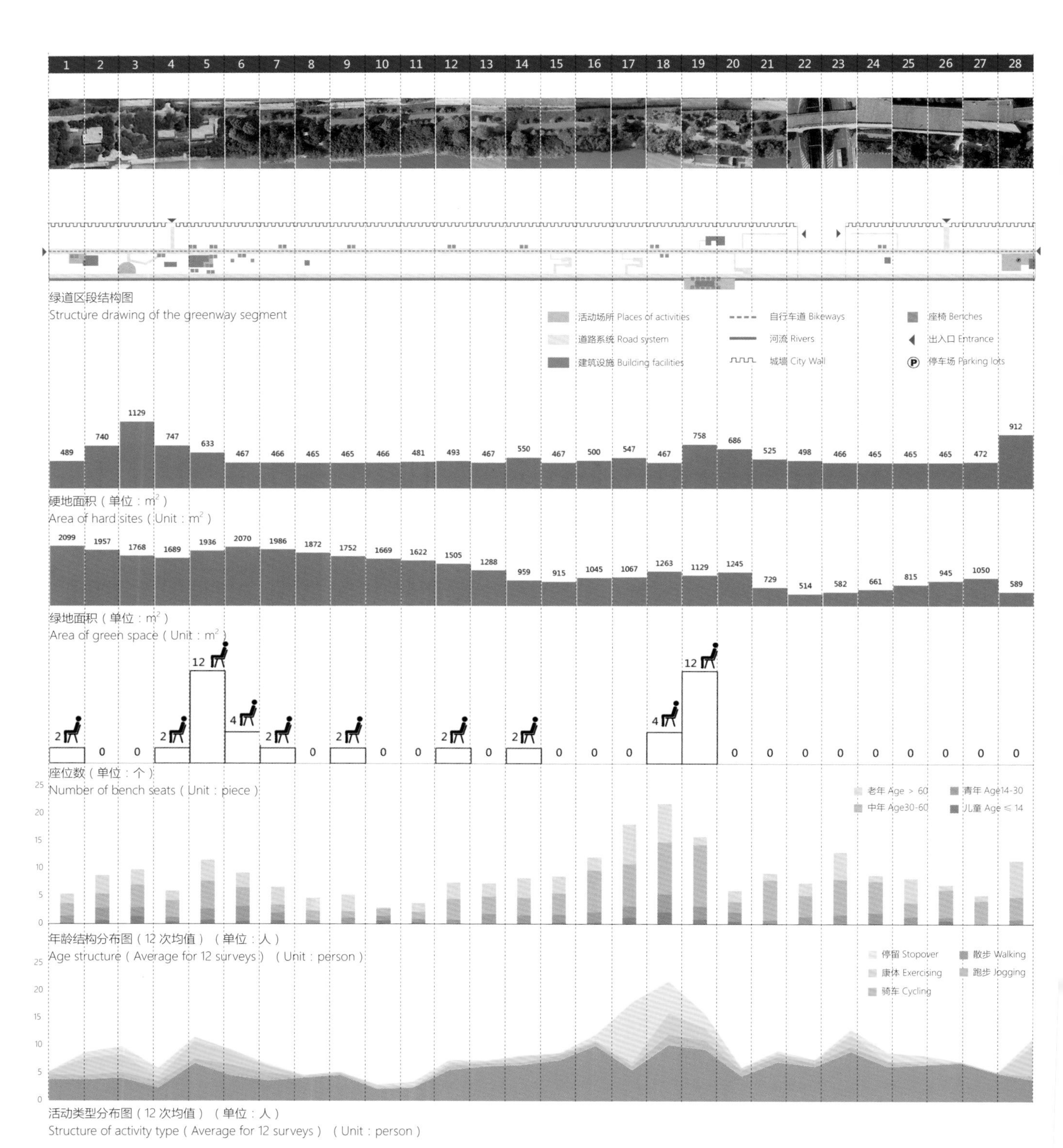
1 2 3 4 5 6 7 8 9 10 11 12 13 14 15 16 17 18 19 20 21 22 23 24 25 26 27 28
绿道区段结构图
Structure drawing of the greenway segment
活动场所 Places of activities
道路系统 Road system
建筑设施 Building facilities
自行车道 Bikeways
河流 Rivers
城墙 City Wall
座椅 Benches
出入口 Entrance
停车场 Parking lots
489 740 1129 747 633 467 466 465 465 466 481 493 467 550 467 500 547 467 758 686 525 498 466 465 465 465 472 912
硬地面积（单位：m²）
Area of hard sites（Unit：m²）
2099 1957 1768 1689 1936 2070 1986 1872 1752 1669 1622 1505 1288 959 915 1045 1067 1263 1129 1245 729 514 582 661 815 945 1050 589
绿地面积（单位：m²）
Area of green space（Unit：m²）
2 0 0 2 12 4 2 0 2 0 0 2 0 2 0 0 0 4 12 0 0 0 0 0 0 0 0 0
座位数（单位：个）
Number of bench seats（Unit：piece）
老年 Age > 60
中年 Age30-60
青年 Age14-30
儿童 Age ≤ 14
年龄结构分布图（12 次均值）（单位：人）
Age structure（Average for 12 surveys）（Unit：person）
停留 Stopover
康体 Exercising
骑车 Cycling
散步 Walking
跑步 Jogging
活动类型分布图（12 次均值）（单位：人）
Structure of activity type（Average for 12 surveys）（Unit：person）

内部空间特征总体分析

区段 5（凤台路—中华门）全长约 1.4km，共等分为 28 段切片。该区段共有 6 个入口，分别位于切片 1、切片 4、切片 22、切片 23、切片 26、切片 28。绿道的内部空间可分为靠墙、滨水两个层次，并在切片 2-5、切片 17-19、切片 27-28 形成较明显的节点活动空间。

从使用人群数据来看，使用人数最多的是切片 18，切片 8-11 使用人群相对较少。这三处集中活动场地能够吸引较多的中老年人停留，主要进行康体活动以及停留休息。其他切片使用人群年龄层次相对均衡，以散步为主。

整个区段切片间硬地面积相差不大，切片 2 和切片 28 硬地面积相对较大。与之对比，绿地面积的波动幅度相对较大：面积最大的切片 1 和最小的切片 22 相差近 3 倍。

在座椅分布上，切片 5、切片 19 的座椅座位数最多，均为 12 个，其他切片均相对较少，一半以上的切片未布置座椅。

Analysis of the characteristics of the interior space

S5 FTR-ZHG is about 1.4km long, equally divided into 28 slices. There are six entrances to this segment, which are located on slices 1, 4, 22, 23, 26, and 28. The interior space of the greenway is divided into two layers, namely, the space leaning against the wall and that at the waterfront. Three main node areas for recreational activities of this segment are sections constituted by slices 2-5, slices 17-19, and slices 27-28, with a large area of hard sites inside for each.

In terms of user data, slice 18 attracts the largest number of users, while the data of slices 8-11 are the smallest. The three nodes areas can attract a lot of middle-aged and senior people to stop, who mainly stay for card or Chinese chess playing and watching. The age structures on the other slices are relatively balanced. The people who are found there mainly go for a walk.

There is not much difference in the area of hard sites in all segment slices. Slice 2 and slice 28 have larger areas of hard sites. In contrast, the area of green space fluctuates greatly. Slice 1, with the largest area of green space, is nearly three times slice 22, which has the smallest.

In terms of distribution of bench seats, slices 5 and 19 are equipped with twelve bench seats the largest number of bench seats in the segment. The number of bench seats drops sharply on other slices, more than half of which have no bench seats.

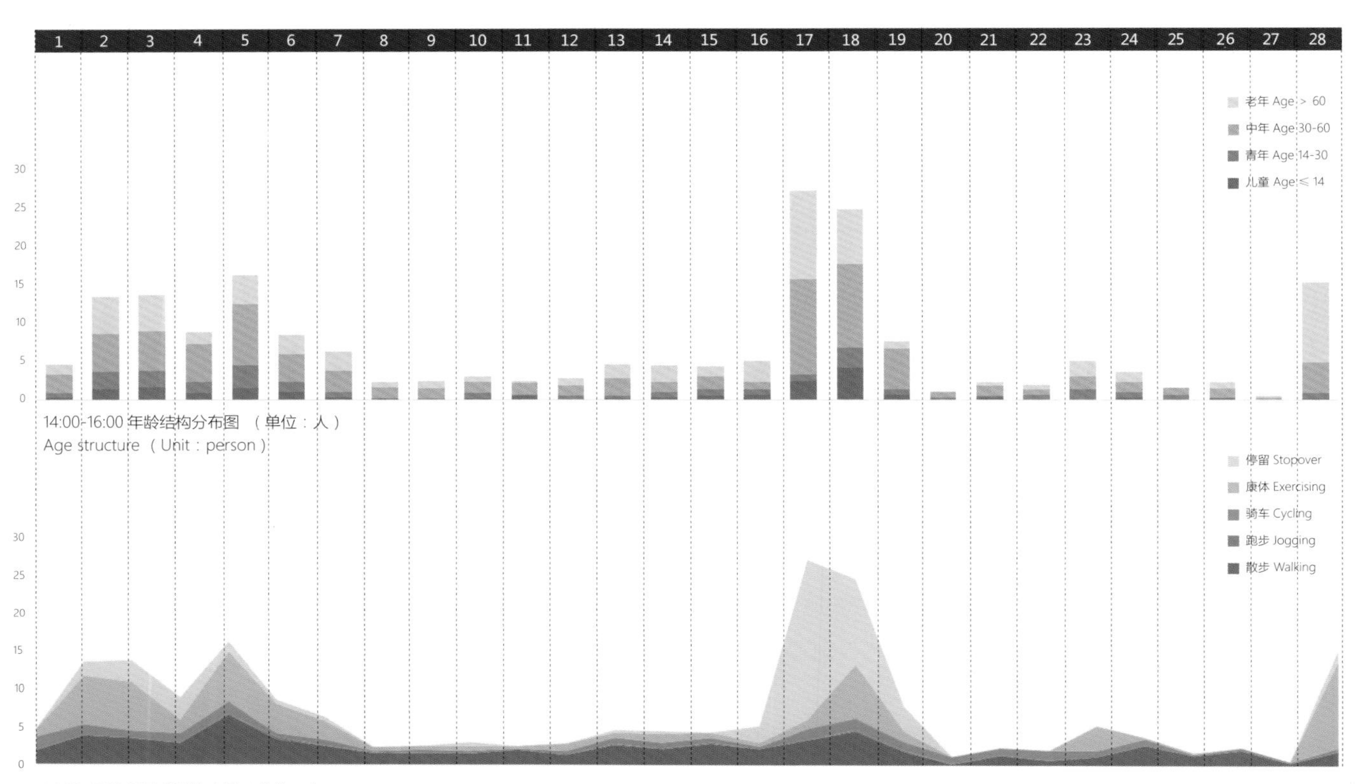

14:00-16:00 年龄结构分布图 （单位：人）
Age structure （Unit：person）

14:00-16:00 活动类型分布图（单位：人）
Structure of activity type（Unit：person）

两个时段的服务绩效分析

周末下午（14:00-16:00）的使用人群在切片 2-5、切片 17-18 和切片 28 三个地段明显集中。与其他切片比较，该地段中老年人明显增多，切片 17-18 以停留休息为主，切片 2-5 以及切片 28 以康体活动为主。

Analysis of service performance of two time periods

On the afternoon of a weekend (14:00-16:00), users are mostly gathered around on slices 2-5, slices 17-18, and slice 28. Compared to the other slices, the number of middle-aged and senior people noticeably increases. Users mainly stay on a seat on slices 17-18, while they go to slices 2-5 and slice 28 mainly for the purpose of exercising.

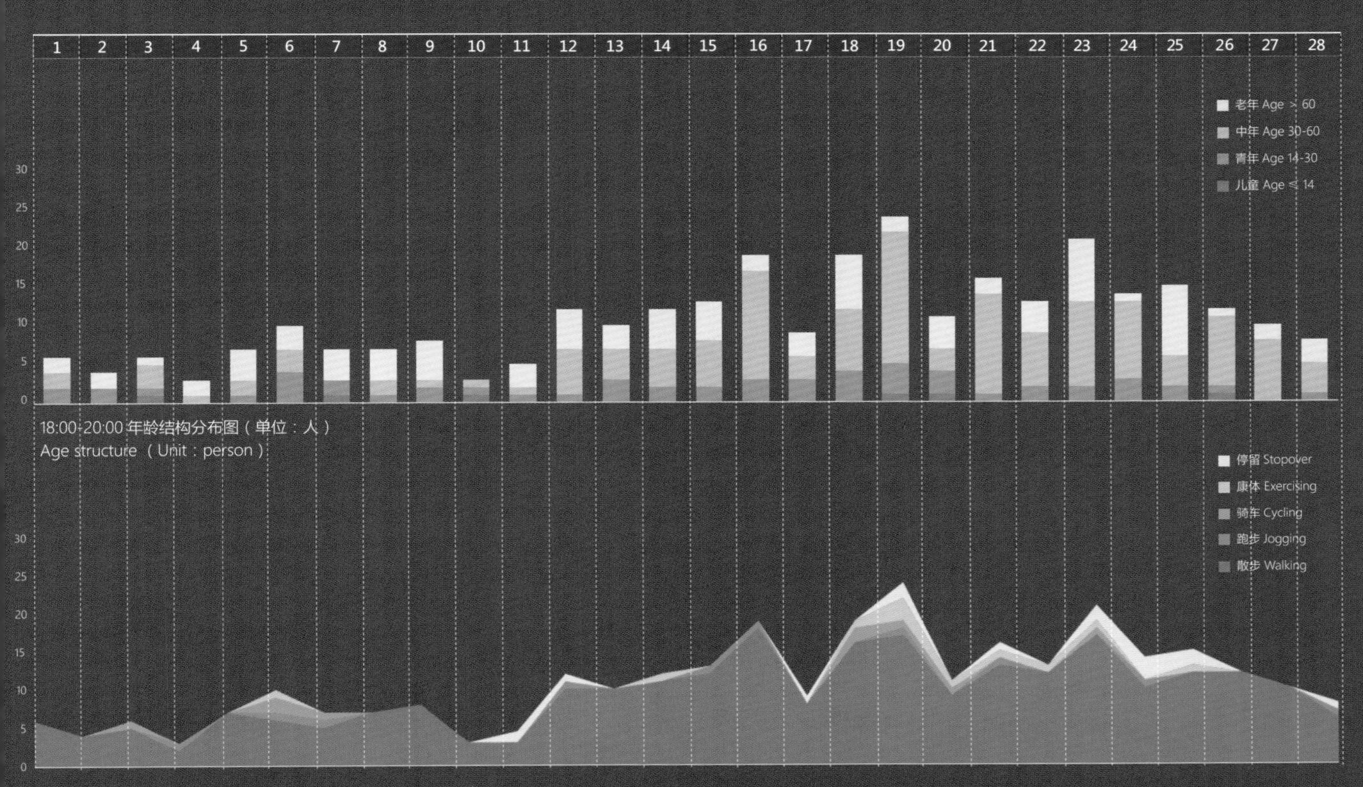

18:00-20:00 年龄结构分布图（单位：人）
Age structure（Unit：person）

18:00-20:00 活动类型分布图（单位：人）
Structure of activity type（Unit：person）

日常傍晚（18:00-20:00）的总体使用人数较之周末傍晚要多，且在该区段空间上的分布更加均衡。使用特征以中老年人散步为主，其中在切片 12-28 内形成使用人群持续高位区间。切片 19 是日常傍晚使用人数最高的切片，但与切片 16、切片 18、切片 23 的使用人数相差并不明显，切片 1-11 使用人数则出现明显下降。

Every day at nightfall (18:00-20:00), the total number of users is larger than that on a weekend afternoon, and spatial distribution on this segment is more balanced. Middle-aged and senior people who go walking mainly use it. Slices 12-28 form a continual peak range of the number of users. Slice 19 is where the number of users is the largest during this time period. However, the number of users among slices 16, 18, and 23 is close to the peak slice, while the number of users on slices 1-11 decreases drastically.

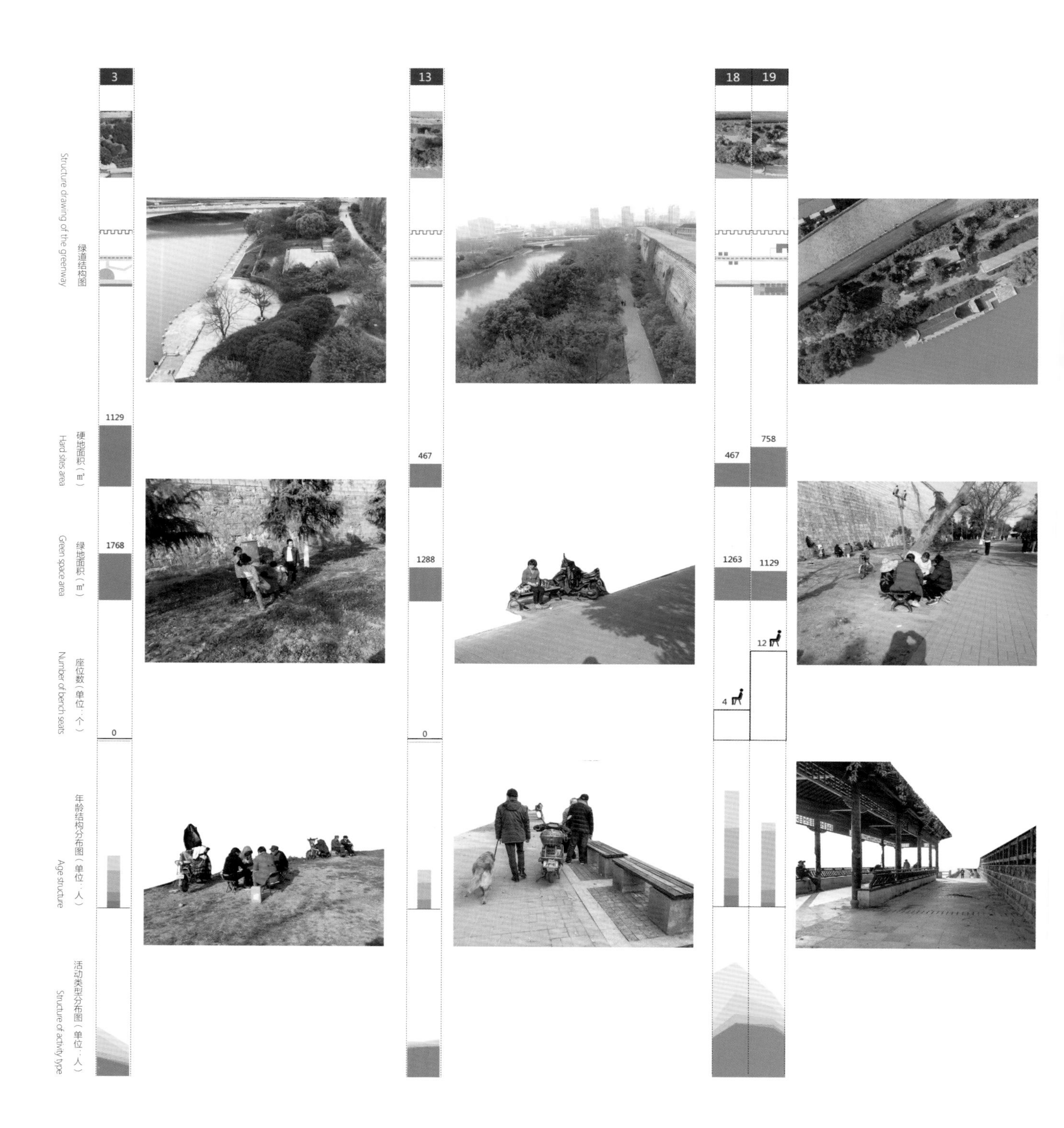
3
13
18
19
绿道结构图
Structure drawing of the greenway
硬地面积（㎡）
Hard sites area
1129
467
467
758
绿地面积（㎡）
Green space area
1768
1288
1263
1129
座位数（单位：个）
Number of bench seats
0
0
4
12
年龄结构分布图（单位：人）
Age structure
活动类型分布图（单位：人）
Structure of activity type

典型切片分析

切片 3 是区段 5（凤台路—中华门）空间较开敞、使用人数较高的切片，该地段活动类型较丰富，虽没有座椅，但有集中的硬质活动场地和开敞草坪，仍然能吸引人群停留。

以切片 13 为代表的地段是该区段最典型的空间类型。切片内没有明显的大面积活动场地，主导活动类型是散步，仅有少数使用者停留休息。切片内部空间较为单调，缺乏吸引大量人群停留活动的条件和亮点。

切片 18 是该区段使用人数最高的地段，该切片座椅设施条件在该区段内最好。周末下午和日常傍晚均有较多人群聚集，周末下午以停留休息和康体活动为主，出现许多中老年人以公共座椅当桌子，并自带板凳打牌等类型活动。日常傍晚则是居民散步休息之所。

Analysis of typical slices

Slice 3 is a wide-open space with the largest number of users on S5 FTR-ZHG. Although no bench seat is set here, with centralized hard sites and spacious lawns, the space on this slice can still attract people to gather around and act diversely.

As a representative, the space of slice 13 is the most typical pattern on S5 FTR-ZHG. There is not any evident wide-open site for recreational activities along the trail. Walking is the dominant activity here. Only a few users prefer to stay around. The experience of space on this slice is tedious, lacking highlights to attract a large number of people to gather.

Slice 18 is where the number of users is the largest. The data of bench seats and the conditions of facilities are the best of the entire segment. A lot of people gather around here both on a weekend afternoon and every day at nightfall. On the afternoon of a weekend, people mainly come here to stop over or exercise. A lot of middle-aged and senior people turn the public bench seats into tables at which they can play Chinese chess or cards in groups and sit around on the stools they brought. In the evening, more nearby residents prefer to come here for a walk.

区段 6（中华门—雨花门 ）抽象图
Abstractive drawing of S6（ZHG – YHG ）

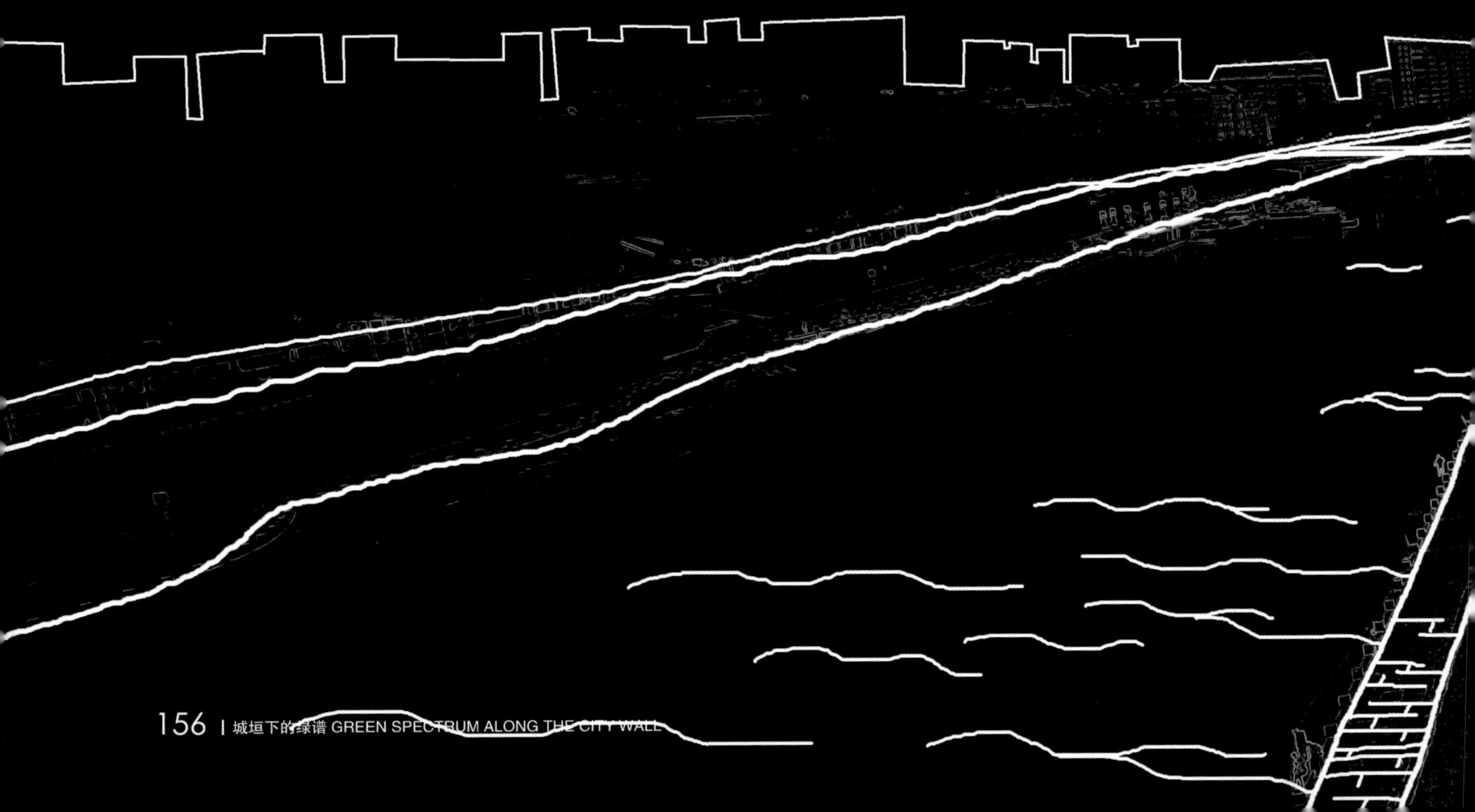

区段 6：中华门—雨花门
Segment 6: Zhonghua Gate-Yuhua Gate

内部空间特征概述

区段 6（中华门—雨花门）位于南京老城的东南角，西起中华门，东至雨花门。该区段绿道经常在传统节日举办灯会等民俗活动，来自南京各处的市民都会来此观赏游玩，同时平日里该区段也是附近居民的重要游憩场所。周末下午该区段场地围满了打牌聊天的附近居民和游客；日常傍晚则有大量居民到此散步，络绎不绝。

Overview of the characteristics of the interior space

S6 SZG-YHG is located at the southeast corner of the old city of Nanjing. It starts from Zhonghua Gate and spreads to Yuhua Gate on the east. At traditional festivals, people often hold a lantern show and other folk events on the greenway of this segment. People come here from all over Nanjing to appreciate the scenery and have fun. On ordinary days, this segment is an important place for daily recreation for surrounding residents. On a weekend afternoon, this segment is crowded with visitors and nearby residents who are fond of playing cards and chatting in small groups. Every day at nightfall, a lot of residents can be found here for daily fitness walking in an endless stream.

设施类别 Type of facilities	具体设施 Specific facilities	数据 Data
游径及交通设施 Trails and transportation facilities	步行道 Footpaths	4.2m
	自行车道 Bikeways	4.2m
	停车场 Parking lot	400m^2
	无障碍道 Barrier-free roads	√
游憩服务设施 Recreational service facilities	游乐器具 Amusement equipment	X
	休息设施 Rest facilities	98 个
	文体活动场地 Cultural and sports field	2403m^2
	无线网 Wi-Fi	X
管理服务设施 Management service facilities	管理中心 Management Center	X
	游客服务中心 Tourist Service Center	X
	信息咨询点 Information consultant	X
商业服务设施 Commercial service facilities	售卖点 Selling	X
	自行车租赁点 Bike rentals	X
	自动售货机 Vendingmachines	X
科普教育设施 Popular science education facilities	科普宣传、展览设施 Popularization and propaganda of science、Exhibition facilities	X
	解说设施 Commentary facilities	X
环境卫生设施 Environmental hygiene facilities	公厕 Restrooms	√
	垃圾桶 Trash cans	√
安全保障设施 Safety & security facilities	照明 Lighting	√
	治安消防 Public security & fire control	√
标识信息设施 Signs	信息、指路、规章标识等 Information, directions, signs, etc.	√

区段6（中华门—雨花门）的主体结构与凤台路—中华门段相同，靠墙一侧有较宽的综合慢行道，滨水地段有独立的亲水游步道。但该区段台地竖向的联系较多样，有通过大面积硬质活动场地与主路连接，也有与服务建筑相连，还有以坡地游步道相连等。整个区段在西部和东部各有一组集中的硬质活动场地，中部以服务建筑为中心也形成了一个较大节点。

The main structure of S6 SZG-YHG is similar to that of S5 FTR-ZHG. The part leaning against the wall has a very wide comprehensive slow lane, while the part at the waterfront is equipped with an independent hydrophilic footpath. However, the forms of vertical connection in this segment are more diverse than that in S5 FTR-ZHG. Open plaza, service building steps, and slope are all used to solve the elevation difference between terraces at the riverbank area. Both the west side and east side of the segment have one open site series for recreational activities. In addition, another large node space is in the middle, with a big service building in the center of the space.

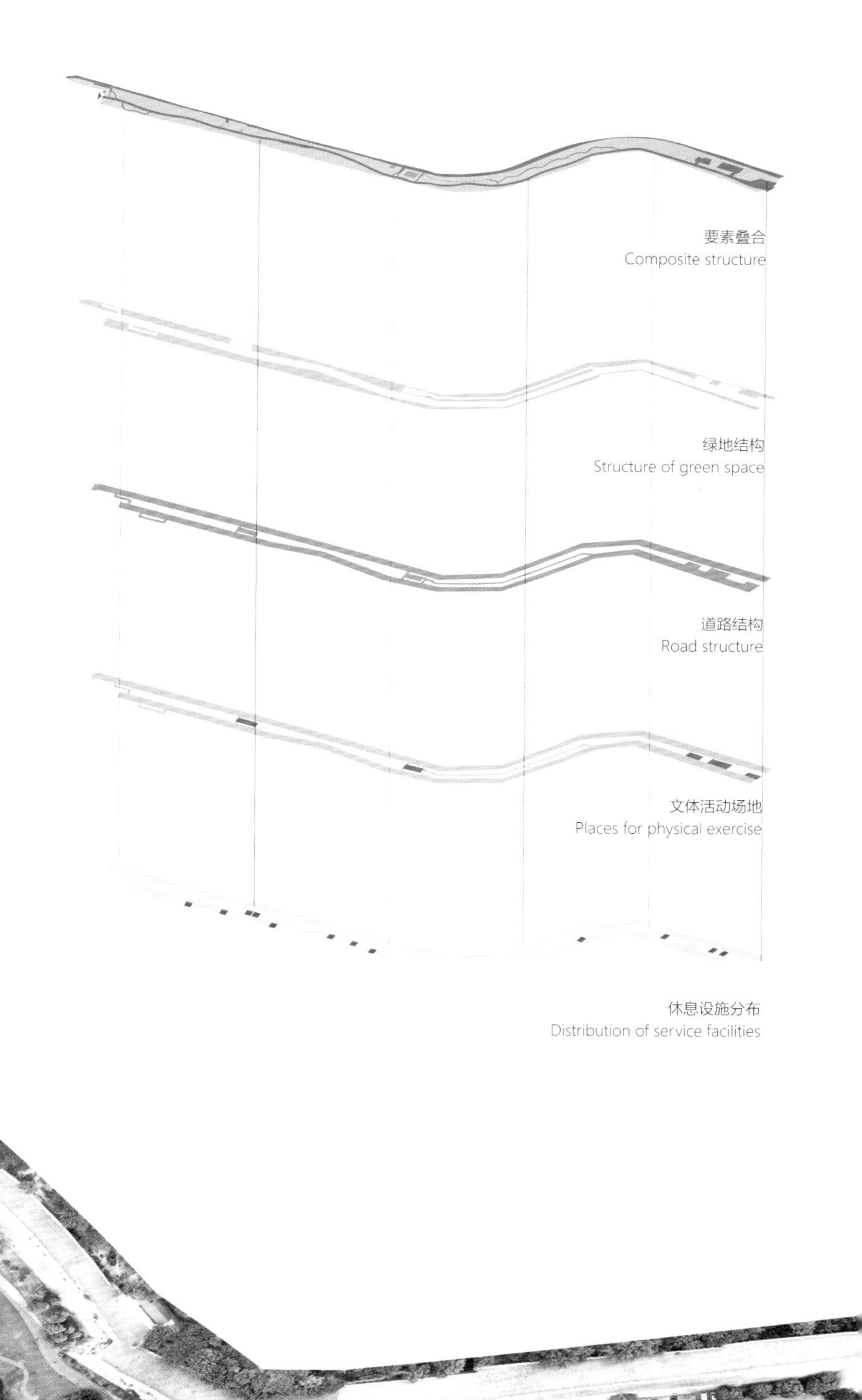

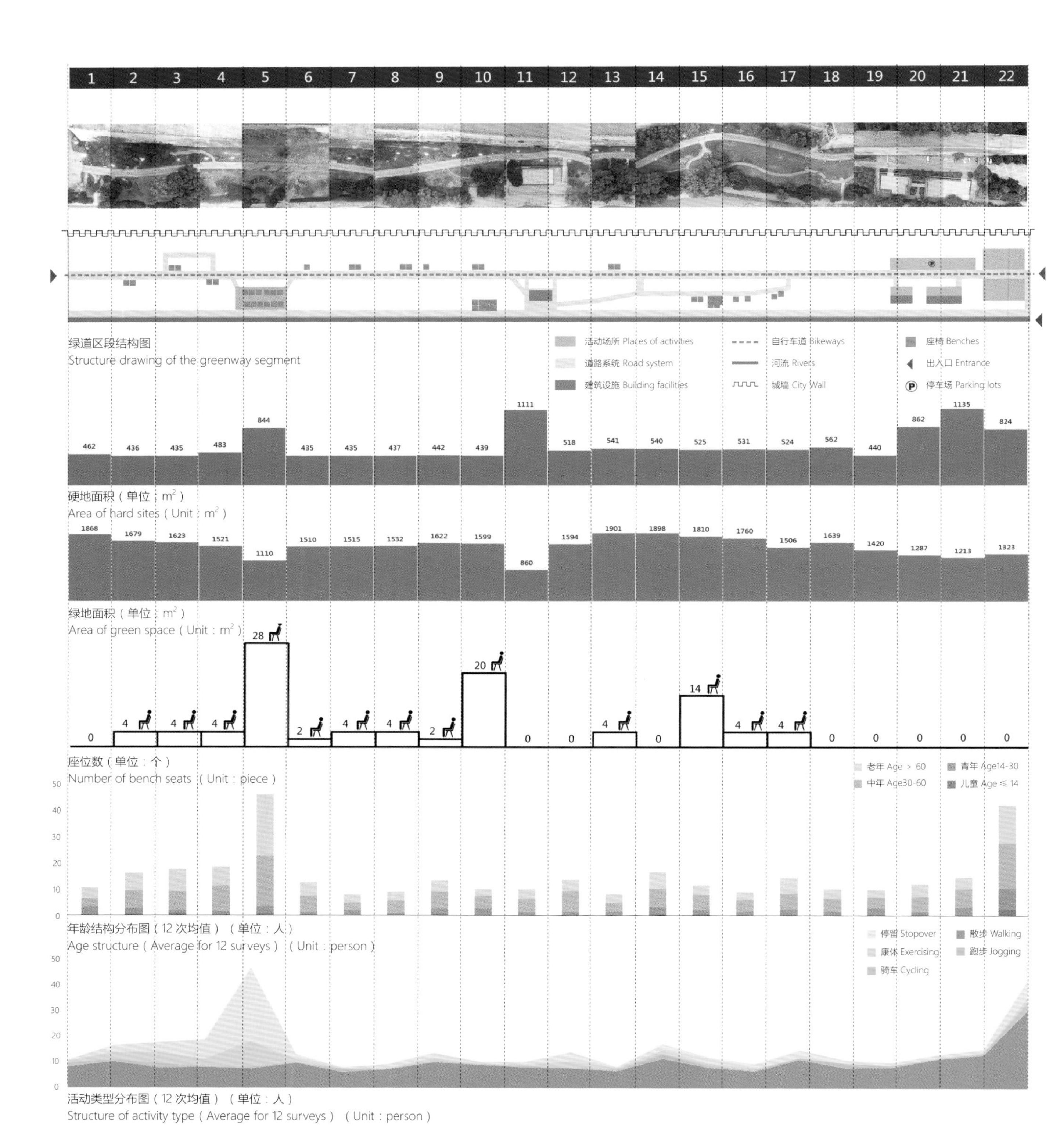
1
2
3
4
5
6
7
8
9
10
11
12
13
14
15
16
17
18
19
20
21
22
绿道区段结构图
Structure drawing of the greenway segment
活动场所 Places of activities
道路系统 Road system
建筑设施 Building facilities
自行车道 Bikeways
河流 Rivers
城墙 City Wall
座椅 Benches
出入口 Entrance
停车场 Parking lots
462 436 435 483 844 435 435 437 442 439 1111 518 541 540 525 531 524 562 440 862 1135 824
硬地面积（单位：m²）
Area of hard sites（Unit：m²）
1868 1679 1623 1521 1110 1510 1515 1532 1622 1599 860 1594 1901 1898 1810 1760 1506 1639 1420 1287 1213 1323
绿地面积（单位：m²）
Area of green space（Unit：m²）
0 4 4 4 28 2 4 4 2 20 0 0 4 0 14 4 4 0 0 0 0 0
座位数（单位：个）
Number of bench seats（Unit：piece）
老年 Age＞60
中年 Age30-60
青年 Age14-30
儿童 Age≤14
年龄结构分布图（12 次均值）（单位：人）
Age structure（Average for 12 surveys）（Unit：person）
停留 Stopover
康体 Exercising
骑车 Cycling
散步 Walking
跑步 Jogging
活动类型分布图（12 次均值）（单位：人）
Structure of activity type（Average for 12 surveys）（Unit：person）

内部空间特征总体分析

区段 6（中华门—雨花门）全长约 1.1km，共等分为 22 段切片。该区段城墙一侧没有出入口，其在城墙与秦淮河间东西向延展，游径由一条综合慢行道与一条亲水步道构成。

该区段内使用人群较集中的有切片 5 和切片 22，这两个切片单元布有该区段最大的两处硬质活动场地。其中，在切片 5 停留休息和康体活动的中老年人较多，切片 22 则多为散步等活动的流动人群，年龄结构更加多样。整个区段中，切片 20-22 拥有较集中的硬地，每个切片硬地面积都超过 800m^2。而绿地在切片 14-17 面积较大，主要是由于城墙在该处转折造成空间放大，从而形成大面积的开敞草坪。

切片 5、切片 10 和切片 15 座椅座位数较多，分别达到 28 个、20 个和 14 个，其余切片座椅座位数均未超过 4 个。

Analysis of the characteristics of the interior space

S6 SZG-YHG is about 1.1km long, equally divided into 22 slices. The segment spreads from east to west between the City Wall and the Qinhuai River, without any entrances on the City Wall side. The trails consist of a comprehensive slow lane and a hydrophilic footpath at the riverside.

Users are mostly gathered on slices 5 and 22 of this segment. These two slice units are equipped with two of the largest hard sites in the entire segment. Among them, most people who stop over or exercise on slice 5 are mainly middle-aged and senior people, while users on slice 22 are usually walking, with a more balanced age structure. Large hard sites are equipped mainly on the area of slices 20-22, with the area on each slice exceeding 800m^2. Meanwhile, an enlarged and wide-open green space is located mainly at the area of slices 14-17, due to the turning of the City Wall.

The number of bench seats is significantly bigger in slices 5, 10, and 15, with the number of seats reaching 28, 20, and 14, respectively. None of the other slices have more than four bench seats.

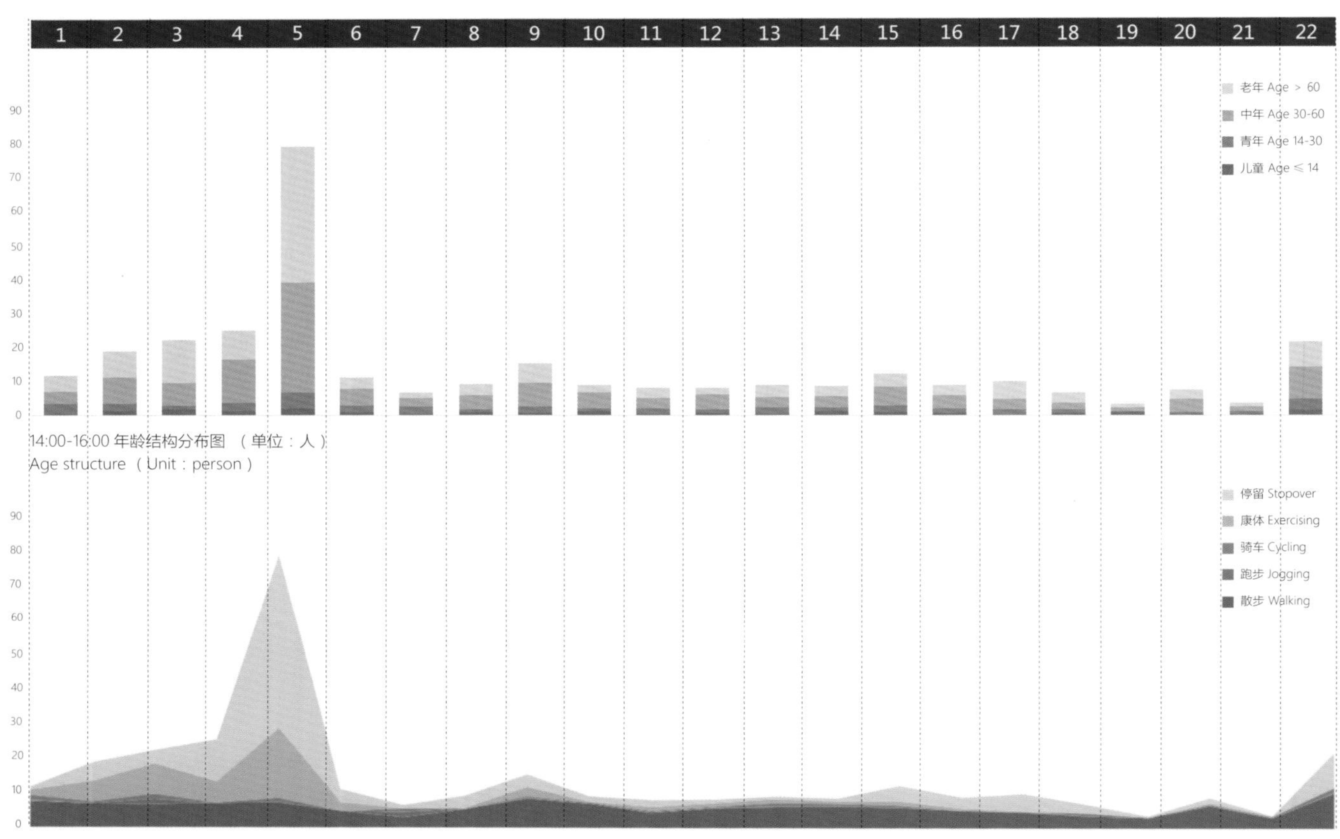

14:00-16:00 年龄结构分布图 （单位：人）
Age structure （Unit：person）

14:00-16:00 活动类型分布图（单位：人）
Structure of activity type（Unit：person）

两个时段的服务绩效分析

周末下午（14:00-16:00）停留休息和康体活动的使用人群集中于切片 5，其中以中老年人为主。从整体的使用活动类型来看，切片 1-5 以停留休息和康体活动为主，其余切片以散步为主。

Analysis of service performance of two time periods

On a weekend afternoon (14:00-16:00), many middle-aged and senior users prefer to gather around at the area of slice 1-5, playing Chinese chess or cards and doing exercising. The number of users reaches its peak in slice 5 in this time period. The user numbers decrease sharply on the slices after slice 5 with the dominant activity changing to walking.

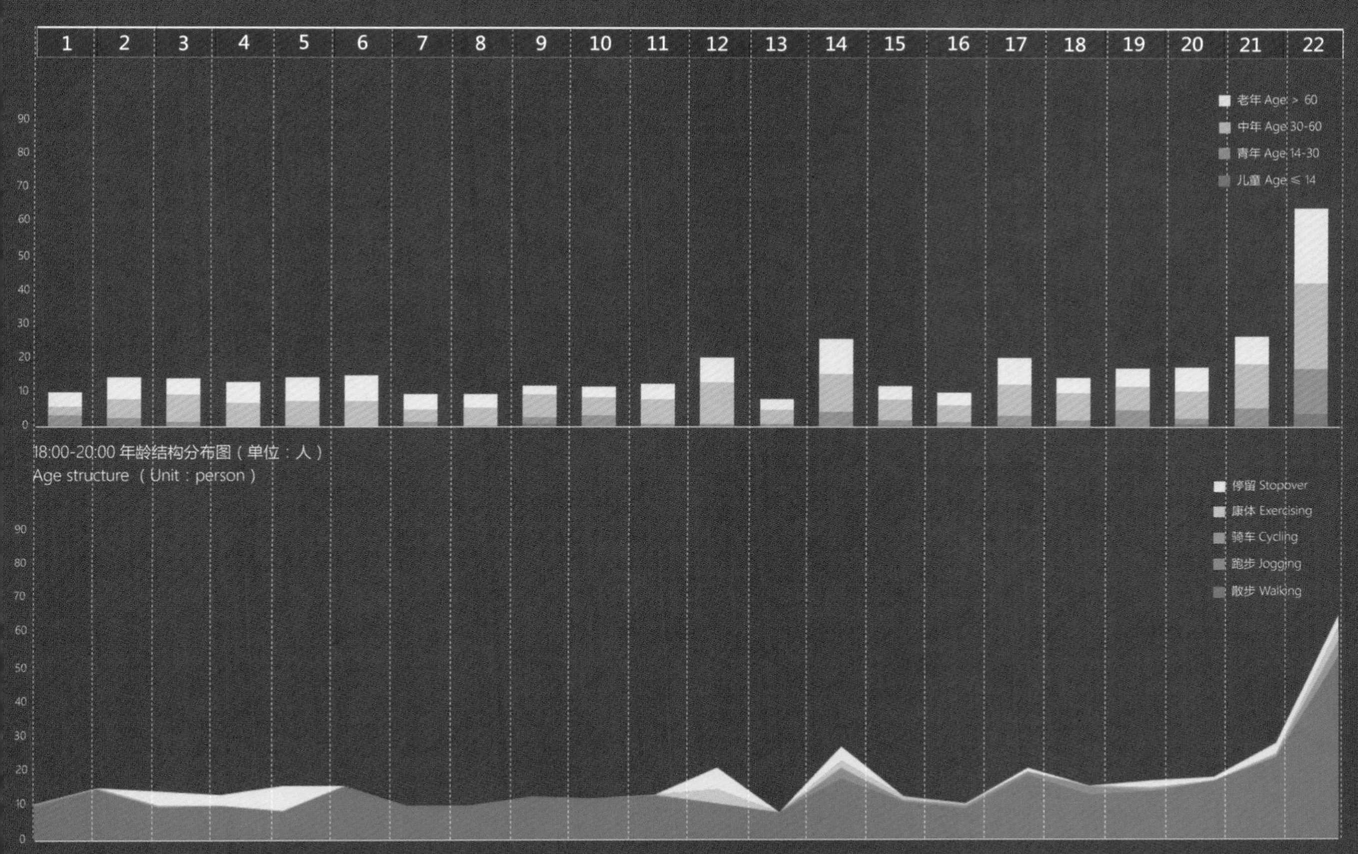

18:00-20:00 年龄结构分布图（单位：人）
Age structure （Unit：person）

18:00-20:00 活动类型分布图（单位：人）
Structure of activity type（Unit：person）

日常傍晚（18:00-20:00）使用人群集中于切片 22。在切片 2-5、切片 12 和切片 14 有少量停留休息和康体活动的人群，其他地段使用活动类型以散步为主。

Every day at nightfall (18:00-20:00), the slice with the biggest number of users moves from slice 5 to slice 22. Although the number is small, a few people can still be found gathering around on slices 2-5, slice 12, and slice 14 to stop over or exercise, but the dominant activity on all the slices in this time period changes to walking.

切片 5 Slice 5

5

绿道结构图 Structure drawing of the greenway

844
硬地面积（㎡） Hard sites area

1110
绿地面积（㎡） Green space area

28
座位数（单位：个） Number of bench seats

年龄结构分布图（单位：人） Age structure

活动类型分布图（单位：人） Structure of activity type

典型切片分析 1

切片 5 是整个区段周末下午使用人数最多的地段，从指标来看，该地段比周边地段的硬地面积有显著提升，并且座椅数目座椅座位数目也是整个区段中最多的。从布局来看，该地段是整个区段为数不多的集中硬地之一，并且视线开敞，与下层亲水步道和上层绿道主游径均有便捷的联系，便于使用人群进入。

在内部，该场地被台地弧形花坛、小花坛和条石座椅划分成多组小型停留空间，也为周末下午周边居民聚集打牌、下棋和围观创造了条件。但也由于该场地被划分得太多太碎，导致在日常傍晚难以满足周边中老年居民聚集进行广场舞等集体健身活动的需求，从而使该场地在两个时段的利用率出现较大反差。

Analysis 1 of typical slices

Slice 5 is where the number of users is the largest on a weekend afternoon. In terms of indicators, the area of hard sites of this slice is noticeably increased compared to the surroundings, and the number of bench seats on this slice is the largest of the entire segment. Making a review on the spatial layout, we find that this slice is one of the few that have centralized areas of hard sites with a wide-open sight in the entire segment. It is conveniently linked to both the hydrophilic trails at the lower level and the main trails of the greenway at the upper level close to the City Wall, so as to facilitate the access of users.

Inside, this area is divided by a big terracing flower bed with an arc border, several small flower bed units in the center, and several stone bench seats into several groups of small spaces, so as to create good conditions for surrounding residents to gather around in small groups to play cards and chess or just watch on a weekend afternoon. However, since this area is divided into too many pieces by the facilities in the middle which are fixed and immovable, it has difficulty meeting the demands for a centralized big space from surrounding middle-aged and senior residents to exercise in a group and engage in activities such square dancing every day at nightfall. As a result, it is not surprising to find a huge difference in utilization ratio between the two time periods on this slice.

切片 11　Slice 11
切片 12　Slice 12

	11	12
绿道结构图 Structure drawing of the greenway		
硬地面积（㎡） Hard sites area	1111	518
绿地面积（㎡） Green space area	860	1594
座位数（个） Number of bench seats	0	0
年龄结构分布图（单位：人） Age structure		
活动类型分布图（单位：人） Structure of activity type		

切片 10

Slice 10

切片 11

Slice 11

切片 12

Slice 12

典型切片分析 2

切片 11-12 同样有着较高的硬地面积，但该地段的使用率却并没有显著提升。其中这两个切片的主体是服务建筑（公厕）及其室外平台和下级亲水平台。与其他大面积活动场地相比，切片 11 的场地并不能吸引太多主游径游客停留休息。

解析该地段的空间特征可以发现，使用人群主要在高程较高的主游径上活动，但由于该地段的公厕以及边上茂密的植物在很大程度挡住主游径上行人的视线，行人难以感知下层亲水平台的存在。同时，公厕南侧的平台场地过大，其实也远远超出实际建筑出入的功能需求，且会对下层亲水平台场地造成压迫。另外，该地段没有任何的座椅设施和可坐空间也在一定程度上降低了使用人群停留的可能性。

Analysis 2 of typical slices

Slices 11-12 are also equipped with large areas of hard sites. The main parts that constitute the space of the two slices are service buildings (public restrooms), outdoor platforms, and hydrophilic platforms at the lower level. If compared to the other slices with equivalent areas of hard sites, slices 11-12 perform very poorly in attracting users from the main trails to stop.

According to the analysis of characteristics of this area, we can find that users are mainly gathered around on the main trails at the upper level. Because the visions of the pedestrians on the main trails are blocked by the public restrooms of this slice and the surrounding leafy trees, the people who walk can hardly perceive the hydrophilic platform at the lower level. Meanwhile, the platform area to the south of the restrooms is so large as to exceed the actual architectural functional requirements and exert vertical pressure from the upper level on the lower hydrophilic platform. Besides, this slice does not have any bench seats or seating spaces, which lowers the possibility of people coming to stop over to a further degree.

	20	21	22
绿道结构图 Structure drawing of the greenway			
硬地面积（㎡） Hard sites area	862	1135	824
绿地面积（㎡） Green space area	1287	1213	1323
座位数（单位：个） Number of bench seats	2	0	0
年龄结构分布图（单位：人） Age structure			
活动类型分布图（单位：人） Structure of activity type			

停车场
Parking lot

切片 20
Slice 20

切片 21
Slice 21

切片 22
Slice 22

典型切片分析 3

切片 20-22 是区段 6（中华门—雨花门）硬地面积最集中的地段，总计约 2800㎡。靠城墙一侧是停车场，切片 20 和切片 21 分别是管理用房和公厕的屋顶活动平台，切片 22 则是绿道出入口的集散广场。

对比集中硬质场地面积，切片 21 面积最大，但是切片 22 的使用人数最多。这是因为，一方面，切片 22 作为东部绿道连接城市道路的出入口，场地相对更加集中开敞，在日常傍晚能吸引周边居民在此集中开展康体活动。另一方面，切片 22 还是小型停车场，不少到周边购物、吃饭的人群将车停至此处，因而部分切片上的活动人群并非都是绿道真正的使用人群。这些因素共同造成切片 20 和切片 21 的使用人数大幅度下降。同时，切片 20 和切片 21 的服务建筑体量较大且周边植物茂密，使得该出入口地段空间有一定收窄，并且在很大程度上遮挡了主游径上人群的视线，这也在一定程度上影响了切片 22 使用人群向西和向南行进的意愿。

Analysis 3 of typical slices

Slices 20-22 form an area with a centralized big area of hard sites on S6 SZG-YHG, with a total area of about 2,800m^2. The site leaning against the City Wall is a parking lot. Slices 20 and 21 are rooftop platforms of management rooms and public restrooms. Slice 22 is an evacuation square to the entrance of the greenway.

In contrast to the area of centralized hard sites, slice 21 has the largest area, but the number of users is the largest on slice 22. According to the survey, since the entrance where the eastern greenway is linked to urban roads, slice 22 has more centralized and wide-open space so as to attract surrounding residents to come and exercise every day at nightfall. On the other hand, slice 22 also functions as a small-size parking lot where a lot of people park their cars before going shopping and dining in the surrounding commercial areas. Therefore, not all the people who go to this part of the slice are the real users of the greenway. All those factors contribute to a drastic decrease in the number of users on slices 20 and 21 nearby. At the same time, since the sizes of the service buildings are too big on slices 20 and 21 and leafy trees grow on the surroundings, both of which block the vision of the people in the entrance to a large degree, the surrounding space linked to the open entrance area in the greenway is experienced to be narrowed down sharply. This also lessens the willingness of users to continuously visit the other sections of the greenway.

切片 20 Slice 20
切片 21 Slice 21
切片 22 Slice 22

	S1:YJG-HYGG	S2:DHG-CCG	S3:CCG-QLG	S4:SXG-JQG	S5:FTR-ZHG	S6:ZHG-YHG
	16	10	17	2	18	5
切片航拍图 Aerial photos of slices						
切片结构图 Structure drawings of slices						
区段排名 & 切片硬地面积（单位：m^2）Rankings of areas of hard sites of slices in the segment (Unit : m^2)	673 4/26	595 18/20	2020 5/30	830 2/24	467 20/28	844 4/22
区段排名 & 切片绿地面积（单位：m^2）Rankings of areas of green space of slices in the segment (Unit : m^2)	1740 10/26	642 17/20	3143 2/30	835 5/24	1263 14/28	1110 21/22
区段排名 & 切片座位数（单位：个）Rankings of numbers of benches on the slices in the segment (Unit : piece)	18 2/26	0 20/20	0 30/30	6 12/24	4 4/28	52 1/22

内部空间特征分析小结

Summary of the analysis of the characteristics of the interior space

使用人群规模最大的切片特征解析

为进一步梳理绿道内部使用人群的空间分布特征，我们分别将 6 个区段使用人数最多的切片提取出来，并将其空间要素量化指标及其在区段切片中的排位信息进行比对分析。在 6 个使用人数最多的切片里，硬地面积排位在区段前 5 位的切片有 4 个；绿地面积排位在区段前 5 位的仅有 2 个；座椅座位数目排位在前 5 位的有 3 个。如果单纯从指标的排序情况来看，硬地面积大的切片上使用人群聚集的可能性较大，其次是座椅座位数，再次是绿地面积。

在 6 个切片中，硬地面积、绿地面积和座椅座位数 3 个指标中 2 个排在前 5 位的切片有 4 个，一个指标排在前 5 位的切片有 1 个，还有 1 个切片 3 个指标均排在后 5 位。不可否认的是空间要素的布局和组织方式将对绿道空间品质起到决定性作用，但审视 6 个使用人群最多的切片中的 3 个空间要素规模指标的排序情况发现，设施规模和容量的支撑同样是绿道服务绩效保障中不可或缺的重要基础。

Characteristics of the slices with the largest numbers of users

To further sort out the characteristics of spatial distribution of the users inside the greenway, we extract from the six segments the slices where the number of users is the largest, and compare and analyze the quantitative indicators of their spatial elements and rankings among the slices of the same segments. Among the six slices with the largest numbers of users, four slices are among the top five in terms of the area of hard sites; only two slices are among the top five of their segments in terms of the area of green space; three slices are among the top five in terms of the number of bench seats. If simply judging from the rankings of the indicators, we find that the slices with the largest areas of hard sites are more likely to have more people gathering around, the next being the number of bench seats, and the third the areas of green space.

Among the six slices, four slices rank top five for two of the three indicators - the area of hard sites, the area of green space, and the number of bench seats; one slice ranks top five for one indicator; another slice ranks last five for all three indicators. It is undeniable that the layout and organization of elements will play a decisive role in the quality of the space of the greenway. However, judging from the rankings of all three indicators of the spatial elements of the slices with the largest numbers of users, we can see that the size of facilities and the capacity to support are equally important foundations necessary for guaranteeing the service performance of the greenway.

	S1:YJG-HYGG	S2:DHG-CCG	S3:CCG-QLG	S4:SXG-JQG	S5:FTR-ZHG	S6:ZHG-YHG
	1	1	9	24	10	7
切片航拍图 Aerial photos of slices	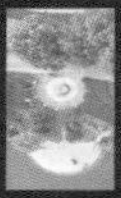		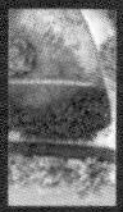	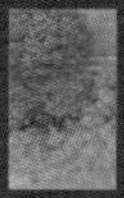	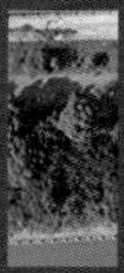	
切片结构图 Structure drawings of slices						
区段排名 & 切片硬地面积（单位：m^2） Rankings of areas of hard sites of slices in the segment (Unit : m^2)	442 16/26	802 15/20	623 27/30	344 22/24	466 23/28	435 22/22
区段排名 & 切片绿地面积（单位：m^2） Rankings of areas of green space of slices in the segment (Unit : m^2)	389 26/26	905 10/20	806 22/30	517 15/24	1669 10/28	1515 14/22
区段排名 & 切片座位数（单位：个） Rankings of numbers of benches on the slices in the segment (Unit : piece)	0 26/26	0 20/20	0 30/30	0 24/24	0 28/28	2 15/22

使用人群规模最小的切片特征解析

同样，我们也将 6 个区段使用人数最少的切片提取出来进行同样的定量数据对比。我们发现，硬地面积排位在区段后 5 位的有 3 个，还有 2 个排在后 6 位；绿地面积排在后 5 位的有 1 个；座椅座位数排在后 5 位的有 5 个，均为 0 个座椅座位。仅从指标的排序情况来看，座位不足和硬地面积不足的切片上人群聚集的可能性最低，绿地面积的大小对于人群聚集的影响力较弱。这与前面使用人数最多的 6 个切片的分析结论是一致的。

在 6 个使用人数最少的切片中，硬地面积、绿地面积和座椅座位数 3 个指标中 2 个排在后 5 位的切片有 3 个，1 个排后 5 位的切片有 3 个（其中 2 个切片还有 1 个指标排在后 6 位）。其反映出来的人群活动设施规模的重要性与之前使用人数最多的切片的分析结果也基本一致。

Characteristics of the slices with the smallest number of users

Similarly, we extract the slices with the smallest number of users from the six segments for similar comparison and contrast of quantitative data. We find that three slices rank last five in terms of area of hard sites, and another two rank last six; one slice ranks last five in terms of area of green space; five slices rank last five in terms of the number of bench seats, all of which are zero. Only judging from the ranking of indicators, we can see that slices with a lack of bench seats and without enough areas of hard sites are the least likely to have users gathered around and that the size of the area of green space does not have much influence on the number of users. This is consistent with the aforementioned conclusion concerning the six slices with the largest numbers of users.

Among the six slices with the smallest numbers of users, three slices rank last five for two of the three indicators - the area of hard sites, the area of green space, and the number of bench seats. Three slices rank last five for one indicator (one of the two slices rank last six for one indicator). The importance of the size of the facilities for human activities is basically consistent with the previous analytical result of the slices with the largest numbers of users.

使用人群总体空间分布特征分析

在切片研究过程中，区段2（定淮门—草场门）空间较均质，使用人群分布也较均匀，波动较小。除此区段外，其他5个区段使用人群在空间上均有明显波动。其中，使用人群波动与绿道总体空间结构收放变化趋势较一致的有区段1（挹江门—华严岗门）和区段3（草场门—清凉门）。这两个区段均是由于城墙走向的变化而形成了绿道空间的收放变化，为使用人群的聚集创造了有利的空间条件。

使用人数波动与绿道总体空间结构收放变化不一致的主要有区段4（水西门—集庆门）。该区段空间整体较狭窄，但在中部空间适当放开。而人群则主要集中在南部较开敞的硬地和北部线性亲水平台，中部较开敞的滨水活动场地人群反而较少。

区段5（凤台路—中华门）和区段6（中华门—雨花门）两个区段整体空间收放变化相对较小，但人群数量波动明显。区段5（凤台路—中华门）使用人群主要在饮马桥西侧大量聚集，区段6（中华门—雨花门）则主要在西部和东部两个较集中的滨水活动场地上聚集了较多使用人群。

Characteristics of general spatial distribution of users

During the study on the slices, we find that S2 DHG-CCG has homogeneous space, with even user distribution in space that does not fluctuate too much. The numbers of users on the other five segments all fluctuate noticeably. Among them, S1 YJM-HYGG and S3 CCG-QLG are where the fluctuation of the number of users is consistent with the change of scaling of the greenway's overall spatial structure. In both segments, the space of the greenway changes in scaling due to a change of direction of the City Wall, which creates positive conditions of space for users to gather around.

On S4 SXG-JQG, the fluctuation of the number of users is inconsistent with the change of scaling of the overall spatial structure of the greenway. The overall space of this segment is narrow, but that in the middle is properly open. The crowds are mainly gathered around on the wide-open hard sites on the south and on the linear hydrophilic platform on the north, while the wide-open place of activities at the waterfront in the middle does not attract too many people.

On S5 FTR-ZHG and S6 SZG-YHG, the change of scaling of the overall space is relatively small, but the number of users fluctuates noticeably. The users of S5 FTR-ZHG mainly gather around on the west side of Yinma Bridge in large quantities, while a lot of users mainly gather around on two central places of activities at the waterfront of both the west and east of S6 SZG-YHG.

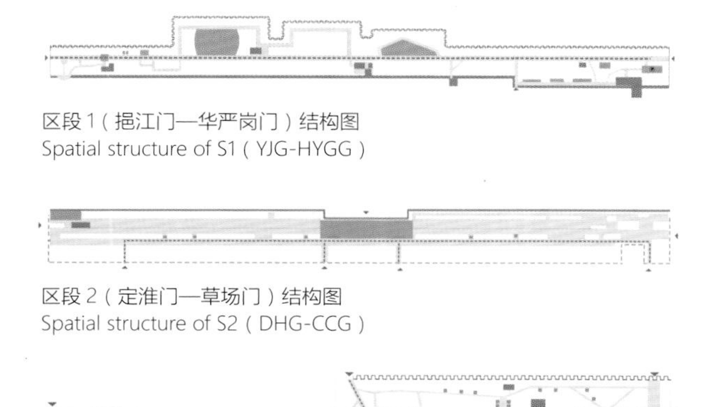

区段1（挹江门—华严岗门）结构图
Spatial structure of S1（YJG-HYGG）

区段2（定淮门—草场门）结构图
Spatial structure of S2（DHG-CCG）

区段3（草场门—清凉门）结构图
Spatial structure of S3（CCG-QLG）

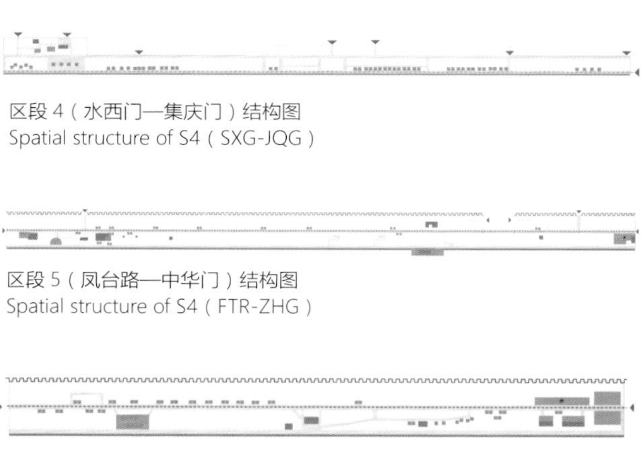

区段4（水西门—集庆门）结构图
Spatial structure of S4（SXG-JQG）

区段5（凤台路—中华门）结构图
Spatial structure of S4（FTR-ZHG）

区段6（中华门—雨花门）结构图
Spatial structure of S6（ZHG-YHG）

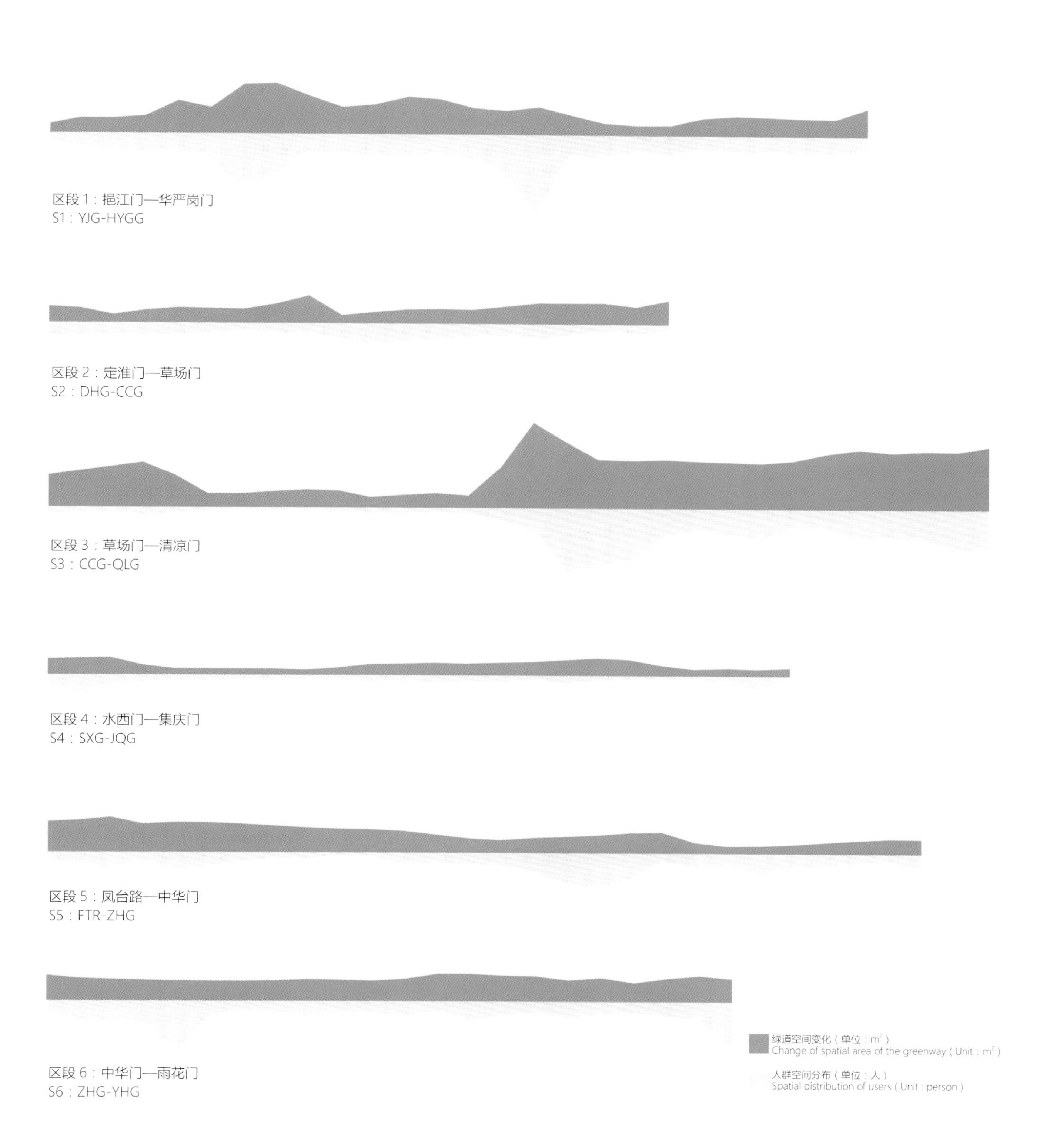
区段 1：挹江门—华严岗门
S1：YJG-HYGG
区段 2：定淮门—草场门
S2：DHG-CCG
区段 3：草场门—清凉门
S3：CCG-QLG
区段 4：水西门—集庆门
S4：SXG-JQG
区段 5：凤台路—中华门
S5：FTR-ZHG
区段 6：中华门—雨花门
S6：ZHG-YHG
绿道空间变化（单位：m²）
Change of spatial area of the greenway (Unit : m²)
人群空间分布（单位：人）
Spatial distribution of users (Unit : person)

内部空间要素的影响总体分析

由于绿道整体空间的收放变化较小，区段 4（水西门—集庆门）、区段 5（凤台路—中华门）和区段 6（中华门—雨花门）3 个区段使用人群的空间分布将与绿道内部空间要素和空间品质特征关联更加紧密，这些在具体的切片分析中已有所反映。

从区段 4（水西门—集庆门）、区段 5（凤台路—中华门）以及区段 6（中华门—雨花门）的使用人群聚集规律来看，使用人群聚集的切片也是停留休息和康体活动开展最密集的切片单元，而该类单元普遍的特征是拥有较大面积的集中活动场地（包括硬地和绿地），但却不总是座椅等休闲服务设施密度最高的地段。根据这一现象可以引出两个可能的结论：一方面，当绿道线性空间相对均质时，集中活动场地（含硬地和绿地）对人群停留休息和开展康体活动的吸引力要远高于游憩服务设施的吸引力。另一方面，调查发现绿道沿线固定的游憩服务设施，例如座椅，如摆放得当能为人群停留休息创造条件，但如摆放不当，实际上对于空间的刚性限定作用过于强烈，限制了停留休息以外其他活动的开展，例如在区段 5（凤台路—中华门）和区段 6（中华门—雨花门）两个区段上经常见到的围坐打牌、聊天等活动实际上就对座椅摆放形式和使用灵活性有着较高要求。

Analysis of the influence of internal elements

Since the overall space of the greenway does not change much in scale, the spatial distributions of the users on S4 SXG-JQG, S5 FTR-ZHG, and S6 SZG-YHG will be more closely linked to both the internal spatial elements of the greenway and the characteristics of spatial structure, which has already been reflected in detailed analyses of the slices.

Judging from how users are gathered around on S4 SXG-JQG, S5 FTR-ZHG, and S6 SZG-YHG, we find that the slices where most people gather are also the slice units where most people stop over and exercise. This type of slice units is an example of what happens when there are large areas of centralized places for activities (including open hard sites and green space), but they are not often the places where bench seats and other service facilities are the most densely distributed. We can draw two possible conclusions from this phenomenon. Firstly, when the linear space of the greenway is relatively homogeneous, centralized places of activities are far more attractive to people who stop over or exercise than recreational service facilities. Secondly, we found that permanent recreational service facilities like bench seats along the greenway will facilitate the activity of stopover only when they are properly installed. If not, they would probably limit space too rigidly, which sets a limit to the possibilities and opportunities of types of activities other than sitting down for a break. For instance, we see a lot of people sit around, play cards, chat, and do other kinds of activities that need more flexibility of the use of the benches on S5 FTR-ZHG and S6 SZG-YHG.

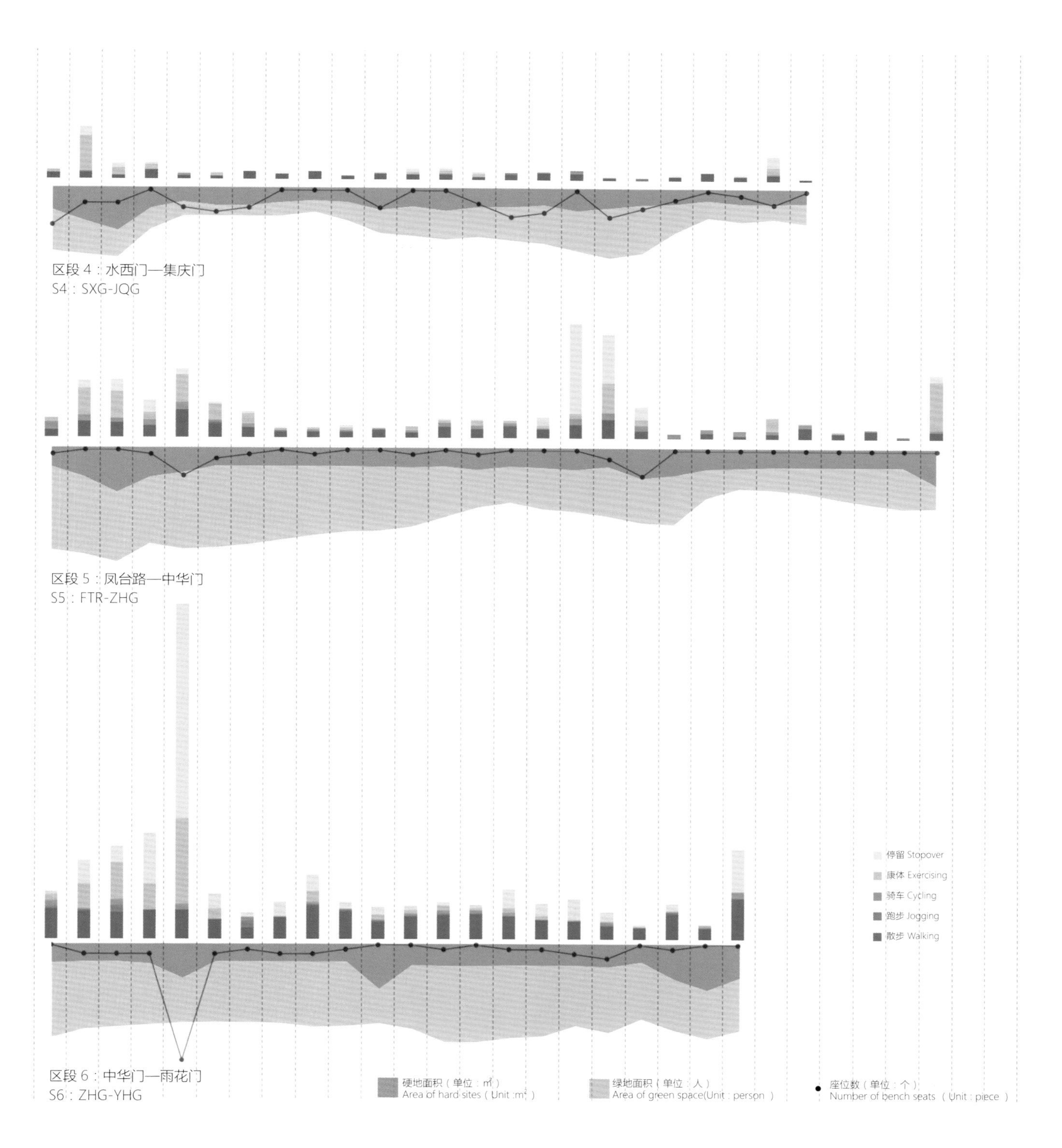
区段 4：水西门—集庆门
S4：SXG-JQG
区段 5：凤台路—中华门
S5：FTR-ZHG
区段 6：中华门—雨花门
S6：ZHG-YHG
停留 Stopover
康体 Exercising
骑车 Cycling
跑步 Jogging
散步 Walking
硬地面积（单位：m²）
Area of hard sites（Unit :m²）
绿地面积（单位：人）
Area of green space(Unit：person）
座位数（单位：个）
Number of bench seats（Unit：piece）

两个时段的服务绩效差异分析

我们发现几乎每个区段周末下午和日常傍晚使用人数最多的切片单元均不相同，其中使用人群聚集位置有明显转移的区段有区段 1（挹江门—华严岗门）、区段 2（草场门—清凉门）和区段 6（中华门—雨花门）。调查发现，与其他区段不同的是，区段 1（挹江门—华严岗门）和区段 3（草场门—清凉门）周末下午和日常傍晚使用人群的年龄结构存在明显差异，周末下午通常青幼年人群比重较大，而日常傍晚中老年人群比重较大，因此造成不同时段人群聚集位置转移的主要原因是不同时段拥有不同年龄的主导使用人群，而不同人群的活动类型、空间需求均不相同。但区段 6（中华门—雨花门）两个时间段主导使用人群的年龄结构并无明显差异，调查发现造成该区段不同时段人群聚集差异的原因主要是同一类使用人群两个时段的主导活动类型差异及其对空间需求的不同。

其他三个区段周末下午和日常傍晚主导使用人群均以中老年人为主，虽然部分区段使用人群最多的切片单元会有所差异，但在空间上位置基本邻近的切片单元没有出现聚集区域大范围转移的情况。部分区段两个时段人群聚集切片的主导使用活动类型会有明显差异，例如区段 4（水西门—集庆门）、区段 5（凤台路—中华门）均在日常傍晚，切片单元会出现停留休息和进行康体活动的人群明显下降的迹象，而在区段 2（定淮门—草场门）则相对稳定。经调查和分析发现，造成该现象的原因在一定程度上是由于使用人群在下午和傍晚两个时段的活动类型差异，但更大程度的原因是该类切片单元场地的照明条件和开敞程度难以满足中老年人傍晚聚集停留休息和进行康体活动的需求。

Variances of the service performance in the two time periods

In the meantime, we find that on almost every segment the slice with the largest number of users on a weekend afternoon is not the same as every day at nightfall. S1 YJM-HYGG, S3 CCG-QLG, and S6 SZG-YHG are where the location of gathered crowds shows noticeable transfer. Through study, we find that what is different from the other segments is that the age structure of users on a weekend afternoon is different from that on a day at nightfall both on S1 YJM-HYGG and S3 CCG-QLG. On weekend afternoons, young people and children are often the dominant user groups, while middle-aged and senior people are the dominant user groups every day at nightfall. As a result, the reason for location transfer of users in two time periods is that different age groups dominate each time periods and that different groups of people differ in preferred recreational activity with a specific spatial demand. However, on S6 SZG-YHG, there is not much noticeable difference in age structure of dominant groups in both time periods. Through study, we find that the main reason for the locations transfer where users gathering around in this segment in different time periods is that the same group of users differs in preferred recreational activities in the two time periods with the corresponding difference in the demands on spatial characteristics.

In the other three segments, the dominant user groups on both a weekend afternoon and every day at nightfall are mainly middle-aged and senior people. Although in some segments there may be a difference in slice units with the largest number of users, in terms of space, the locations of the slice units with the largest number of users are basically close to one another without any significant transfer in large scale of aggregated areas. In some segments, in both time periods, the dominant activities on slices where the most people are gathered may differ greatly. For example, on S4 SXG-JQG and S5 FTR-ZHG, the number of users who stop over and exercise on the slice units every day at nightfall decreases noticeably, while the number on S2 DHG-CCG is relatively stable. After the review and analysis, we found that the cause of this phenomenon is, to a certain extent, the difference in activities occurring in the two time periods. However, to a greater extent, the reason is that the lighting conditions and the openness of the space on these slice units can hardly meet the demand of middle-aged and senior people who want to gather around, stop over, and exercise at nightfall.

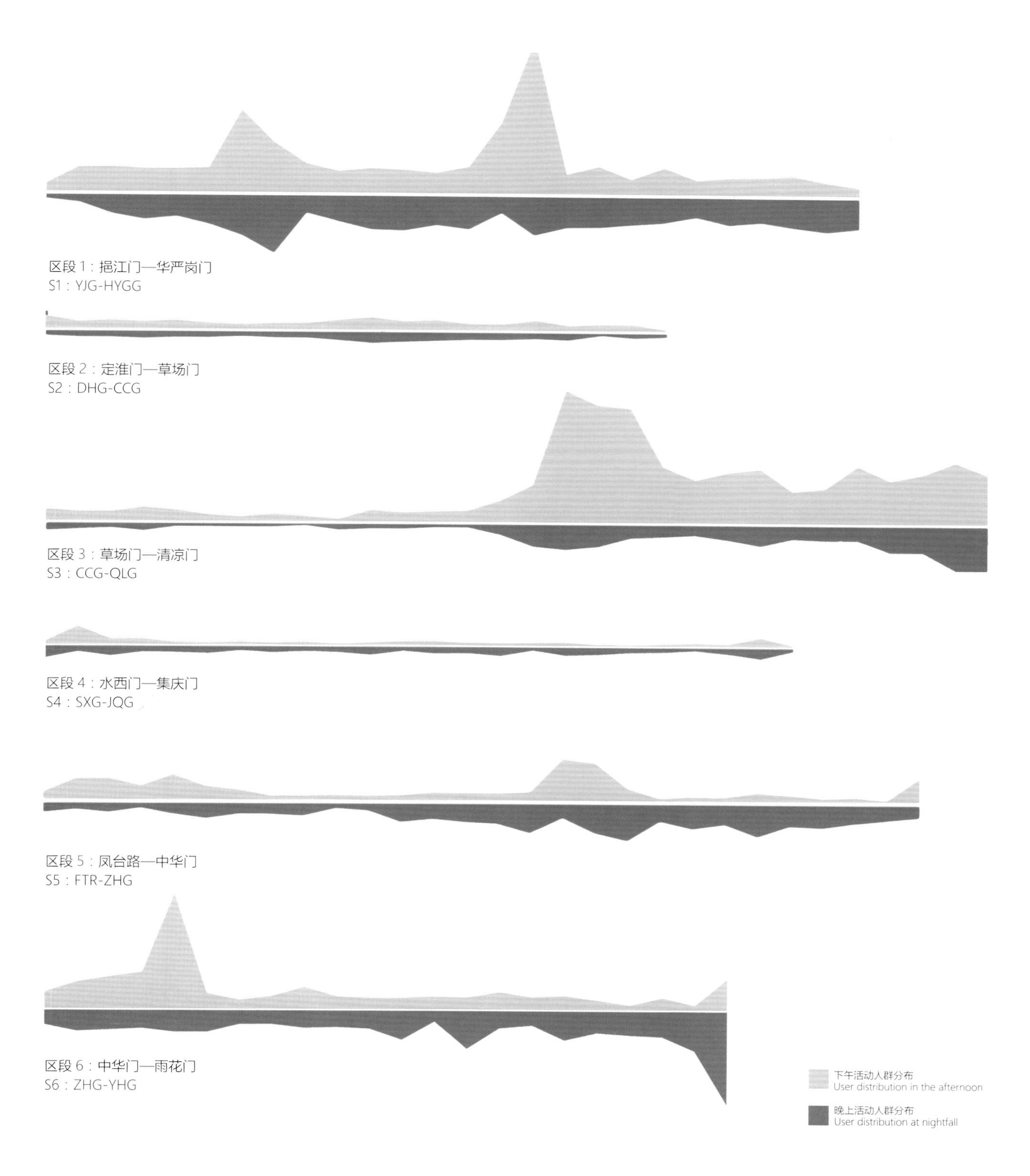
区段 1：挹江门—华严岗门
S1：YJG-HYGG
区段 2：定淮门—草场门
S2：DHG-CCG
区段 3：草场门—清凉门
S3：CCG-QLG
区段 4：水西门—集庆门
S4：SXG-JQG
区段 5：凤台路—中华门
S5：FTR-ZHG
区段 6：中华门—雨花门
S6：ZHG-YHG
下午活动人群分布
User distribution in the afternoon
晚上活动人群分布
User distribution at nightfall

使用人群的年龄结构与活动类型特征分析

从 6 个调查区段的人群年龄结构来看，中老年人是目前明城墙绿道的核心使用群体，而且随着中国城市人口老龄化趋势的进一步发展，在建成环境绿道里中老年人逐渐成为主导使用人群的特征很可能会进一步强化。绿道的使用人群结构将在很大程度上决定其使用活动类型的结构，在本次调查中，以散步、广场舞、棋牌等为代表的中国城市中老年群体的主要游憩行为活动类型已在绿道上充分显现。这也是为何目前在绿道游径使用上，与中老年人为主体的散步人数相比，青年人占主体的骑行人数微不足道，甚至可以忽略不计的重要原因。

Characteristics of age structure and activity type of users

Judging from the age structure of user groups in the six segments of the survey, we found that middle-aged and senior people are currently the core group of users of the Nanjing Ming Dynasty City Wall Greenway. With further aging of the Chinese population in urban areas, the trend of middle-aged and senior people being the dominant user group on the greenway in its built environment would likely continue. The user structure of the greenway will determine the types of activities that take place in it. In this survey, in Chinese cities represented by walking, square dancing, Chinese chess and card playing, the main recreational activities highly popular in the groups of middle-aged and senior people have fully shown themselves on the greenway. This is why in terms of the current use of the greenway the number of cyclists, who are mostly young people, is insignificant and even ignored, compared to that of middle-aged and senior people who mainly walk.

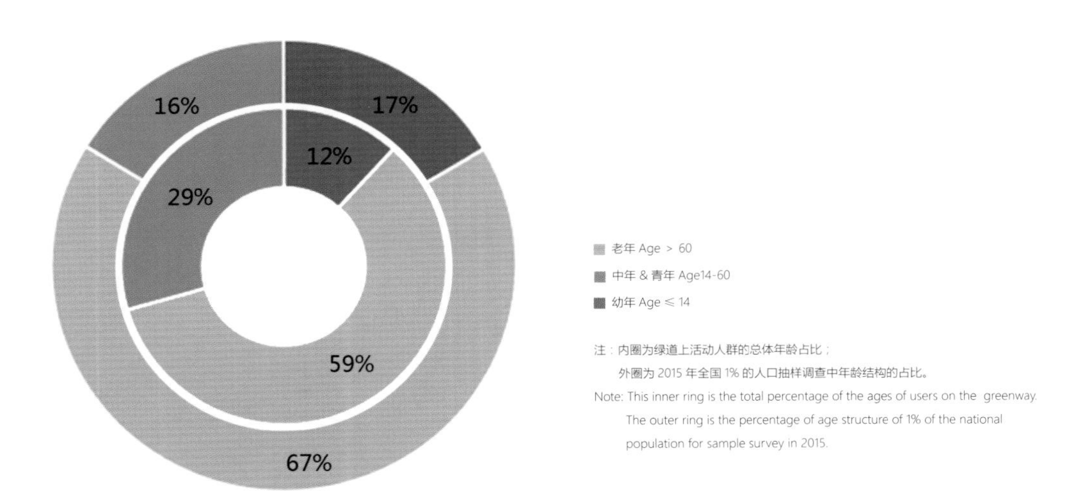

注：内圈为绿道上活动人群的总体年龄占比；
外圈为 2015 年全国 1% 的人口抽样调查中年龄结构的占比。
Note: This inner ring is the total percentage of the ages of users on the greenway.
The outer ring is the percentage of age structure of 1% of the national population for sample survey in 2015.

S1：YJG-HYGG
S2：DHG-CCG
S3：CCG-QLG
S4：SXG-JQG
S5：FTR-ZHG
S6：ZHG-YHG
S1：YJG-HYGG
S2：DHG-CCG
S3：CCG-QLG
S4：SXG-JQG
S5：FTR-ZHG
S6：ZHG-YHG
停留 Stopover
康体 Exercising
骑车 Cycling
跑步 Jogging
散步 Walking
老年 Age > 60
中年 Age 30-60
青年 Age 14-30
儿童 Age ≤ 14

石
头
城

内部空间要素服务绩效分析

Analysis of service performance of key elements of the greenway interior space

在对绿道内部空间特征与服务绩效进行切片分析的基础上，本书还针对座椅、硬地、绿地和游径四大空间要素及其服务绩效进行了专门分析。

其中，对于座椅、硬地和绿地三类要素，本书采用了相似的分析路径，即对6个区段要素的位置和布局进行梳理，将其规模和密度等指标进行量化，并与要素服务绩效（例如使用人数规模、密度以及占比等）指标进行关联分析，讨论前者对后者的影响方式和规律。同时，结合典型区段的要素空间特征及其服务效果，对要素积极与消极的布局模式进行探索。

由于游径较之其他三类要素稍显特殊，本书首先在这6个区段游径组合模式和宽度进行量化分析的基础上，将其分别与周末下午和日常傍晚游径上的散步、跑步和骑车人群密度指标进行关联分析，并对每个区段游径的断面特征进行比对，从而针对游径空间特征对绿道服务绩效的影响进行探讨。

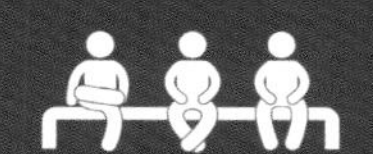

Besides the slice analysis of every single greenway segment, another analysis on the service performance of the key elements (including bench seats, hard sites, green space, and trails), which integrates and compares the corresponding datas of six segments, is also made in the study.

Among the four types of elements, a similar pathway is employed in the analysis of bench seats, hard sites and green space. In the process of analysis, the location and layout of the elements and the indicators of number and density of the elements are classified or quantified at first. Then, the results of classification and quanification are analyzed correlatively together with the indicators of service performance (such as the number, density, and proportion of users, etc.), by which the connections and influences in between are discussed respectively. Moreover, a review of the spatial characteristics and service effect of the elements was done in one or two typical segments, so as to make a further discussion on the layout patterns of the element.

Due to the difference from the other three elements, the trails of six segments are firstly classified on its modes of inner structure and the widths. Next, together with the comparison of the cross-section densities of each trail , the classification results are correlatively analyzed with the density of pedestrians, joggers and cyclers on each greenway on a weekend afternoon and every day at nightfall. A discussion is made in the end to explore the influence from the trails to the service performance of the greenway.

空间要素分析——座椅
Analysis of key elements - bench seats

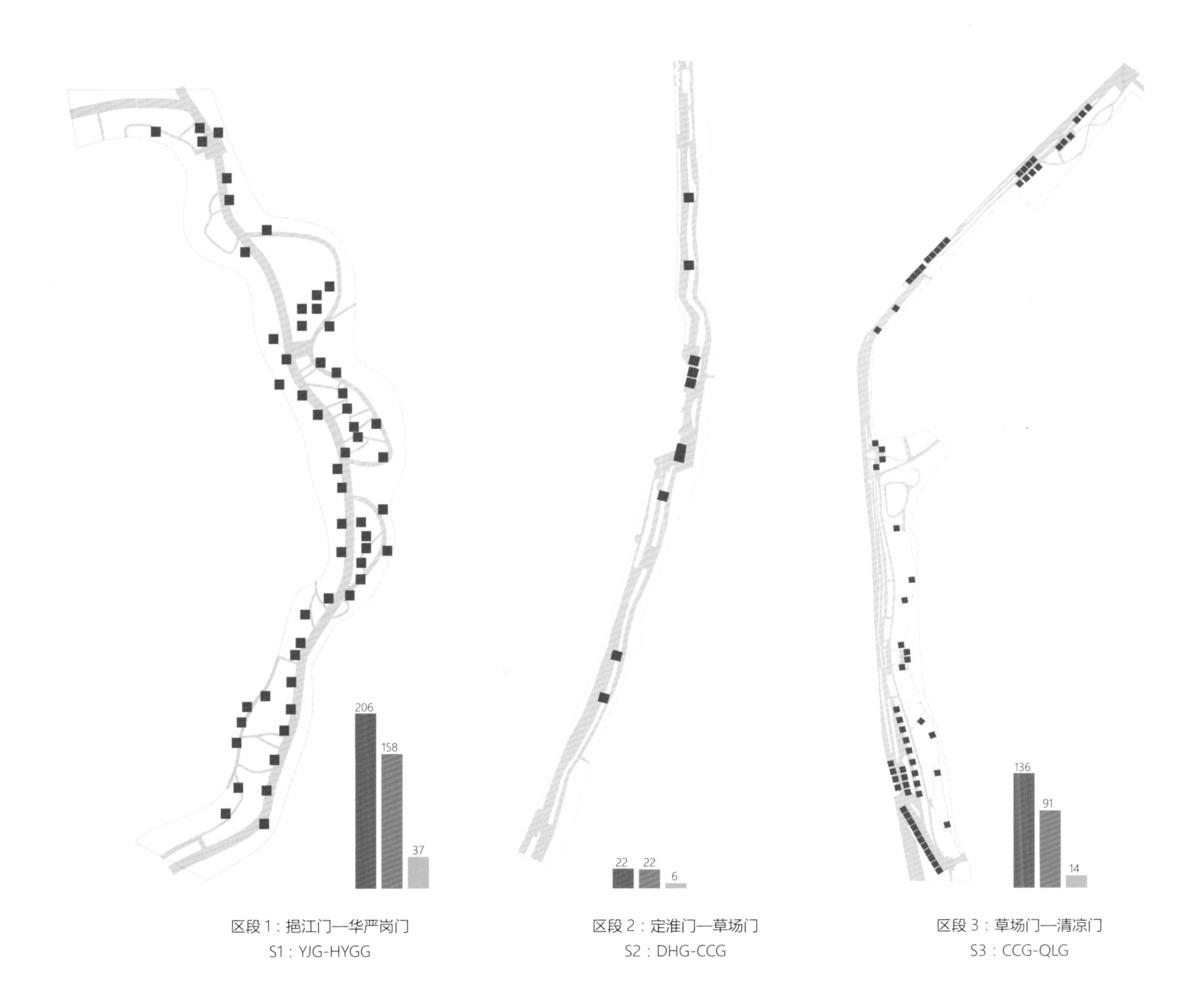

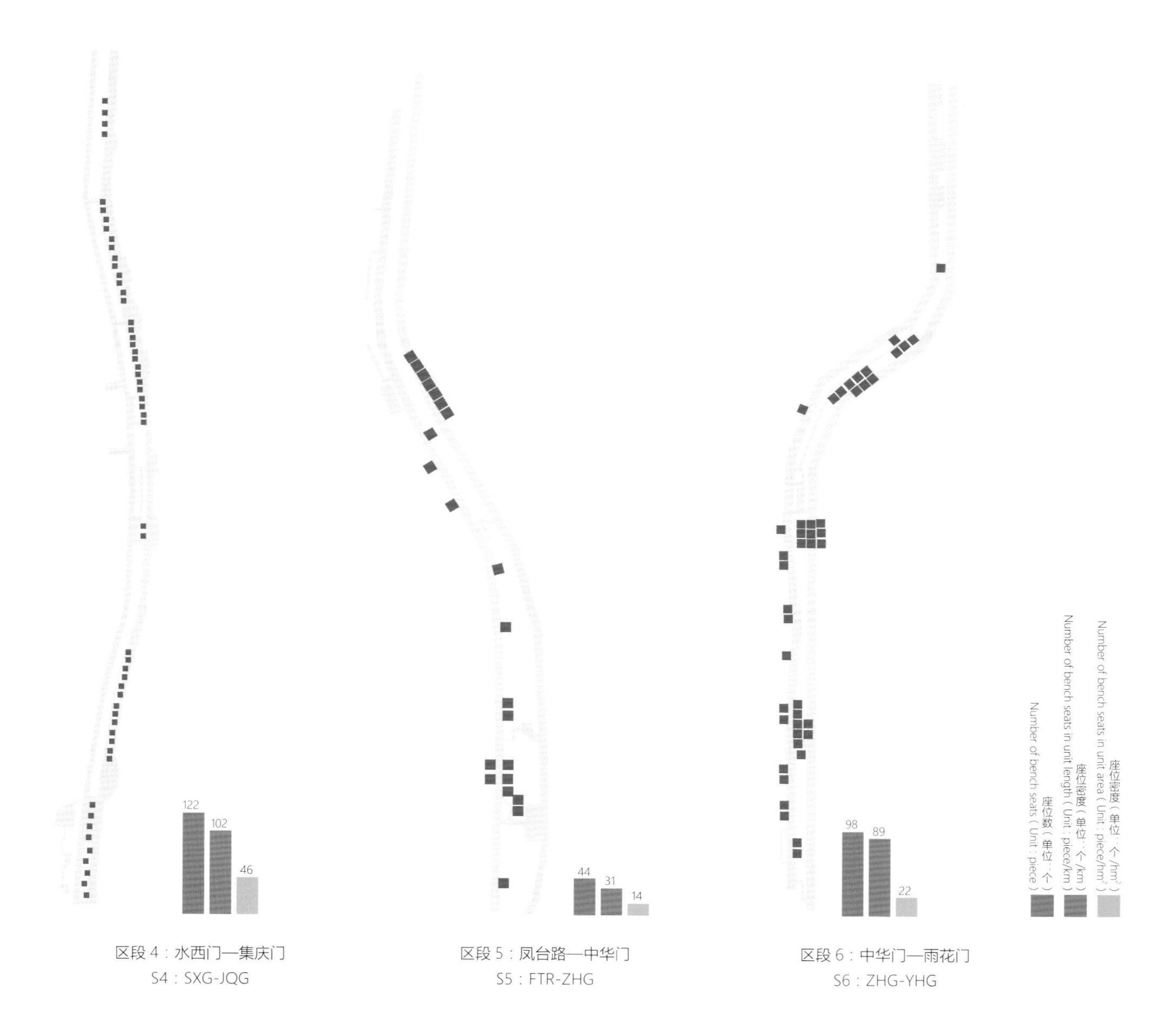
122
102
46
区段 4：水西门—集庆门
S4：SXG-JQG
44
31
14
区段 5：凤台路—中华门
S5：FTR-ZHG
98
89
22
区段 6：中华门—雨花门
S6：ZHG-YHG
座位数（单位：个）
Number of bench seats (Unit : piece)
座位密度（单位：个/km）
Number of bench seats in unit length (Unit : piece/km)
座位密度（单位：个/hm²）
Number of bench seats in unit area (Unit : piece/hm²)

25
118
57
S1：YJG-HYGG
8
29
14
15
17
77
S2：DHG-CCG
13
10
45
24
172
126
S3：CCG-QLG
11
29
21
17
11
9
S4：SXG-JQG
10
8
7
30
59
134
S5：FTR-ZHG
5
15
34
37
117
119
S6：ZHG-YHG
9
48
49
下午停留人数占比（单位：%）
Percentage of people who stop over in the afternoon（Unit：%）
下午停留人数（单位：人）
People who stop over in the afternoon（Unit：person）
下午停座比（单位：%）
Potential user-bench ratio in the afternoon（Unit：%）
傍晚停留人数占比（单位：%）
Percentage of people who stop over at nightfall（Unit：%）
傍晚停留人数（单位：人）
People who stop over at nightfall（Unit：person）
傍晚停座比（单位：%）
Potential user-bench ratio at nightfall（Unit：%）

座椅使用特征总体分析

左图将 6 个区段座椅的使用情况进行了归纳，分为停留人数、停留人数占绿道使用人数比例（简称停留人数占比）和停留人数占座椅座位总数比例（简称停座比）三项指标，并进行下午以及傍晚时段数据的对比分析。

从平均停留人数来看，6 个区段周末下午停留数都远大于日常傍晚，不同区段差异较大，下午差异尤为明显。区段 3（草场门—清凉门）(2, 3, 4) 下午停留人数最多，而区段 4（水西门—集庆门）(3, 2, 1) 停留人数无论是周末下午还是日常傍晚都是 6 个区段中最少的。从停留人数占比来看，周末下午停留人数比重也远大于日常傍晚。区段 6（中华门—雨花门）(4, 4, 3) 和区段 5（风台路—中华门 ）(5, 5, 4) 周末下午停留人数占比较大，区段 2（定淮门—草场门）(6, 6, 6) 和区段 4（水西门—集庆门）占比较小。但是到了傍晚，区段 2（定淮门—草场门）成了停留人数占比最大的区段，而该区段下午占比是 6 区段中最小的。从停座比来看，下午的停座比也远大于傍晚。各区段停座比差异悬殊，区段 5（风台路—中华门）、区段 3（草场门—清凉门）和区段 6（中华门—雨花门）下午的停座比都超过了 100%，但是区段 4（水西门—集庆门）无论是下午还是傍晚停座比都小于 10%，存在着大量座椅的闲置浪费。

指标显示，座椅利用情况与座椅座位数和密度指标出现反差最大的是区段 6（中华门—雨花门）和区段 4（水西门—集庆门），前者在座椅座位数和密度指标排位靠后的情况下仍能吸引较多人群停留休息，并拥有较大的停座比，而后者正好相反。在解析该反差所产生的原因时，我们发现这 2 个区段的座椅空间布置形式存在较大差异。其中，区段 6（中华门—雨花门）的座椅在滨水和沿路空间布置，并且呈现出组团及行列等多种形式，能满足不同停留人群的多样化需求。而区段 4（水西门—集庆门）由于空间所限，仅有的形式就是在滨水地段进行行列布置，这在很大程度上限制了座椅的使用方式，并大大降低了其被不同人群使用的可能性。

注：括号内数字依次为座位数排序、单位长度座位数排序、单位面积座位数排序。

The characteristics of the use of bench seats

The figure on the left divides how the bench seats in the six segments are used into three categories: the number of people who stop over, the percentage of the number of people who stop over in the total number of users of the greenway segment, and the percentage of the number of people who stop over in the total number of bench seats (potential user-bench ratio). The data from weekend afternoons and every day at nightfall can be compared and analyzed.

Judging from the average number of people who stop over, we find that the number of people who stop over on a weekend afternoon is much larger than that of every day at nightfall in the six segments. Different segments differ greatly, and the difference in the afternoon is especially evident. S3 CCG-QLG (2,3,4) has the largest number of people who stop over in the afternoon, while S4 SXG-JQG (3,2,1) has the smallest number of people who stop over in all six segments, whether it is on a weekend afternoon or every day at nightfall. Judging from the percentage of the number of people who stop over, we find that the number of people who stop over on a weekend afternoon is also much larger than that of every day at nightfall. S6 ZHG-YHG (4,4,3) and S5 FTR-ZHG (5,5,4) both have a large percentage of the number of people who stop over on weekend afternoons, while the percentages on S2 DHG-CCG (6,6,6) and S4 SXG-JQG are relatively small. However, at nightfall, S2 DHG-CCG becomes the segment with the largest percentage of the people who stop over, while that percentage on the weekend afternoons is the smallest among all six segments. In terms of potential user-bench ratio, the ratio on weekend afternoons is much higher than that on every day at nightfall. There is a huge difference in potential user-bench ratio among all the segments. The ratio on weekend afternoons exceeds 100% for S5 FTR-ZHG, S3 CCG-QLG, and S6 ZHG-YHG, but the ratio in S4 SXG-JQG, where many benches remain empty, is below 10% whether for a weekend afternoon or every day at nightfall.

According to the indicators, the sharpest contrast between the use of bench seats and the number and density of bench seats appears on S6 ZHG-YHG and S4 SXG-JQG. Although the number and density of bench seats both rank among the last few, the former can still attract a lot of people to stop over and keep a high potential user-bench ratio, which is quite the opposite for the latter. In analyzing the cause of such a contrast, we find out that these two segments differ greatly in the arrangement of bench space. The bench seats on S6 ZHG-YHG are installed at the waterfront and along the trails in the form of groups, processions, and other formations, so as to satisfy the diverse demands of different groups. However, since S4 SXG-JQG is limited by space, the only form of bench arrangement is processions at the waterfront, which greatly reduces the potential use for the seats.

Note: The numbers in the parentheses are (the ranking of the number of bench seats in all 6 segments, the ranking of the number of bench seats in unit length in all 6 segments, and the ranking of the number of bench seats in unit area in all 6 segments).

1 | 2 | 3 / 4

典型区段座椅使用特征解析

Characteristic analysis of the use of bench seats on typical slices

区段 4（水西门—集庆门）(3, 2, 1) 的座椅座位数和密度在 6 个区段中排名较靠前，但是实际停留人数较少和停座比较低。实际调研发现，该段是 6 区段中最为狭窄的一段，自行车道和步行道的宽度都不超过 3m，绿化空间更加局促，整个区段呈现出明显的通行空间特征，并不适宜人停留。另外，该区段的多数座椅较简陋，部分直接将绿化台边沿铺上木板改造成座椅，却未做妥善管理和维护，座椅脏乱不堪，很少能吸引人坐下停留。

S4 SXG-JQG is the segment with the greatest number of bench seats and greatest bench seat density. But there is a low number of people who stop over and the segment's potential user-bench ratio is extremely small. According to the on-spot survey, we found that this segment is the narrowest of all six segments; the widths of either bikeways or footpaths are less than 3m; the green space is more cramped; the entire segment is characteristic of being noticeably used for traffic, thus unsuitable for people to stopover. Besides, most of the bench seats on this segment are simple and crude, some of which are directly made of planks and paved on the border of greening bed without any proper maintenance. As a result, those bench seats are in a poor, dirty condition and fail to attract people to sit.

区段6（中华门—雨花门）(4, 4, 3) 是6个调研区段中下午停留人数占比最大的区段，也是下午的停座比超过100%的区段之一。实际调研发现，该区段靠近南京老城南居民区，许多老人都会自带小马扎来到绿道成群结队地下棋和打牌，还吸引不少人站在一旁围观，沿步行道两侧的座椅和休憩亭几乎全部被该类活动占用。由于此类人群数目远远超过了绿道座椅座位数目，因此很多人群自带马扎或坐在花坛边缘、石块等上面。

Of all six segments, S6 ZHG-YHG (4,4,3) has the largest percentage of people who stop over on weekend afternoons. It is also one of the segments where the potential user-bench ratio exceeds 100% on weekend afternoons. According to the on-spot survey, we find that since this segment is close to the residential area of the old city of Nanjing, a lot of seniors bring their own campstools to the greenway and play cards or chess in groups, which attracts a lot of people to stand around and watch. The bench seats and pavilions along both sides of the footpaths are almost occupied by people doing such activities in small groups. People bring their own stools or sit on rocks and flower bed edges since the number of people greatly exceeds the number of bench seats on the greenway.

1
2
3
4

POSITIVE 积极

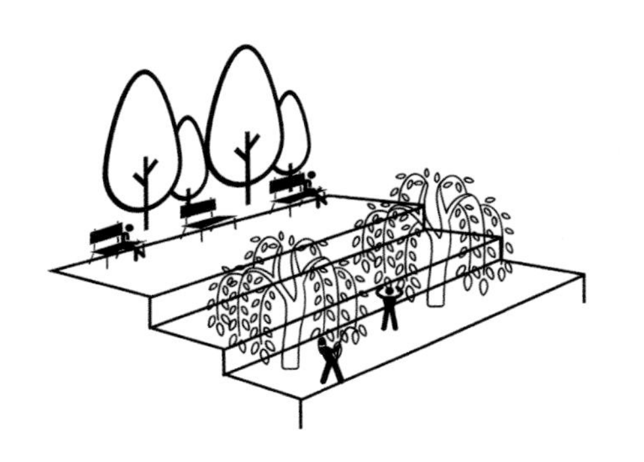

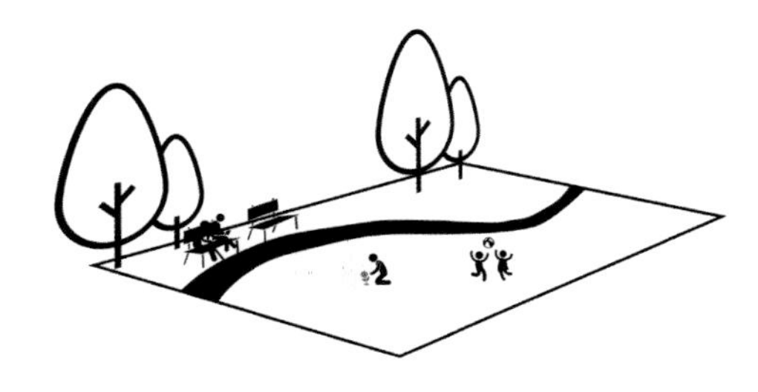

- 原型：区段 3（草场门—清凉门）切片 4
- 位于阶梯状平台最高处
- 视野良好
- 可以远眺水面，近看过往人流

· Prototype：S3（CCG-QLG）slice 4
· At the highest point of the stair-step platform
· Wide vision
· Distant view of the surface of the water and close look at crowds of people who pass by

- 原型：区段 6（中华门—雨花门）切片 15
- 位于草地上，小路一侧
- 绿化景观良好
- 可以看草地上活动的人群

· Prototype：S6（ZHG-YHG）slice 15
· Located on the grass on one side of the footpath
· Proper greenery landscape is good
· Be able to see the people who do activities

- 原型：区段 3（草场门—清凉门）切片 29
- 位于开敞的硬质广场一侧
- 视野良好
- 可以看广场上活动的人群

· Prototype：S3（CCG-QLG）slice 29
· Located on one side of the wide-open hard plaza
· Wide view
· Be able to see the people who do activities on the plaza

消极 NEGATIVE

- 原型：区段 1（挹江门—华严岗门）切片 8
- 围绕树池做一圈座椅
- 视野较差
- 形式重复单调

· Prototype：S1（YJG-HHGG）slice 8
· Benches are installed around the planting pool
· Limited and blocked vision
· Repeated andmonotonous form

- 原型：区段 4（水西门—集庆门）切片 16
- 位于狭窄道路一侧
- 与过往行人距离过近
- 空间过于急促

· Prototype：S4（SXG-JQG）slice 16
· Located on one side of the narrow road
· Too close to the people who pass by
· Space is too cramped

- 原型：区段 3（草场门—清凉门）切片 8
- 背靠高墙，面对灌木墙
- 灌木遮挡视野
- 整个空间较为封闭

· Prototype：S3（CCG-QLG）slice 8
· It leans against the high city wall and faces the quickset hedge
· Vision blocked by shrubs
· The entire space is sealed off

空间要素分析 —— 硬地
Analysis of spatial elements - hard sites

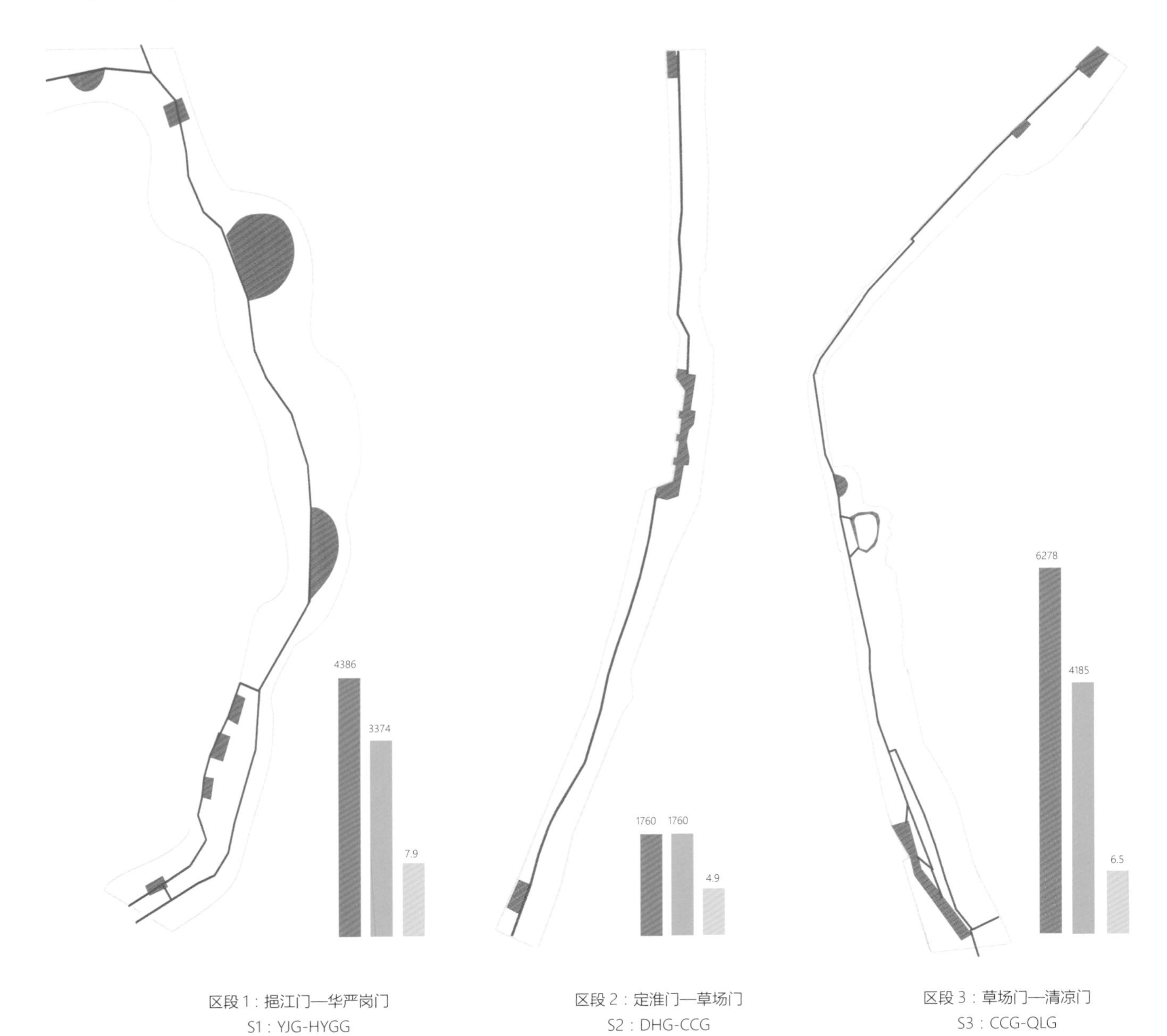

区段 1：挹江门—华严岗门
S1：YJG-HYGG

区段 2：定淮门—草场门
S2：DHG-CCG

区段 3：草场门—清凉门
S3：CCG-QLG

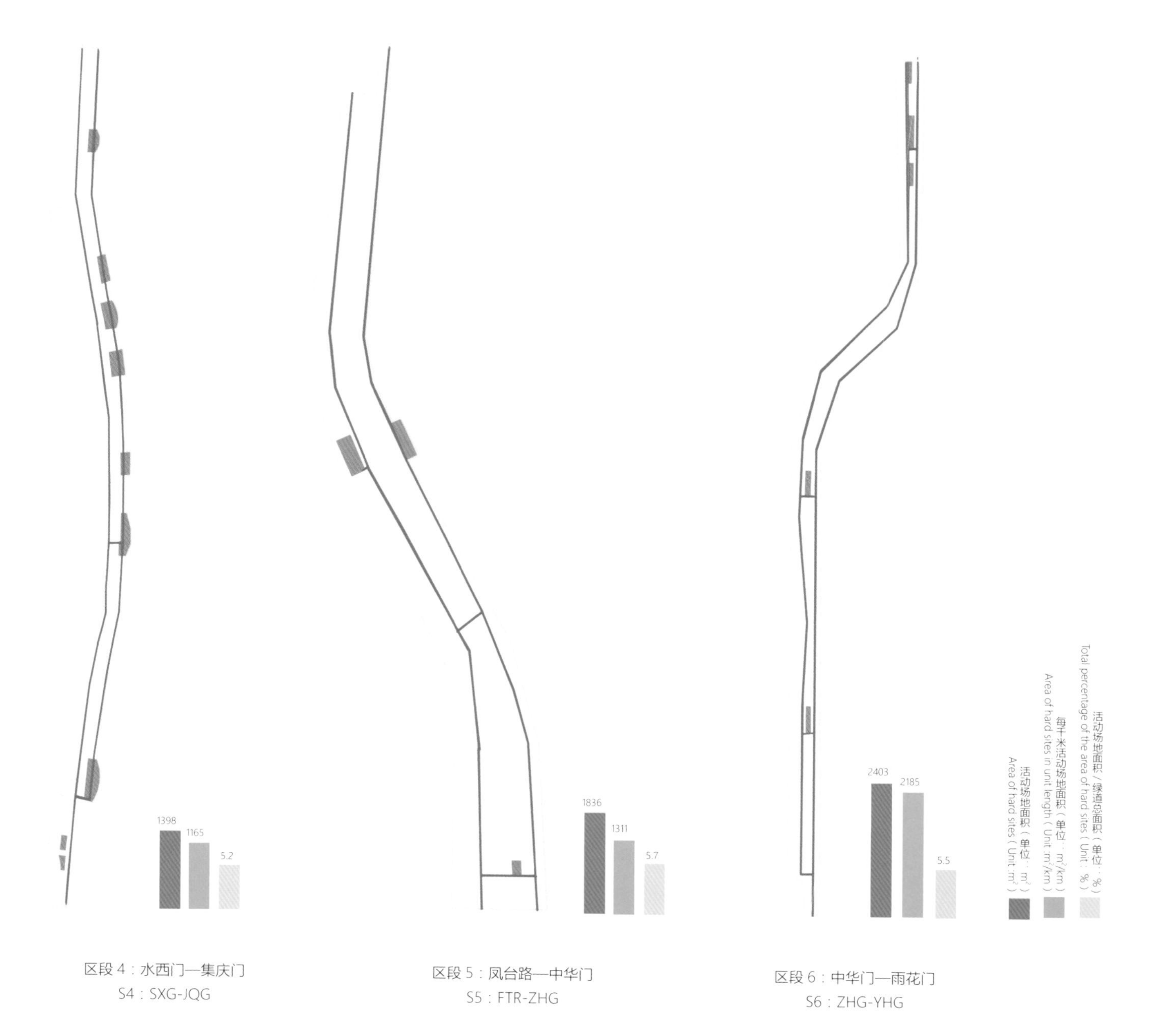

区段 4：水西门—集庆门
S4：SXG-JQG

区段 5：凤台路—中华门
S5：FTR-ZHG

区段 6：中华门—雨花门
S6：ZHG-YHG

硬地使用特征总体分析

集中硬质场地通常能吸引人群聚集、停留，进行各种康体活动。为更客观地把控人群在绿道硬地上的活动特征，我们选取周末下午与日常傍晚康体活动人数、康体活动人数占绿道使用人数比例（简称康体活动人数占比）和康体活动人数占活动场地面积比例（简称康体活动密度）数据进行对比分析。从整体指标对比可发现硬地面积和密度标高的区段更容易吸引康体活动人群聚集。

在周末下午时段，区段3（1，1，2）的各组指标值均最大，区段1（2，2，1）康体活动人数和密度次之，但该区段康体活动人数占比仅排第4位。区段2（5，4，4）各指标值均排序最低，区段4（6，6，5）虽然康体活动人数与密度仅排第5位，但在总人数中的占比排序第2位。

在日常傍晚时段，各区段相关指标值均出现不同程度的下降，排序也出现改变。其中，区段3（草场门—清凉门）康体活动人数占比仍排序最高，但康体活动人数和密度排序最高的区段则变为区段1（挹江门—华严岗门）。区段2（定淮门—草场门）各指标仍排序最低，但指标下降幅度最大的则是区段5（4，5，3）和区段6（3，3，4），区段4（水西门—集庆门）的指标下降幅度相对较小。

从实际调查情况来看，周末下午和日常傍晚康体活动类型和主体均存在较大差异，其中周末下午以青年运动和儿童娱乐为主，而在日常傍晚则以中老年人健身活动为主。日常傍晚时段出现降幅较大的区段所反映的问题其实是，该区段硬地在两个时段开展康体活动的适宜性存在较大差异。这种问题产生的根本原因既有场地空间设计包容性的问题，也有场地环境（如照明条件等）方面的问题。

注：括号内数字表示硬地面积排位、硬地面积密度排位、硬地面积占比排位。

The characteristics of the use of hard sites

The centralized and open hard sites usually attract people to gather around, stop over, and do all kinds of physical exercises. To manage the characteristics of human activities on the greenway objectively, we selected, compared, and analyzed data such as the number and density of people who exercise, percentage of people who exercise out of the total number of users of the greenway segment, and density on both weekend afternoons and every day at nightfall. Judging from the comparison of the indicators, we found that there are more possibilities for the segments to attract the people gathering for exercise, on which the indicators of the area and density of hard sites are greater.

On the afternoon of a weekend, all the indicators of S3 CCG-QLG (1,1,2) are the highest, while the number and density of people who exercise on S1 YJG-HYGG (2,2,1) rank the 2nd, but the percentage of those who exercise on this segment only ranks 4th. All the indicators of S2 DHG-CCG (5,4,4) rank last. Although the number and density of people who exercise and its density of S4 SXG-JQG only rank the 5th, its percentage of the total number of people using the segment ranks the 2nd.

Every day at nightfall, relevant indicators of different segments drop to varying degrees, along with the change of rankings. Among the segments, the percentage of people who exercise in S3 CCG-QLG still ranks 1st, but the number and density of people who exercise in S1 YJG-HYGG ranks 1st. All the indicators of S2 DHG-CCG still rank last, but the segments with the most drastic decrease of indicators are S5 FTR-ZHG (4,5,3) and S6 ZHG-YHG (3,3,4), and the decrease range of all the indicators of S4 SXG-JQG is relatively small.

According to the on-spot survey, the types of physical exercise and those who do those exercises on weekend afternoons and every day at nightfall differ greatly. On a weekend afternoon, most of the users are mainly young people who exercise and playing children, while every day at nightfall most of the users are mainly middle-aged and senior people who exercise. The problem reflected by the big decrease range of indicators every day at nightfall is that the suitability of hard sites in this segment for doing different physical exercises in each time periods differ a lot. The root cause of this problem is not only a lack of flexibility and inclusiveness in site design but also the poor condition necessary facilities (like lighting and so on).

Note: The numbers in the parentheses show (the ranking of the area of hard sites in all 6 segments, the ranking of the area of hard sites in unit length in all 6 segments, and the ranking of the percentage of the area of hard sites in the total area of segment in all 6 segments).

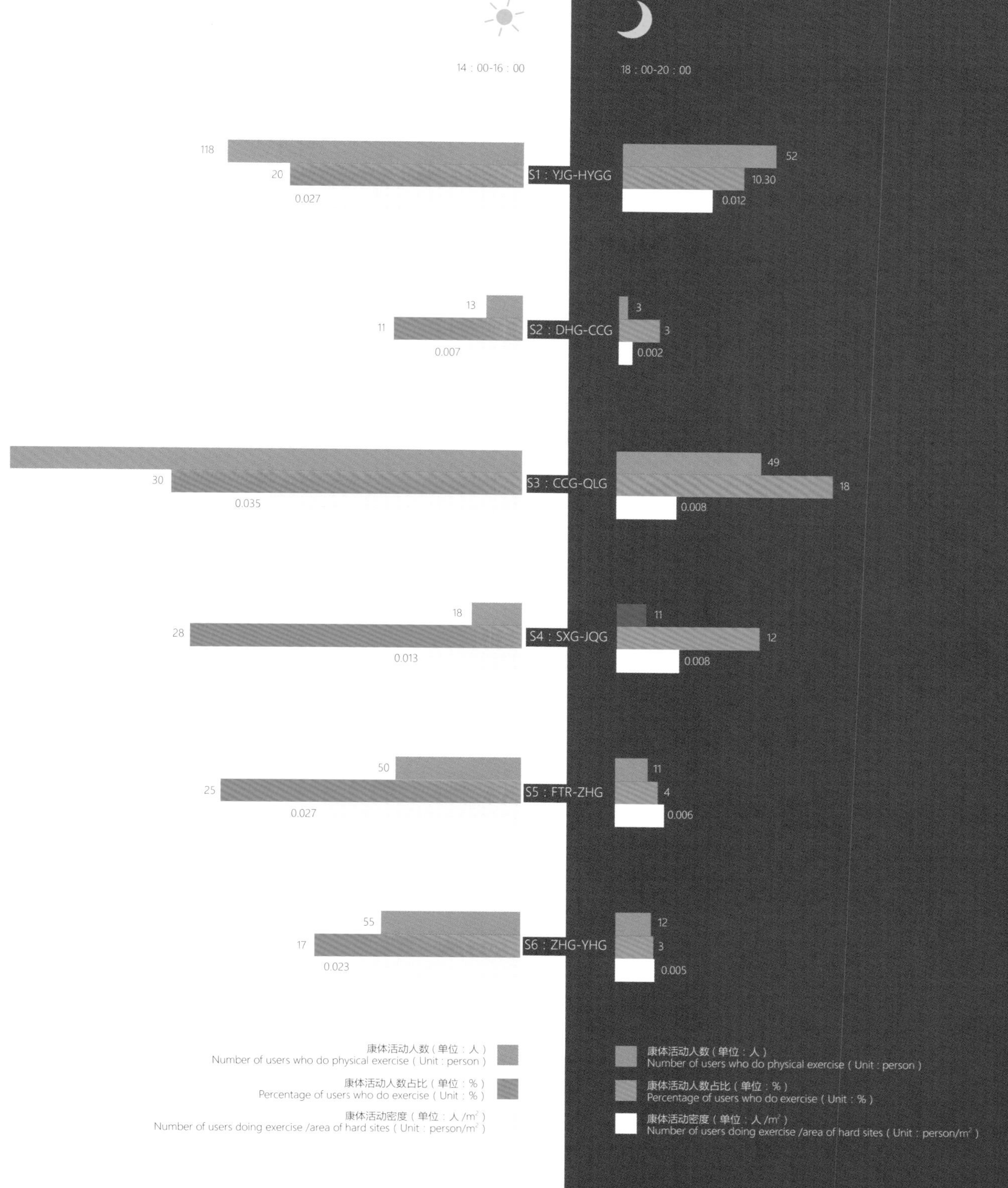
14：00-16：00
18：00-20：00
118
20
0.027
S1：YJG-HYGG
52
10.30
0.012
13
11
0.007
S2：DHG-CCG
3
3
0.002
217
30
0.035
S3：CCG-QLG
49
18
0.008
18
28
0.013
S4：SXG-JQG
11
12
0.008
50
25
0.027
S5：FTR-ZHG
11
4
0.006
55
17
0.023
S6：ZHG-YHG
12
3
0.005
康体活动人数（单位：人）
Number of users who do physical exercise（Unit：person）
康体活动人数占比（单位：%）
Percentage of users who do exercise（Unit：%）
康体活动密度（单位：人/m²）
Number of users doing exercise /area of hard sites（Unit：person/m²）
康体活动人数（单位：人）
Number of users who do physical exercise（Unit：person）
康体活动人数占比（单位：%）
Percentage of users who do exercise（Unit：%）
康体活动密度（单位：人/m²）
Number of users doing exercise /area of hard sites（Unit：person/m²）

区段 1 : 挹江门—华严岗门 (切片 8)
S1 : YJG-HYGG (Slice8)

区段 2 : 定淮门—草场门 (切片 12)
S2 : DHG-CCG (Slice12)

区段 3 : 草场门—清凉门 (切片 28)
S3 : CCG-QLG (Slice)

区段 4 : 水西门—集庆门 (切片 3)
S4 : SXG-JQG (Slice3)

区段 5 : 凤台路—中华门 (切片 3)
S5 : FTR-ZHG (Slice3)

区段 6 : 中华门—雨花门 (切片 21)
S6 : ZHG-YHG (Slice21)

14：00-16：00

18：00-20：00

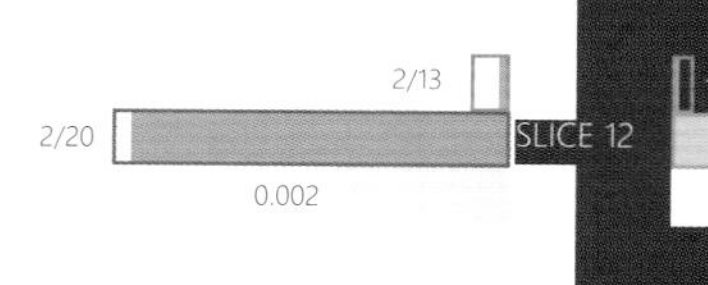

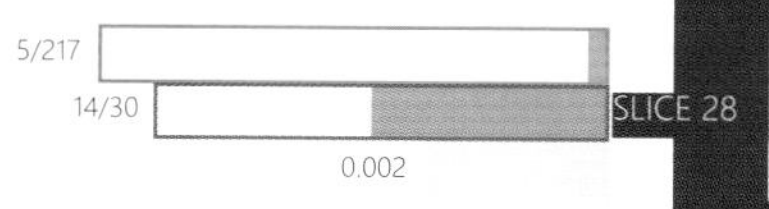

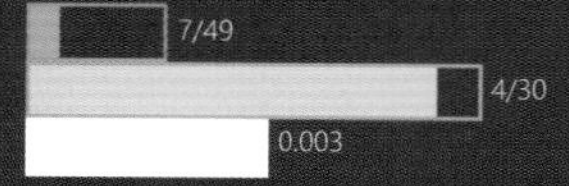

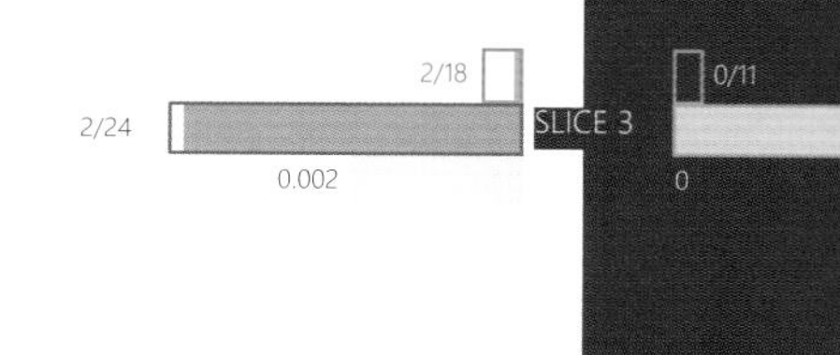

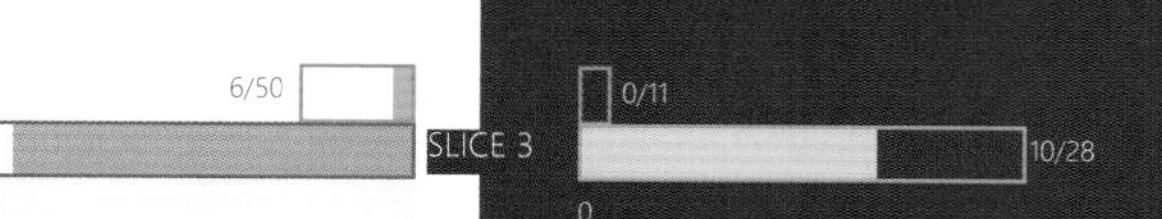

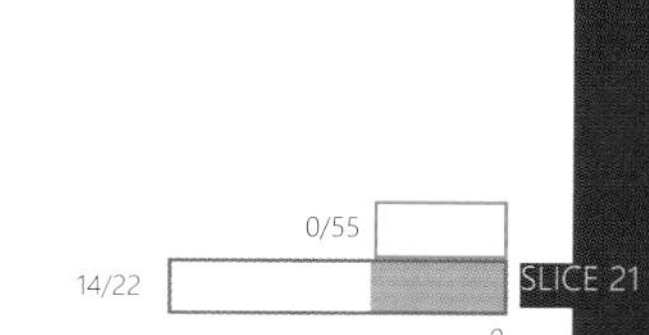

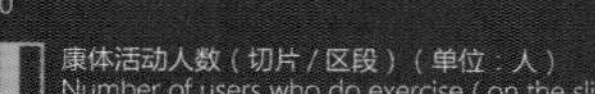

康体活动人数（切片 / 区段）（单位：人）
Number of users who do exercise (on the slice/on the segment) (Unit : person)

切片康体活动人数排序
Ranking of numbers of users who do exercise on slices

切片康体活动人数 / 硬地面积（单位：人 /m²）
Number of users who do exercise on the slice / Area of hard sites on the slice (Unit : person/m²)

典型切片硬地使用特征分析

为进一步解析硬地面积对康体活动的影响，我们将每个区段硬地面积最大的切片提取出来进行对比分析，发现此类切片在周末下午和日常傍晚康体活动人数占到同区段前 5 位的仅略过一半，排在第 1 位的仅有区段 1（挹江门—华严岗门）日常傍晚时段。

由此反映出，除了面积因素外，使用人群康体活动的开展还受制于其他相关因素。周末下午时段的区段 3（草场门—清凉门）以及日常傍晚时段的区段 4（水西门—集庆门）、区段 5（凤台路—中华门）切片上的康体活动人数仅能排在区段中游，与另一时段排位对比均出现较大幅度下降，可见其对不同人群或不同康体活动类型的吸引力出现了差异。而区段 6（中华门—雨花门）在两个时段的康体活动人数排位均处于中游，反映出该切片硬地利用率较低，场地设计可能存在较大问题。

Characteristic analysis of the use of hard sites on typical slices

To further analyze the influence of the area of hard sites on physical exercise, we extracted, compared, and analyzed the slices that have the largest areas of hard sites and discovered that there are only slightly more than half the slices where the number of people who exercise ranks top five in the same segment both on a weekend afternoon and every day at nightfall and that only the slice on S1 YJG-HYGG ranks the 1st every day at nightfall.

This shows that whether or not users do exercise is also subject to other relevant influences aside from available area. The number of people who do exercise on S3 CCG-QLG on weekend afternoons and on both S4 SXG-JQG and S5 FTR-ZHG every day at nightfall only ranks in the middle, but drops drastically, compared to the ranking in another time period. Thus, it can be seen that there is a difference in attraction to these slices on the part of different dominant users or different dominant types of exercise done by people in one age group. In both time periods, the number of people who exercise on S6 ZHG-YHG ranks in the middle, showing a low utilization ratio for hard sites in the segment and there is a high probability that the root of the problems lie in the site design.

1 | 2 | 3 / 4

典型区段硬地使用特征解析

区段 3（草场门—清凉门）是 6 个区段中硬地空间形态和铺装样式最丰富、进行康体活动人数最多的区段。该区段硬地集中分布在北部、中部和南部三个绿道出入口处，北部场地是较宽敞的大平台，设施较少，停留休息与进行康体活动的人数较少，是该区段中使用率较低的活动场地。中部“鬼脸照镜”场地和步行桥相连，周边有植被、景观石、水池，形成了一个集中的滨水活动场地，在周末下午大量家长带着孩子们一起进行亲子活动，是该区段使用率较高、人气较旺的活动场地。南部场地是一块长方形硬地，空间开敞，周末下午有人自带器材来此运动健身，并且在傍晚时能吸引大量中老年人开展康体健身活动。

Characteristic analysis of the use of hard sites on typical segments

In terms of spatial forms and paving styles, the hard sites on S3 CCG-QLG have the largest area among all six segments. This segment also has the largest number of people who exercise. The hard sites of this segment are mainly distributed at greenway entrances, which can be found in the northern, middle, and southern parts of the segment. The northern hard sites are big spacious platforms with few facilities, and the number of people who stop over and exercise is less than one third of the average number.Therefore, they have the lowest utilization ratio of the entire segment. In the middle, the hard site of "Funny Face Mirrors" is connected to a footbridge. The surrounding vegetation, landscape stones, and pools form a centralized waterfront place where a lot of parents bring their children to play on the weekend afternoons, so this area has the highest utilization ratio and is the most popular. The southern area is a rectangular hard site with a wide-open space where people bring their own sports equipment to do various forms of exercise on weekend afternoons. At nightfall, this area can attract a lot of middle-aged and senior people to exercise.

POSITIVE 积极

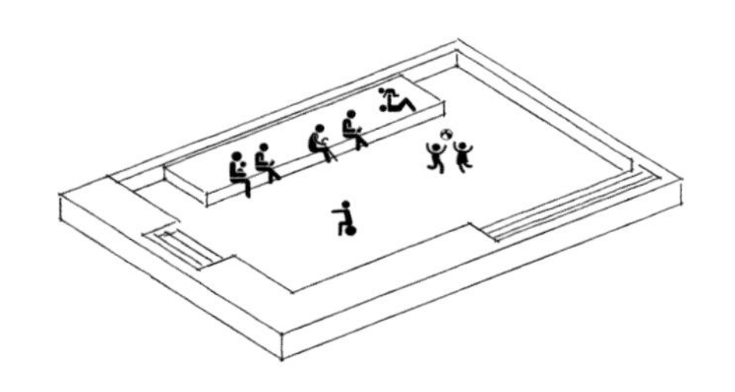

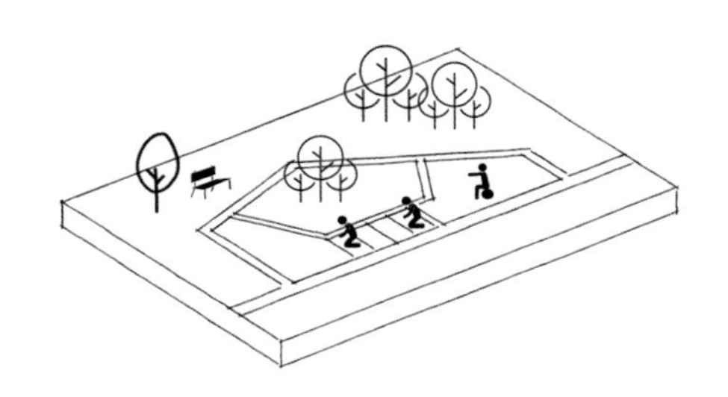

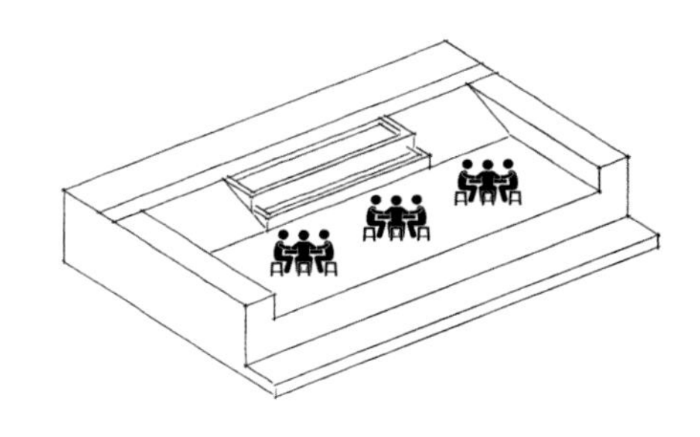

- 原型：区段 1（挹江门—华严岗门）切片 7
- 下沉的小型硬地广场
- 高度适宜的平台提供了大量可坐空间

· Prototype：S1（YJG-HYGG）slice 7
· A small sunken hard plaza
· A platform of a proper height provides a lot of space for users to sit

- 原型：区段 1（挹江门—华严岗门）切片 15-16
- 城墙外凸形成的活动场地
- 有沙池、树篱迷宫以及多种游乐设施

· Prototype：S2（YJG-HYGG）slices 15 -16
· Places of activities formed by outwardly convex city wall
· Sand pit, hedgemaze, and other kinds of entertainment facilities

- 原型：区段 6（中华门—雨花门）切片 5
- 低于综合慢行道的下沉广场
- 场地上的石块提供了聚集的可能

· Prototype：S6（ZHG-YHG）slice 5
· A sunken plaza lower than comprehensive slow lanes
· Rocks on the ground provide opportunities for people to gather, stay, and do all kinds of activities

消极 NEGATIVE

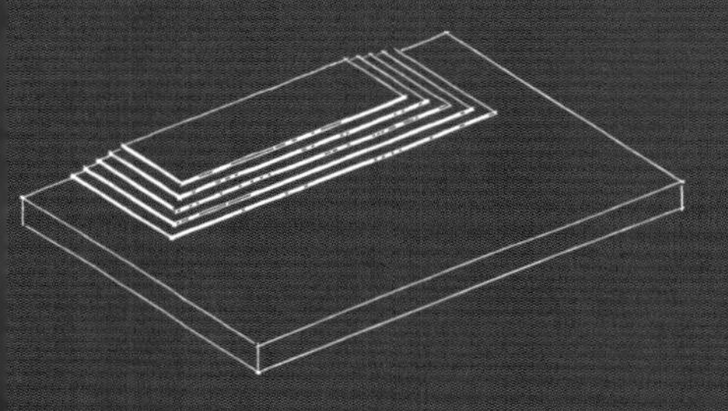

- 原型：区段 3（草场门—清凉门）切片 4
- 滨水平台
- 无任何设施与植被

· Prototype: S3（CCG-QLG） slice 4
· A waterfront platform
· No facilities or vegetation

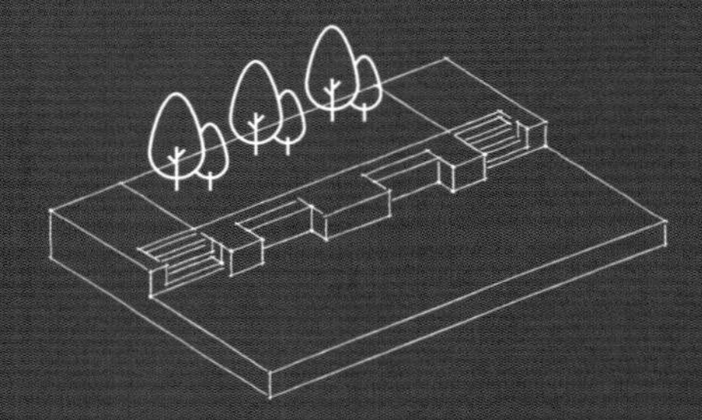

- 原型：区段 4（水西门—集庆门）切片 11
- 滨水平台
- 有内凹的座椅，但无树荫遮蔽

· Prototype: S4（SXG-JQG） slice 11
· A waterfront platform
· Inwardly concave benches, but no shades of trees

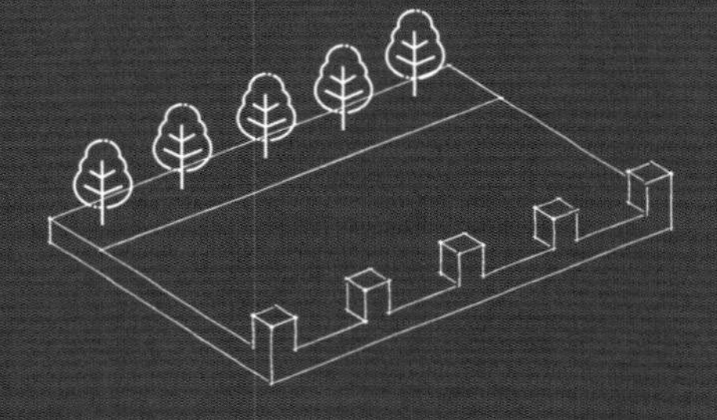

- 原型：区段 4（水西门—集庆门）切片 17
- 滨水平台
- 平台狭窄，无设施

· Prototype: S4（SXG-JQG） slice 17
· A waterfront platform
· Narrow platform. No facilities

空间要素分析——绿地
Analysis of spatial elements - green space

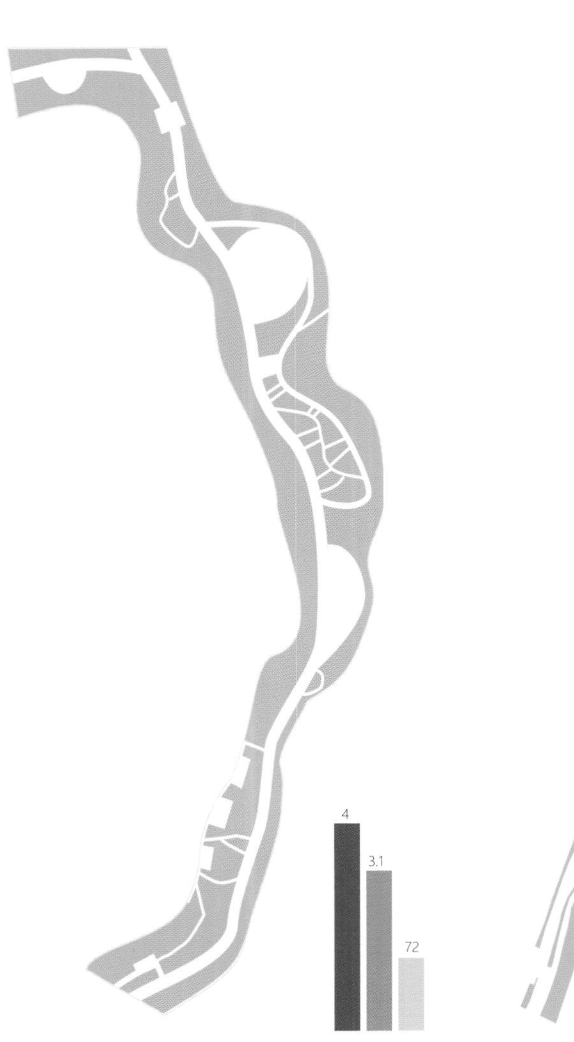

区段 1：挹江门—华严岗门
S1：YJG-HYGG

区段 2：定淮门—草场门
S2：DHG-CCG

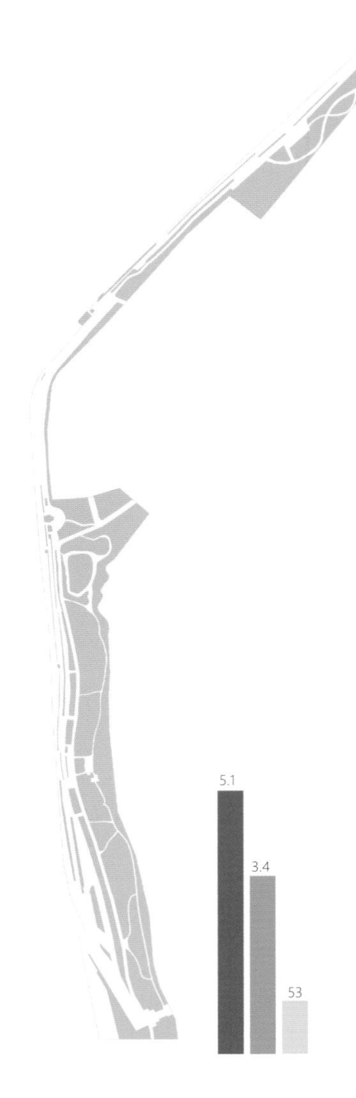

区段 3：草场门—清凉门
S3：CCG-QLG

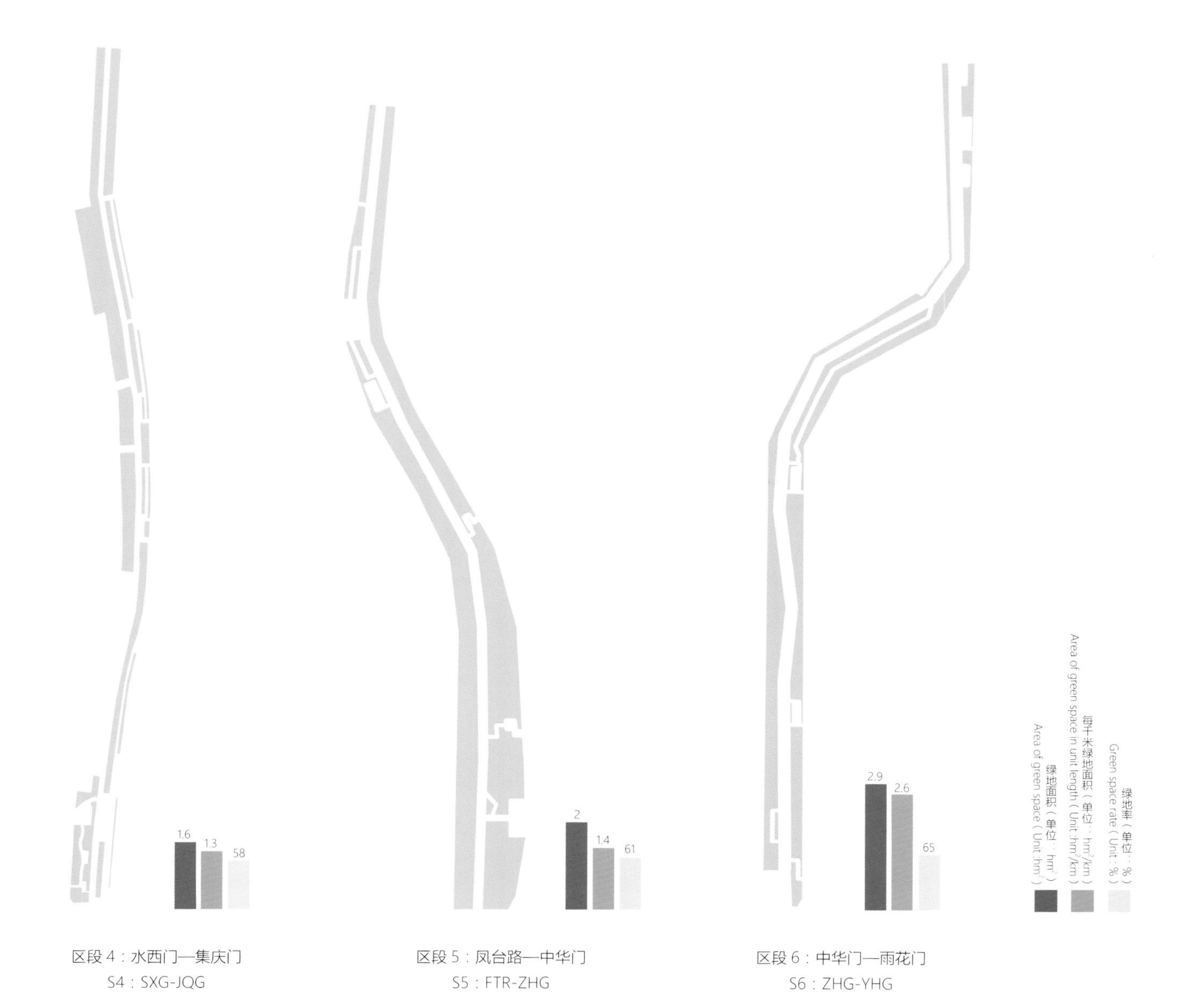

区段 4：水西门—集庆门
S4：SXG-JQG

区段 5：凤台路—中华门
S5：FTR-ZHG

区段 6：中华门—雨花门
S6：ZHG-YHG

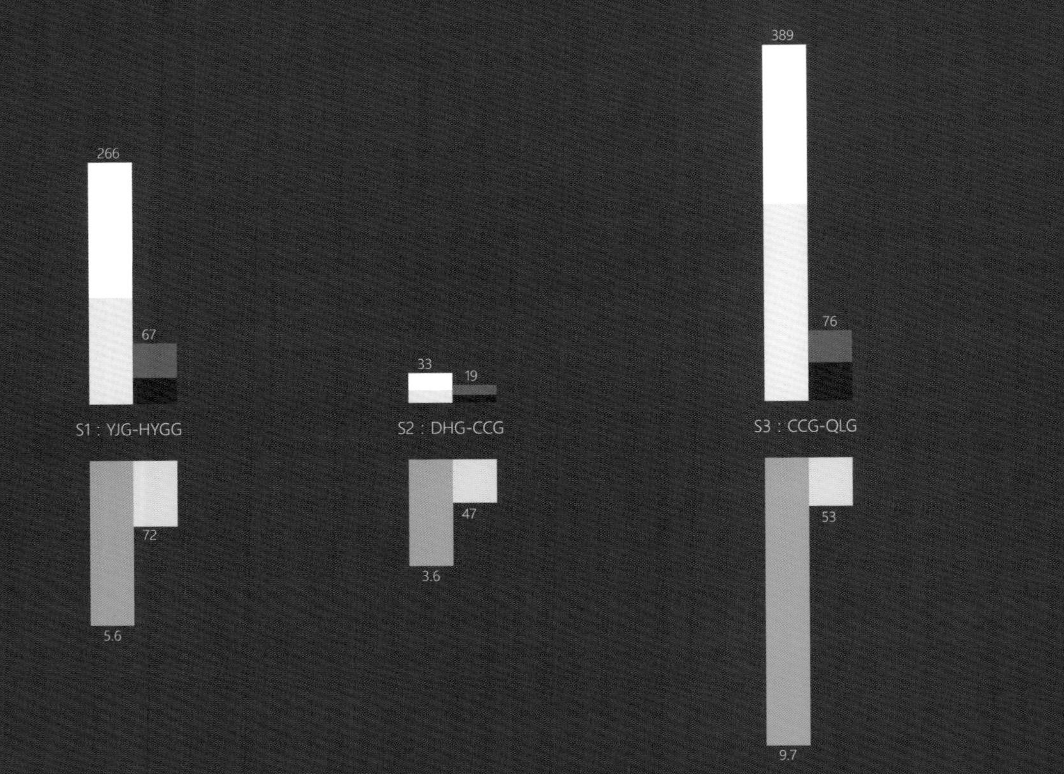

绿地使用特征总体分析

人群对绿地的使用通常集中在下午，而在傍晚的使用则大幅度下降，鉴于此我们选取 6 个区段周末下午与绿地使用关联较紧密的指标，即康体活动人群总数和密度以及停留休息人群总数和密度，并将其与绿地面积和绿地率进行关联对比。

从停留和康体活动人群总数来看，不同区段间差异十分明显。除区段 2（5，4，6）外，其他区段停留和康体活动人群总数基本与绿地面积成正比，其中区段 3（1，1，5）绿地面积最高，停留与康体活动人群总数最高，其次是区段 1（2，2，1），而区段 4（6，6，4）人数最低。周末下午停留和康体活动人群密度指标与人群总数的排名和趋势完全一致，也是区段 3（草场门—清凉门）和区段 1（挹江门—华严岗门）较高，

The characteristics of the use of green space

Green space on the greenway is always used and occupied in the afternoon, while its utilization ratio decreases greatly at nightfall. In view of this, we selected the indicators that are closely linked to the use of green space on weekend afternoons from the six segments-that is, the number and density of people who exercise and the number of people who stop over, and correlate and compare the data with the area of green space and ratio of green space.

Different segments differ noticeably in the total numbers of those who stop over and those who exercise. Except in S2 DHG-CCG (5,4,6), the total number of those who stop over and those who exercise in the other segments is basically in direct proportion to the areas of green space. Among the segments, S3 CCG-QLG (1,1,5) has the largest area of green space and the largest total number of those who stop over and those who exercise. S1 YJG-HYGG (2,2,1) ranks the 2nd, while S4 SXG-JQG (6,6,4) ranks last. On weekend afternoons, the total density of people who stop over and exercise is consistent with the ranking of the number of people who do these activities, with S3 CCG-QLG and S1 YJG-HYGG ranking first and S2 DHG-CCG and S4 SXG-JQG ranking last.
A very interesting phenomenon is that although green spaces are

区段 2（定淮门—草场门）和区段 4（水西门—集庆门）较低。

一个有趣的现象是，虽然绿地并非绿道使用人群停留和康体活动的主要空间，但从人群使用数据与绿地面积关联对比中可看出，使用人数与区段的绿地面积明显呈正比。绿地面积越大，活动人数越多，并且活动人群密度也会上升。在调研中发现，一方面开敞绿地可以为人群提供大量可坐空间，吸引人停留；另一方面绿地面积越大，也能为空间收放、视线开闭、植物造景等方面创造大量的机会，这对间接吸引人群停留和康体活动均能产生积极的促进作用。

注：括号内数字依次为绿地面积排序、绿道单位长度绿地面积排序、绿地率排序。

not the main places where greenway users stop over or exercise, we can see from the correlation and comparison of user data and the area of green space that the number of users is obviously in direct proportion to the area of green space on the segment. The bigger the area of the green space is, the larger the number of users is, and the density of users stopover and doing exercise also increases. During the survey, we find out that wide-open green spaces can attract a lot of people to stop over or play, on the one hand, which is already well realized. On the other hand, instead of being occupied directly by itself, the bigger the area of the green space is, the more opportunities it can create in aspects including enlargement in a continuous homogeneous space, openness of the sight, joyful experience from planting and landscaping and so on. This can play an active role in attracting more people to stop over and exercise indirectly.

Note: The numbers in the parentheses are successively (the ranking of the area of green space in all 6 segments, the ranking of the area of green space in unit length of the greenway in all 6 segments, and the ranking of greening rate in all 6 segments).

典型区段绿地使用特征分析

区段 3（草场门—清凉门）是 6 个区段中绿地面积最大、停留和康体活动人群总数最多的区段。实际调查显示，该区段大量绿地主要集中在南半段，绿地最宽部分可达 80m。该区段绿地空间开敞，要素丰富，为多样化的休闲游憩活动的开展创造了良好条件。如果不算历史文化遗存等景点的吸引力，该区段周末午后集中的开敞绿地比硬质场地对使用人群有更高的吸引力。

Characteristic analysis of the use of green space on typical segments

S3 CCG-QLG is the segment that has the biggest area of green space and the biggest number of users who stop over and exercise among the six segments. The survey shows that a great area of green space is located on the southern half of the segment, with the widest part of the green space reaching 80m. The green space in this segment is wide open with a variety of amenities, and can thus support all kinds of visitors' recreational activities. If we don't take into account the attractiveness of the historic relics and other scenic spots in this segment, the centralized and wide-open green space of this segment on weekend afternoons plays a more important role in attracting users compared to hard sites.

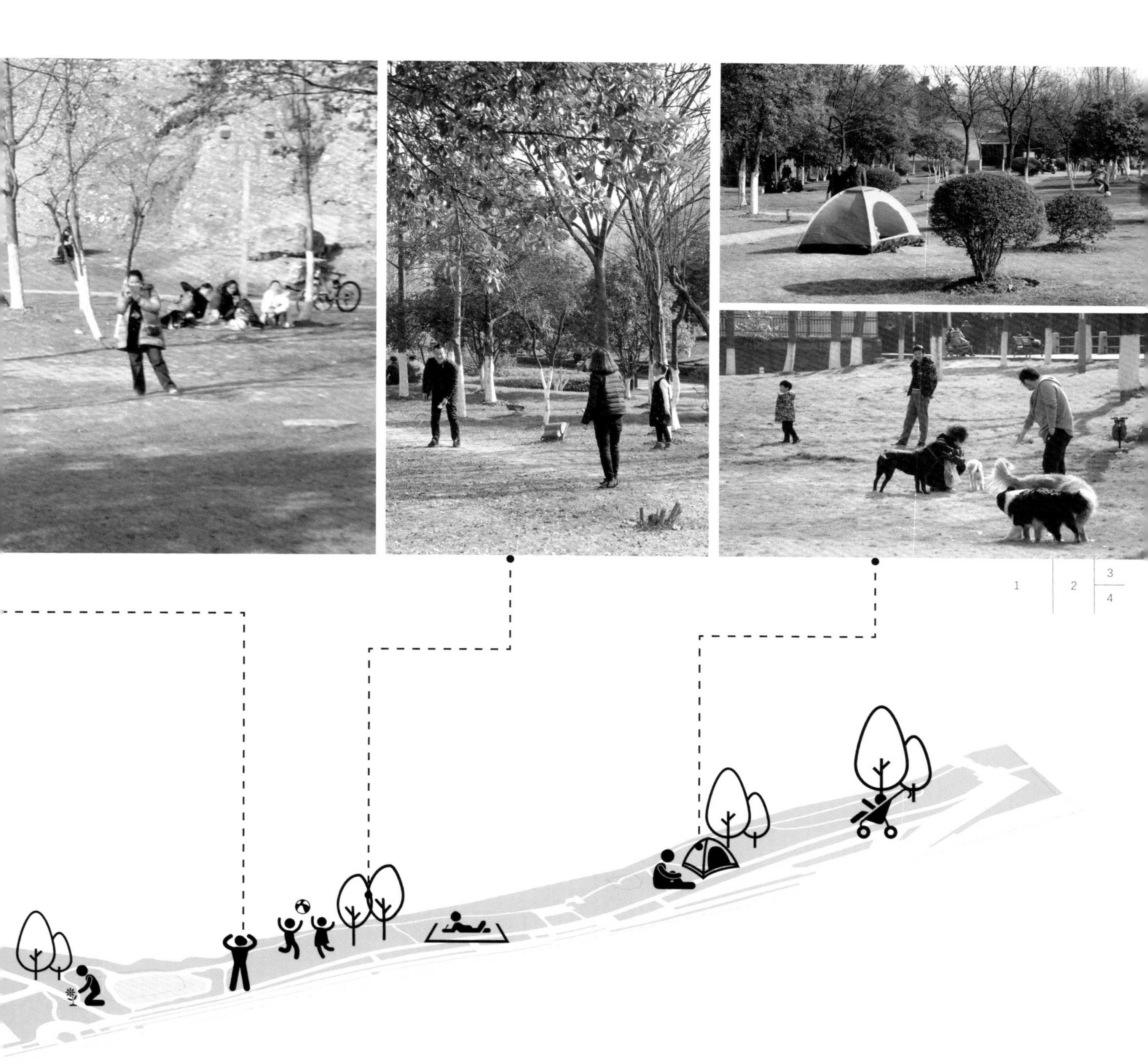
1
2
3
4

区段 4（水西门—集庆门）的绿地面积是 6 个区段中最小的，停留和康体活动人群总数也最少。通过调查分析，该区段是 6 个区段中最狭窄的一段，因此绿地空间非常局促，呈条状布置，其宽度仅能进行空间界定和隔离，完全无法容纳人群活动。另外，该区段绿化种植多以较高的草丛、成片的灌丛以及分叉较低、郁闭度较高的中小乔木为主，这在很大程度上降低了散步通行活动、人群的停留休息和其他使用的可能性。

The area of green space in S4 SXG-JQG is the smallest of the six segments, with the smallest number of people who stop over and exercise. From the survey we found out that this segment is also the narrowest of the six segments. Therefore, the green space of this segment is very cramped and arranged in the shape of a belt, which is only available to be used for spatial definition and isolation with limited width instead of supporting any recreational activities. In addition, since the vegetation of this segment is mostly high grass, bushes, and small-and-medium sized arbors with low branching and high crown density, they actually lower down the possibility of people to sit around and make a different use of this segment in addition to walking.

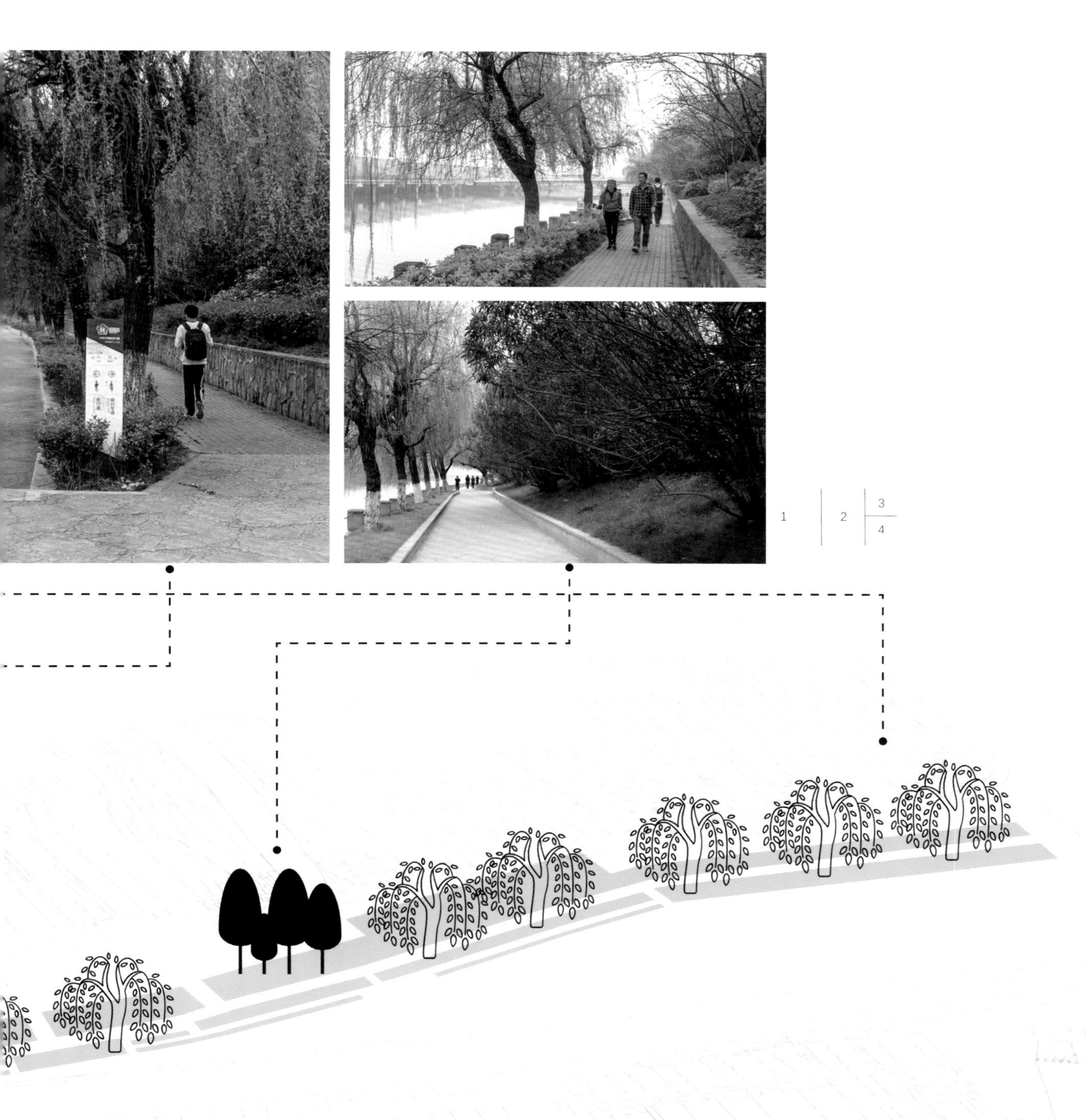
1
2
3
4

POSITIVE 积极

平坦的草地
Flat lawn

灌木丛前的草地
Lawn in front of shrubs

低郁闭度的树林
Woods with low canopy density

积极与消极空间分析

调研发现绿道上的植物种类以及种植方式对人群停留休息和活动开展的影响比较大。经调研分析，我们发现在绿道周边平坦草地、灌木丛前草地或是视线开敞的落叶树下活动的人群占绿地上活动人群的 90% 以上，而较高的草丛、成片的灌丛以及郁闭度较高、视线遮蔽的常绿树下则鲜有人群使用。

消极 NEGATIVE

高于脚背的草地
Lawn higher than the back of foot height

成片的灌丛
Patches of shrubs

高郁闭度的树林
Woods with high canopy density

Positive and negative spatial analysis

From the survey, we found out that the types of vegetation along the greenway and how they are planted have a great influence on people who stop over or for other recreational activities. More than 90% of the people who do activities on the green space are found gathering on the flat lawns surrounding the greenway, on the green area in front of the bushes, and under the deciduous trees with a wide-open view, while few people can be seen on the area which is planted with high grass, bushes, and evergreens with high crown density that block vision.

空间要素分析——游径
Analysis of spatial elements-trails

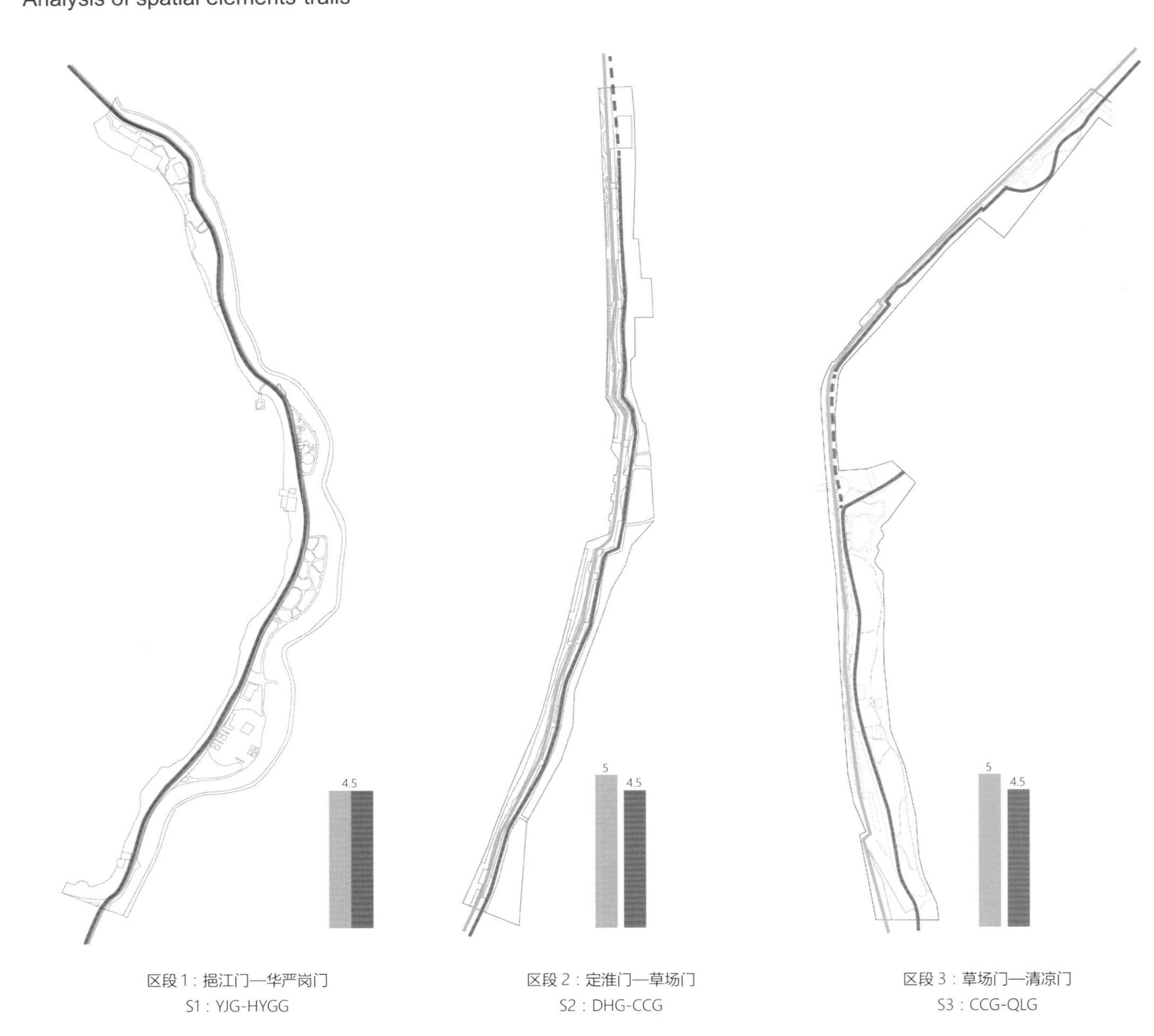

区段 1：挹江门—华严岗门
S1：YJG-HYGG

区段 2：定淮门—草场门
S2：DHG-CCG

区段 3：草场门—清凉门
S3：CCG-QLG

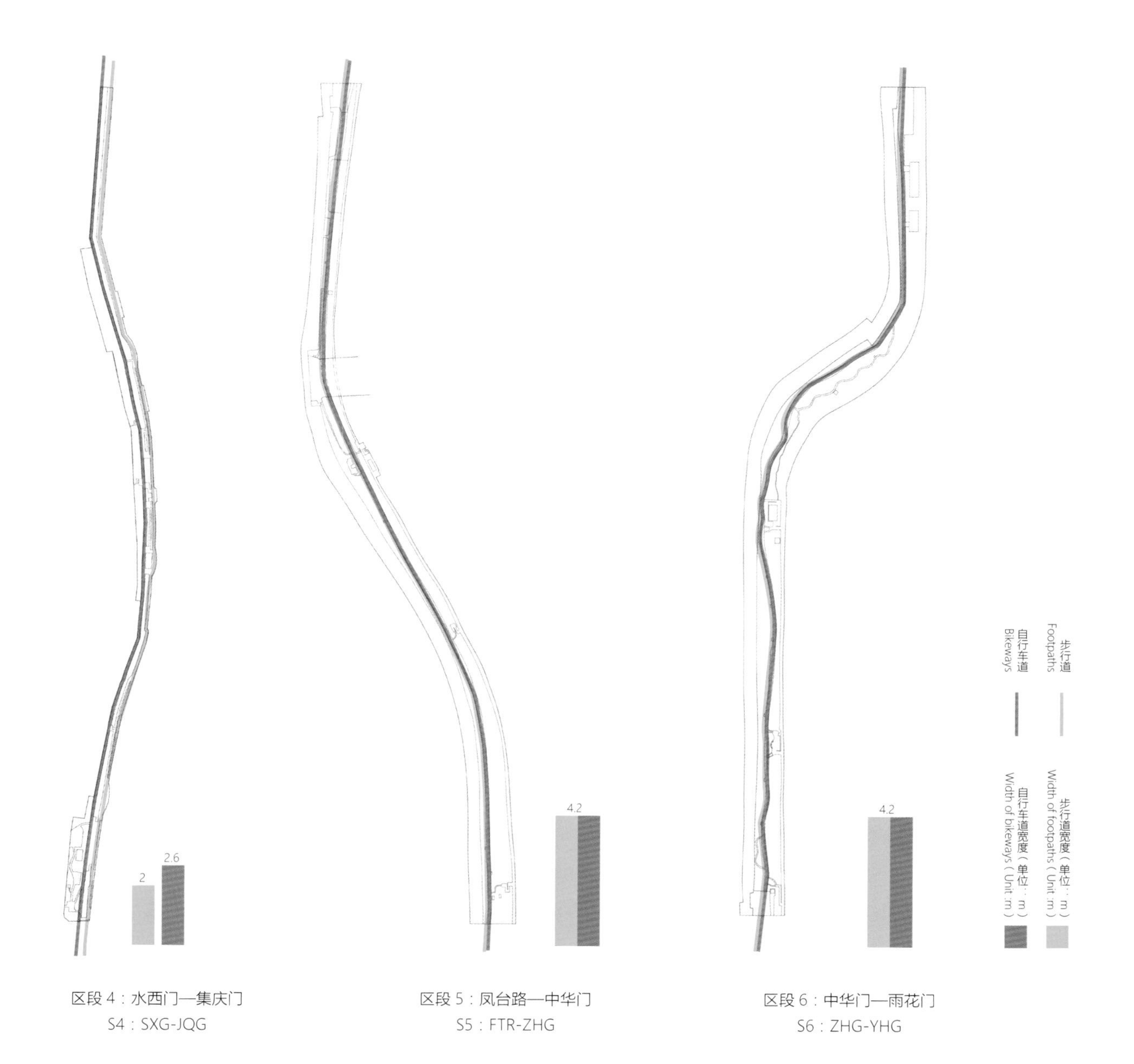

区段 4：水西门—集庆门
S4：SXG-JQG

区段 5：凤台路—中华门
S5：FTR-ZHG

区段 6：中华门—雨花门
S6：ZHG-YHG

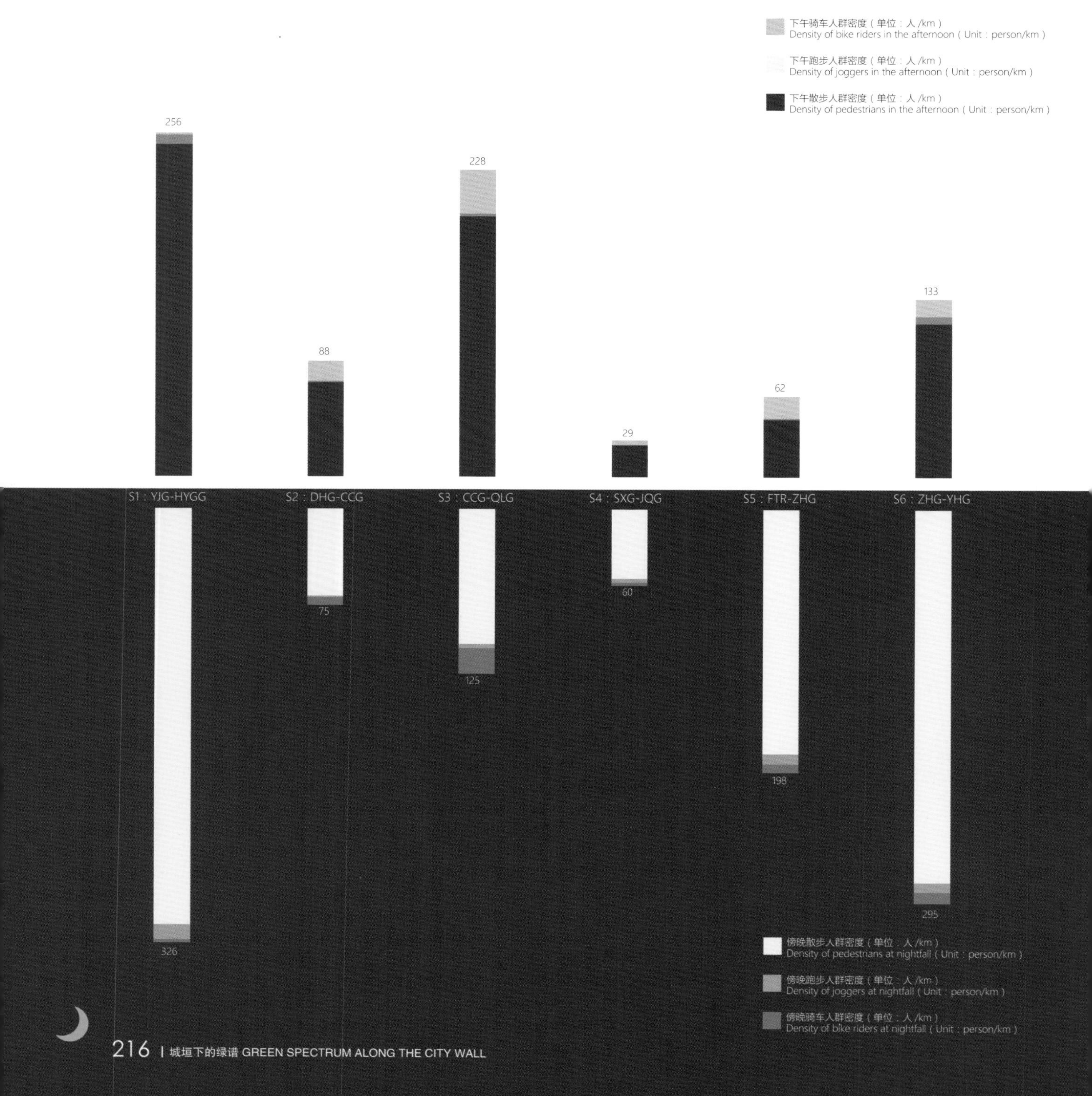
下午骑车人群密度（单位：人/km）
Density of bike riders in the afternoon (Unit : person/km)
下午跑步人群密度（单位：人/km）
Density of joggers in the afternoon (Unit : person/km)
下午散步人群密度（单位：人/km）
Density of pedestrians in the afternoon (Unit : person/km)
256
88
228
29
62
133
S1 : YJG-HYGG
S2 : DHG-CCG
S3 : CCG-QLG
S4 : SXG-JQG
S5 : FTR-ZHG
S6 : ZHG-YHG
326
75
125
60
198
295
傍晚散步人群密度（单位：人/km）
Density of pedestrians at nightfall (Unit : person/km)
傍晚跑步人群密度（单位：人/km）
Density of joggers at nightfall (Unit : person/km)
傍晚骑车人群密度（单位：人/km）
Density of bike riders at nightfall (Unit : person/km)

游径由步行道和自行车道组成，两者在绿道上均呈线性展开。由于6个区段长度各异，因此我们采用密度数据（使用人数/千米）来进行关联对比分析，并主要选择散步、跑步和骑车三个使用类型数据，分周末下午和日常傍晚两个时段进行对应研究。从数据中可以看出，在两个时段内，6个区段散步人数一直占据主导，除区段2（定淮门—草场门）外，各区段日常傍晚散步人群密度均高于周末下午。其中，区段1（挹江门—华严岗门）的散步人群密度在两个时段内均为最高，而区段4（水西门—集庆门）则在两个时段均排位最后。对比两者的主游径结构和布局走向可以发现，区段1采用综合慢行道，并施行鼓励步行、限制骑行的措施，加之绿道空间较开阔、周边景点和活动场地丰富，均在很大程度上促进步行人群的使用。反观，区段4主游径中步行道和自行车道分离，且受到整个绿道空间的限制，步行道宽度较低仅有2m，勉强能支持3人并行，这在一定程度上也限制了大规模散步人群的使用，加之绿道空间均质、景点较少，对散步人群的吸引力也大幅度削弱。

在3个使用活动类型中，骑车人群密度在各区段间差别较悬殊，并且周末下午的使用人群密度远高于日常傍晚。由于所有区段日常傍晚骑行使用人数几乎可以忽略不计，因此在此不专做讨论。在采用综合慢行道型主游径的三个区段中，区段5（风台路—中华门）的骑行密度最高，区段1最低，骑行人数与散步人数成反比，实际上也印证了两种活动在综合慢行道上相互干扰和限制的特点。在区段1就为保障散步人群不受骑行人群干扰，对骑行活动采取了一定的限制措施。在采用自行车道、步行道分离型游径的三个区段，密度最高的是区段3（草场门—清凉门），最低的则是区段4。由于该类型游径两种活动人群相互干扰较少，可以发现骑行人数与区段使用总人数及散步人数基本成正比。

Aggregate analysis of the characteristics of the use of trails

Trails consist of footpaths and bikeways, both of which spread out linearly on the greenway. Due to six segments' different lengths, we adopted density data (number of users/kilometer) to carry out correlation, comparison, and analysis, mainly from walking, jogging, and bike-riding data, and studied the data from both afternoon weekends and every day at nightfall. We can see from the data that in both time periods the number of those who go walking always dominates the six segments and that except for in S2 DHG-CCG, the density of people who go walking every day at nightfall is higher than the number who go walking on weekend afternoons. Moreover, the density of those who go walking on S1 YJG-HYGG is the greatest for both time periods, while S4 SXG-JQG ranks last in both time periods. After comparing the structures and layouts of the main trails of the two, we can see that S1 YJG-HYGG adopts comprehensive slow lanes and takes measures that encourage walking and limit bike-riding. In addition, wide-open greenway space and a variety of surrounding scenic spots and places for diverse recreational activities attract pedestrians to a great extent. In contrast, in S4 SXG-JQG, the footpaths and bikeways of the main trails are separated. Limited by the entire greenway space, the width of the sidewalks is only 2m and is barely enough to allow 3 people to walk shoulder to shoulder, which limits a large number of people from using the trails. In addition, the homogeneity of the greenway space and a lack of scenic spots weaken the attraction of the trails to those who are fond of walking.

For the three kinds of activities on trails, there is a huge difference in the density of bike riders among the six segments, and the density of users weekend afternoons is far higher than every day at nightfall. Since the number of bike riders every day at nightfall in all the segments can be ignored, we do not discuss this aspect here. Of the three segments that adopt comprehensive slow lanes as their main trails, S5 FTR-ZHG has the highest density of bike riders, while S1 YJG-HYGG has the lowest. The number of bike riders is inversely proportional to that of those who go walking, which in fact further proves that the two types of activities interfere with and restrict each other in terms of comprehensive slow lanes. On S1 YJG-HYGG, restrictive measures have been imposed on bike-riding, so as to prevent bike riders from interfering with pedestrians who are considered to be the primary users of this segment. Among the three segments that use the trails to separate bikeways and sidewalks, S3 CCG-QLG has the highest density of bike riders, while S4 SXG-JQG has the lowest. Since the two groups of people who do these two types of activities seldom interfere with each other on this pattern of trails, we can see that the number of bike riders is basically in direct proportion to the total number of users of the segment and the number of those who go walking on the segment.

游径断面空间特征分析

Characteristic analysis of cross - section space of trails

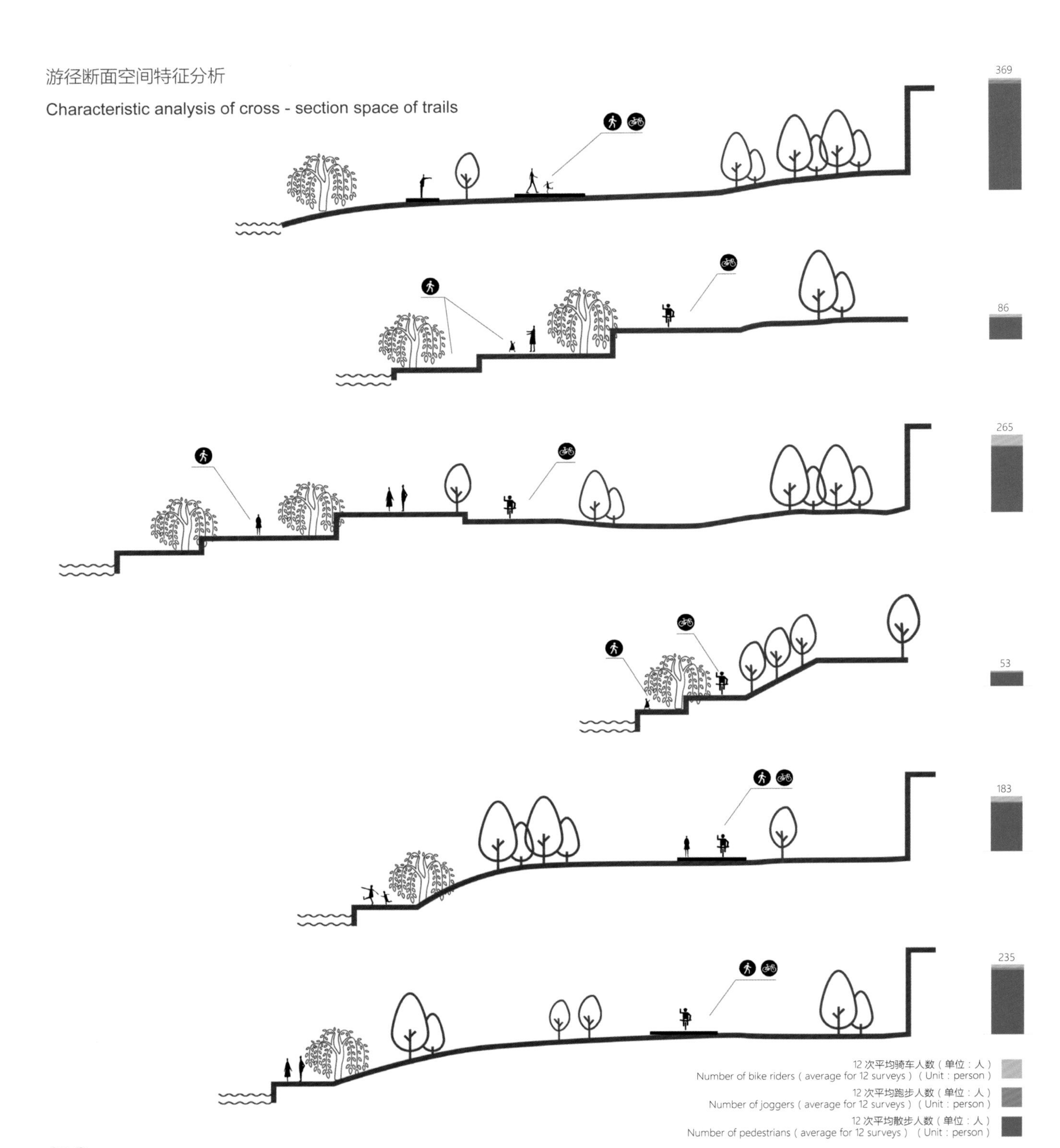

区段 1：挹江门—华严岗门
S1：YJG-HYGG

区段 2：定淮门—草场门
S2：DHG-CCG

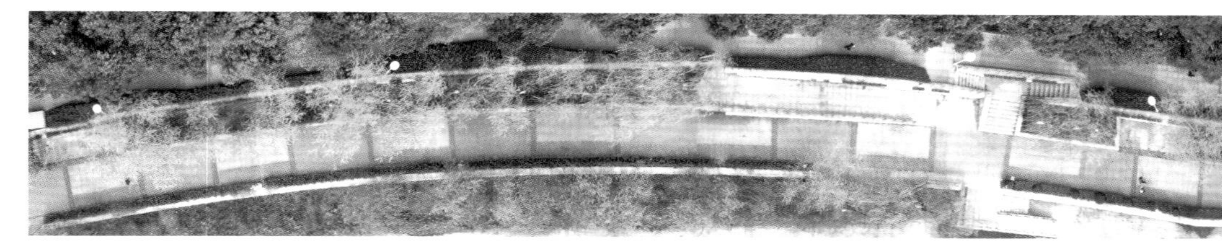

区段 3：草场门—清凉门
S3：CCG-QLG

区段 4：水西门—集庆门
S4：SXG-JQG

区段 5：凤台路—中华门
S5：FTR-ZHG

区段 6：中华门—雨花门
S6：ZHG-YHG

游径使用问题的反思

首先，由于建成环境绿道多是在城市原有绿地和设施基础上改造而成，部分绿道空间限制比较严重，因而并非所有区段都能为游径设置提供充分的空间条件，这也是调查区段有一半不得不采用综合慢行道型主游径的重要原因，甚至一些区段（如区段 1）为了保障绝大多数步行人群的使用，还在一定程度上限制骑行活动。

其次，骑行使用比步行使用对于绿道游径的要求要苛刻得多，但由于建成环境情况较复杂，如果在绿道规划、设计、管理和维护中没有能够完全将各种复杂因素协调统筹到位，也会阻碍游径的骑行使用。例如在调查中发现，区段 2（定淮门—草场门）由于管理维护不周，在绿道最北侧的自行车道被一小型垃圾搜集点阻断；区段 3（草场门—清凉门）中部由于受现状建筑影响，自行车道也发生了中断，要顺利通过只能借道残疾人坡道，将车推到下层步行道通行。

最后，对于中国城市而言，骑行实际上是居民日常出行的重要途径，建成环境绿道除了作为休闲游憩的重要载体外，实际上还应被视为城市慢行系统的重要组成部分进行综合安排和整合。在调查区段中，区段 3（草场门—清凉门）、区段 5（凤台路—中华门）等区段骑行人数较多的重要原因就是这些区段的自行车道与城市路网以及与居民小区都有很好的衔接，并成为居民出行、衔接城市道路系统的重要交通纽带。

Reflections on the use of trails

Since most greenways in built environments are transformed from the original green space and facilities of the city, part of the greenway space is seriously restricted. Therefore, not all the segments can provide enough spatial conditions for the setup of a trail system, which is also an important reason why half of the segments in the survey have to adopt comprehensive slow lanes combining footpaths and bikeways as their main trails. Bike riding is even restricted on some segments (e.g. S1 YJG-HYGG), so as to guarantee that the primary users, who prefer to go for a walk, can use the trails more comfortably without interference from bikes.

Second, bike riding has stricter requirements on the greenway trails than walking. However, due to the complex nature of a built environment, if various kinds of complex factors cannot be fully coordinated and planned in greenway planning, design, and maintenance, the occurrence of bike riding on the trails will be hindered. For example, from the survey we found that in S2 DHG-CCG the bikeway on the northern side of the greenway is cut off by a small garbage collection point that is there due to poor maintenance. The bikeway in the middle of S3 CCG-QLG is also cut off by some existing buildings. If anyone wishes to pass successfully, they must stop riding and push the bike to the pathway at the lower level by passing through a ramp for the disabled.

Moreover, as far as Chinese cities are concerned, bike riding is a very important way for residents to travel on a daily basis. In addition to being a key carrier of recreation, the greenway in a built environment should actually be seen as an essential part of an urban slow traffic system and integral to planning. Among the segments under survey, the key reason why S3 CCG-QLG and S5 FTR-ZHG have a larger number of bike riders is that the bikeways of these segments are well connected to the city traffic network and surrounding residential quarters so they function as key transportation links for residents to travel around, and they also function as an adaptor between the trail system in communities and the road system of the city.

绿道
GREENWAY
汉中门至集庆门段
GW001C / c02
西
West
东
East
自行车道
步行道
Walking Path

区段1：：挹江门—华严岗门
S1:YJG-HYGG

区段2：：定淮门—草场门
S2:DHG-CCG

区段3：：草场门—清凉门
S3:CCG-QLG

区段4：：水西门—集庆门
S4:SXG-JQG

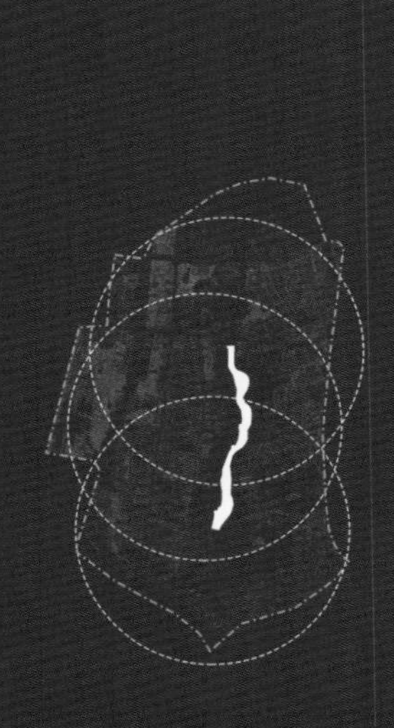

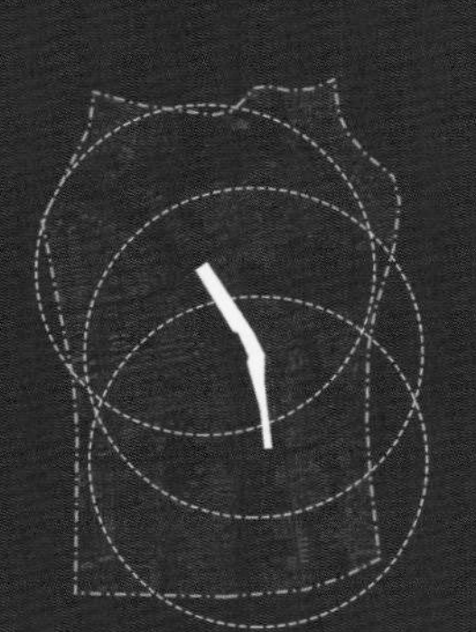

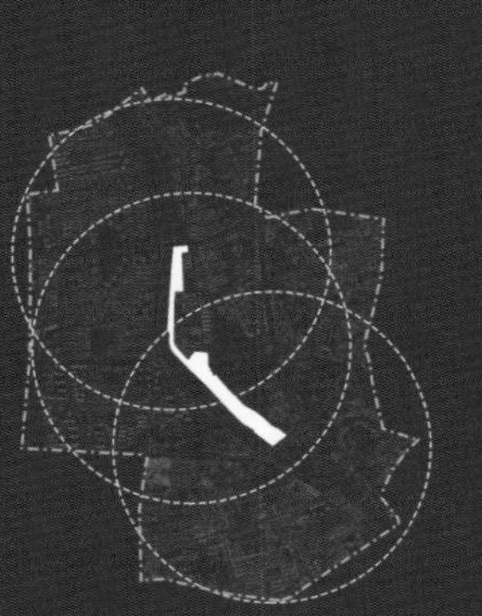

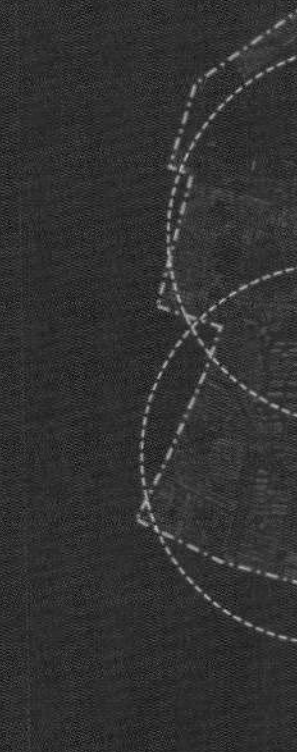

绿道外部空间特征分析

Characteristic Analysis of the Exterior Space of the Greenway

区段5：凤台路—中华门
S5:FTR-ZHG

区段6：中华门—雨花门
S6:ZHG-YHG

什么是绿道外部空间

What is the exterior space of a greenway

本书中所提及的绿道外部空间主要是指绿道周边对绿道使用关联紧密的空间及其内部要素。

为了明确是绿道外部研究空间的范围，我们在绿道出入口处，以 15 分钟步行距离（1000m）为半径划定外部关联空间范围。为了便于研究开展和数据分析，以外部关联空间范围边界最近的街道为界，形成街区完整的研究范围（如图）。

在该研究范围内，我们主要分析其中的用地结构、开发强度和人口密度、交通结构等数据，并探讨其对绿道使用人群规模和使用强度的影响特征。

The exterior space of the greenway mentioned in this study mainly refers to the spatial range where the elements closely linked to the use of greenway in the surroundings of the greenway are located.

To define the range of the exterior space of the greenway in this study, we set the range of correlated exterior space at the entrance of the greenway with a 15-minute-walk radius (1,000m). Meanwhile, to facilitate study development and data analysis, we define the nearest urban road as the border of the range of the correlated exterior space as the boundary to form a complete study scope of the street block.

Within this scope, we mainly analyze data including land use structure, development intensity, and traffic conditions, and discuss their modes of action and influence on the service performance of the greenway.

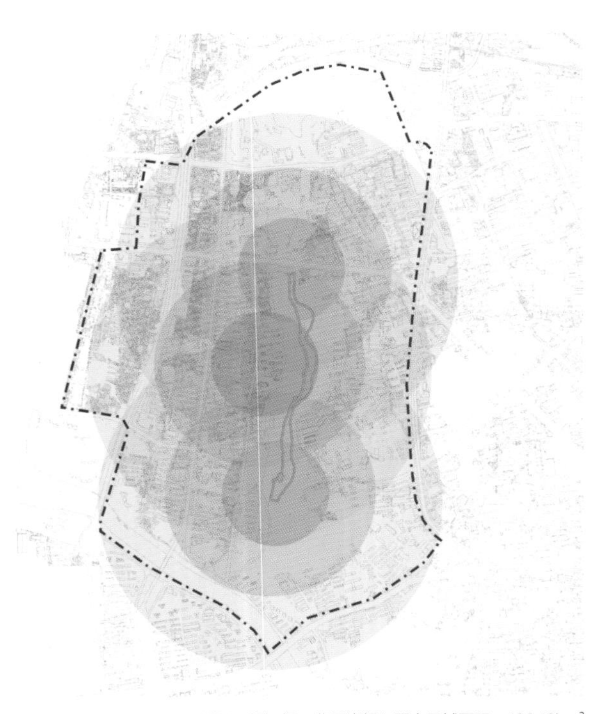

区段 1（挹江门—华严岗门）研究区域面积：486.40hm²

Area of study region of S1 (YJG-HYGG) : 486.40hm²

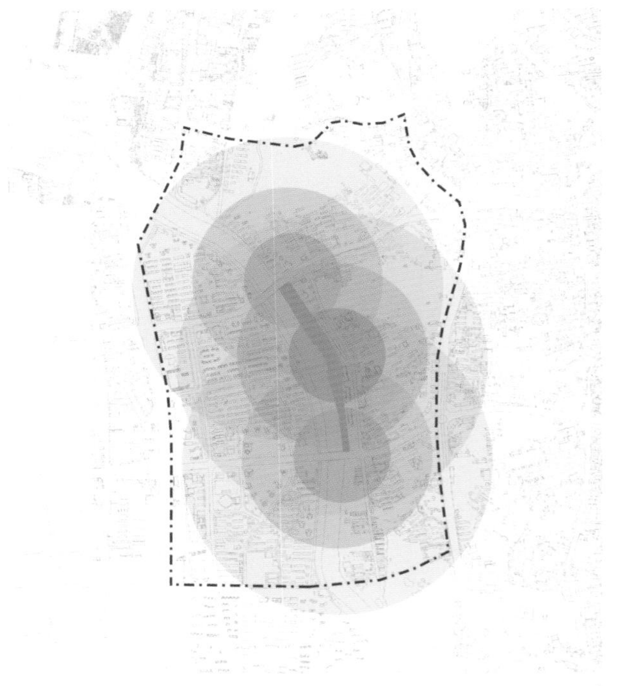

区段 2（定淮门—草场门）研究区域面积：406.91hm²

Area of study region of S2 (DHG-CCG) : 406.91hm²

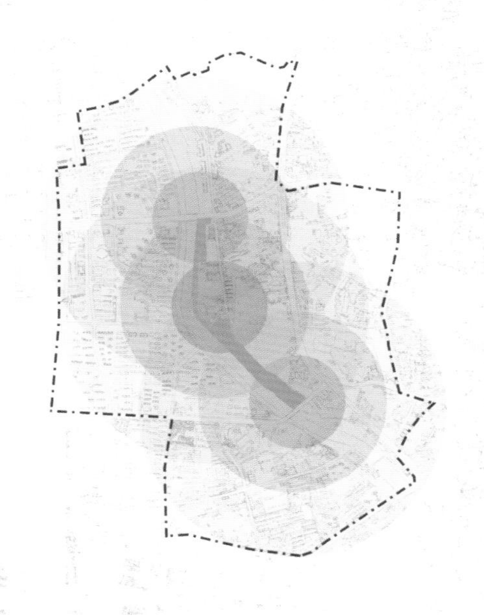

区段 3（草场门—清凉门）研究区域面积：597.30hm²
Area of study region of S3（CCG-QLG）：597.30hm²

区段 4（水西门—集庆门）研究区域面积：533.30hm²
Area of study region of S4（SXG-JQG）：533.30hm²

区段 5（凤台路—中华门）研究区域面积：597.02hm²
Area of study region of S5（FTR-ZHG）：597.02hm²

区段 6（中华门—雨花门）研究区域面积：610.10hm²
Area of study region S6（ZHG-YHG）：610.10hm²

怎样关联分析绿道外部空间特征与服务绩效

How to correlate and analyze the characteristics of greenway exterior space and service performance

建成环境绿道的服务对象源于外部城市空间，同时其布局也需依托城市道路、河流、带状遗址等线性空间要素展开，因而其服务绩效不仅与自身内部空间特质相关，也与绿道周边外部城市空间环境特质紧密关联，例如，城市用地结构、城市路网结构与密度、居民密度、空间环境紧凑度等均将对绿道服务绩效产生直接影响。由于影响因素较多，且作用方式较复杂，要对各影响因素进行梳理和针对性分析并非易事。

鉴于此，我们主要通过定量方法，首先将调查的 6 个绿道区段外部各个空间要素特征进行量化，将其与绿道内部的使用人群总体规模和密度指标进行关联比对，通过分析各区段外部空间特征指标与使用人群规模和密度数据变化的联系规律，尝试探讨各外部空间要素对绿道服务绩效的作用方式。

The service object of the greenway in a built environment comes from surrounding urban space, and its layout also depends on linear spatial elements including urban roads, rivers, and historic sites stripped for development, so its service performance is not only related to its own spatial characteristics but also closely linked to the environmental traits of the urban space surrounding the greenway. For example, urban land use structure, the structure and density of the urban road network, the density of the surrounding residential population, and the compactness of development of the surrounding land all influence directly the service performance of the greenway. Due to many influencing factors and their complex modes of action, it is not easy to put these in order and make a corresponding analysis.

To search for a solution, we mainly rely on quantitative methods in our study. First, we quantify the characteristics of all kinds of spatial elements of the exterior space in all six segments, correlate and compare the total number and density of users inside the greenway, and attempt to discuss the modes of action of the external spatial elements to the service performance of the greenway by analyzing the relationship between the indicators of the characteristics of exterior spaces of the segments and the user data.

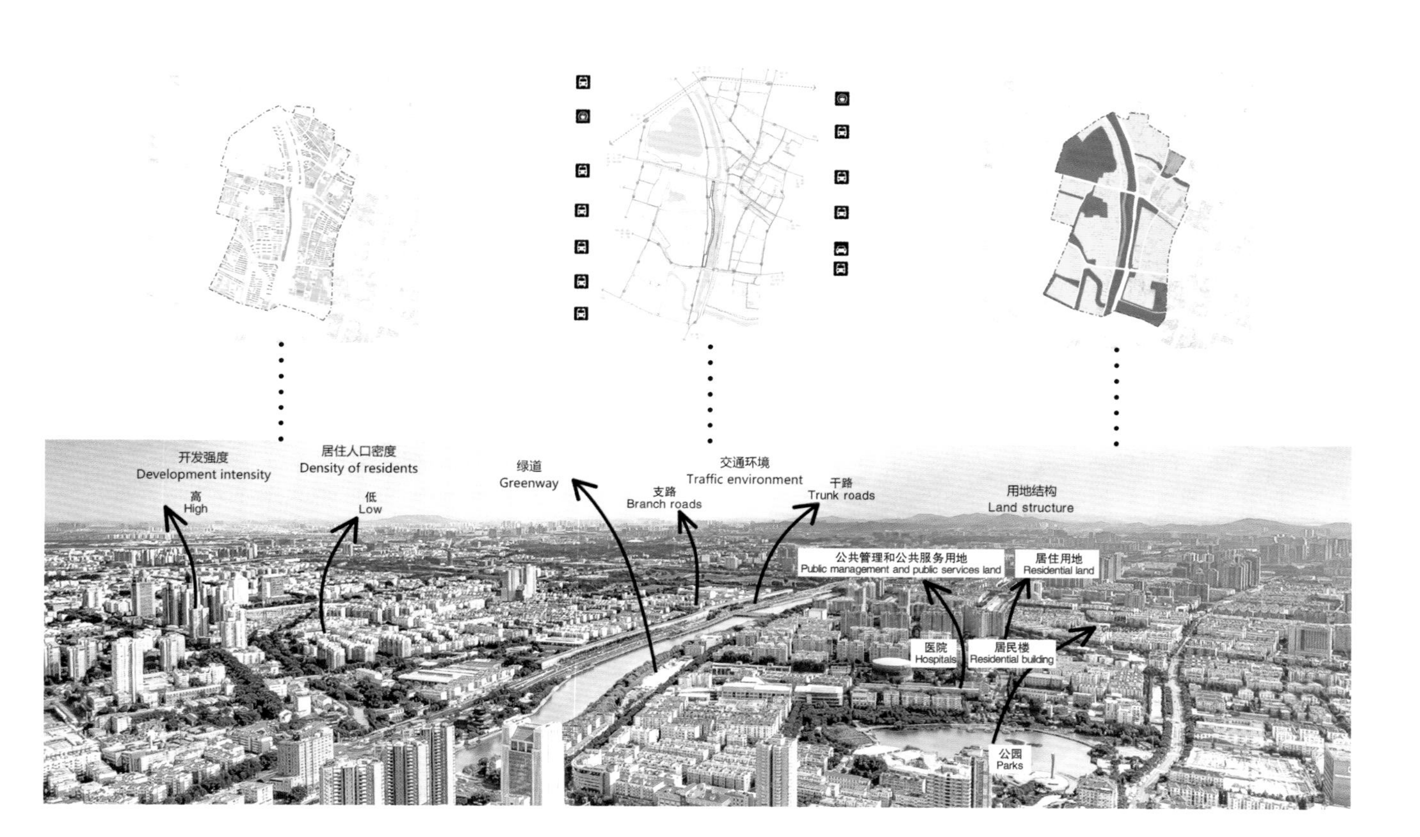
开发强度
Development intensity
高
High
居住人口密度
Density of residents
低
Low
绿道
Greenway
支路
Branch roads
交通环境
Traffic environment
干路
Trunk roads
用地结构
Land structure
公共管理和公共服务用地
Public management and public services land
居住用地
Residential land
医院
Hospitals
居民楼
Residential building
公园
Parks

外部空间特征与服务绩效关联分析

Correlation analysis of characteristics of greenway exterior space and greenway service performance

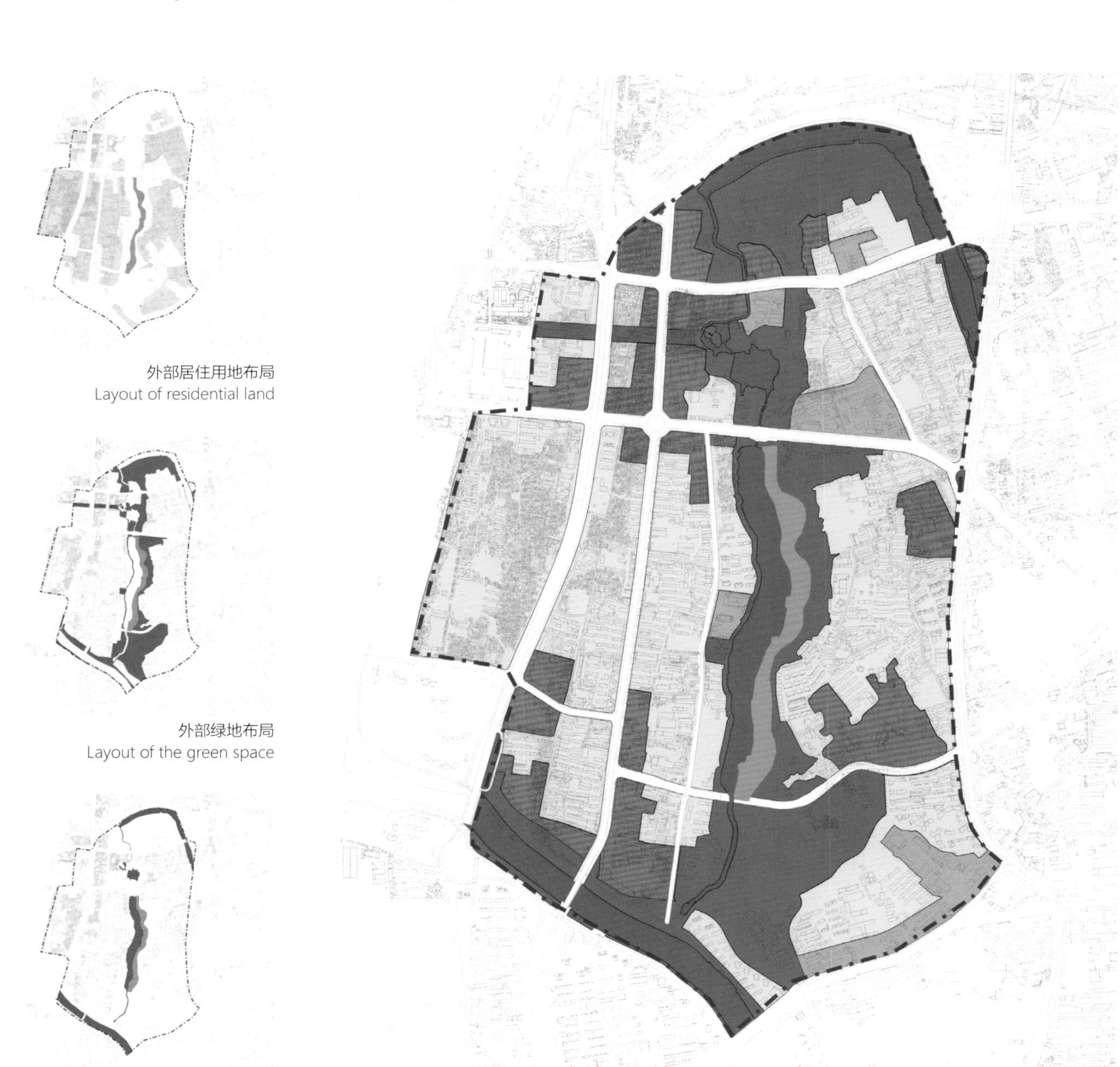

外部居住用地布局
Layout of residential land

外部绿地布局
Layout of the green space

外部水系布局
Layout of the water system

外部用地类型布局
Layout of the landuse of the exterior space

外部空间用地结构：区段 1（艮江门—华严岗门）

该区段所在区域以居住用地为主，占用地总面积的一半。绿地与广场用地（简称“绿地”）面积也较大，占总用地面积约20%。商业服务业设施用地（简称“商业用地”）和公共管理与公共服务用地（简称“公共服务用地”）面积合计占比约 16%。从布局看，绿道东、西侧分别为公园绿地和护城河水系，再外侧为居住用地。商业用地和公共服务用地则主要沿南北向的郑和路、热河路和虎踞路布置，大多离绿道有一定距离。

Landuse structure of the exterior space of S1 (YJG - HYGG)

The region where this segment is located is mainly residential land that covers half of the total land area. Green spaces and square land ("green space" for short) also have a large area, covering 20% of the total land area. Commercial and business facilities land ("commercial land" for short) and administration and public services land ("public services land" for short) cover about 16% of the total land area. In terms of layout, the east and west sides of the greenway are parks and city moat, respectively, and the outside is residential land. The commercial land and public services land are mainly located along Zhenghe Road, Rehe Road, and Huju Road, all of which travel from the south to north, with a certain distance from the greenway.

Note: The names of the land use types are all from *Code for Classification of Urban Land Use and Planning Standards of Development Land* (GB 50137 –2011).

绿地
Green Space

商业用地
Commercial land

公共服务用地
Public services land

居住用地
Residential land

用地类型	比例	面积
其他类型的用地 Other landuse types	19.54%	95.02hm²
绿地 Green Space	19.72%	95.92hm²
商业用地 Commercial land	12.07%	58.72hm²
公共服务用地 Public Service land	4.66%	22.69hm²
居住用地 Residential land	44.01%	214.05hm²

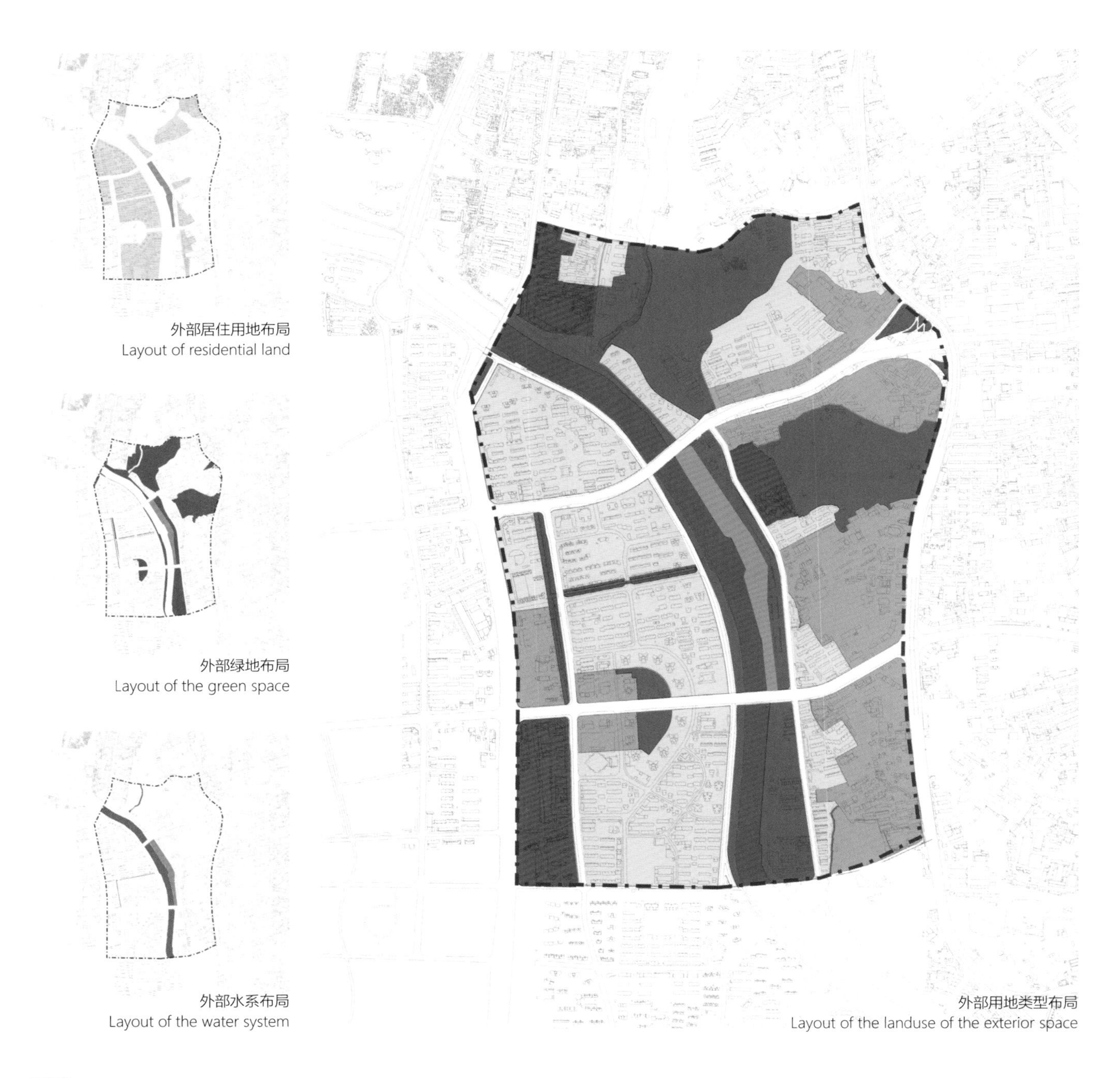

外部居住用地布局
Layout of residential land

外部绿地布局
Layout of the green space

外部水系布局
Layout of the water system

外部用地类型布局
Layout of the landuse of the exterior space

外部空间用地结构：区段 2（定淮门—草场门）

该区段所在区域居住用地约占总面积三分之一，绿地和公共服务用地是区域内占比较大的用地类型，分别约占 21% 和 18%。商业用地面积较小，约占总面积的 6%。从布局上看，绿道西侧、秦淮河对岸是大面积的居住用地，绿道东侧紧挨一片面积较大的公共服务用地和绿地。商业用地则散布在区域内，并有一小片与绿道东北相邻。

Landuse structure of the exterior space of S2 (DHG - CCG)

The residential land in the district where this segment is located is about one third of the total land area. Green space and public services land are the types of land with a bigger coverage of area within the district. The former is 21%, while the latter is 18%. The area of commercial land is smaller, covering about 6% of the total land area. In terms of layout, to the west of the greenway and across the Qinhuai River are large areas of residential land, while the east of the greenway is right next to a large area of public services land and green space. The commercial land is dispersed around the district, with a small part next to the northeast portion of the greenway.

用地类型	比例	面积
其他类型的用地 Other landuse types	20.90%	85.06hm²
绿地 Green Space	21.28%	86.60hm²
商业用地 Commercial land	6.44%	26.19hm²
公共服务用地 Public Service land	17.66%	71.87hm²
居住用地 Residential land	33.72%	137.19hm²

外部居住用地布局
Layout of residential land

外部绿地布局
Layout of the green space

外部水系布局
Layout of the water system

外部用地类型布局
Layout of the landuse of the exterior space

外部空间用地结构：区段 3（草场门—清凉门）

该区段所在区域居住用地约占总面积 40%，绿地和公共服务用地占比相对较大，分别约占 22% 和 19%。商业用地面积较小，约占总面积的 6%。审视用地布局，大面积居住用地分布在绿道西侧；秦淮河对岸，紧邻绿道东侧的是大面积公共服务用地和公园绿地。商业用地分散布置于区域外围。

Landuse structure of the exterior space of S3 (CCG - QLG)

The residential land in the district where this segment is located covers about 40% of the total land area. The percentages of green space and public services land are relatively bigger, at 21% and 19% respectively. The area of commercial land is smaller, covering about 6% of the total area. In terms of the layout of the land, large areas of residential land are distributed along the west of the greenway and across the Qinhuai River. Right next to the east of the greenway are large areas of public services land and parks. The commercial land is dispersed around the outside of the district.

用地类型	比例	面积
其他类型的用地 Other landuse types	13.23%	79.03hm^2
绿地 Green Space	21.77%	130.01hm^2
商业用地 Commercial land	5.60%	33.45hm^2
公共服务用地 Public Service land	19.19%	114.63hm^2
居住用地 Residential land	40.21%	240.18hm^2

外部居住用地布局
Layout of residential land

外部绿地布局
Layout of the green space

外部水系布局
Layout of the water system

外部用地类型布局
Layout of the landuse of the exterior space

外部空间用地结构：区段 4（水西门—集庆门）

该区段外部空间用地类型相对单一，基本由居住用地和绿地组成。居住用地约占用地总面积的 53%，绿地约占 28%。该区域没有商业用地，公共服务用地占比也较小。与前几个区段不同的是，该区段绿道位于秦淮河西岸，东西两侧都是大面积居住用地，仅有的一小块公共服务用地集中分布在区域东北边缘。

Landuse structure of the exterior space of S4 (SXG - JQG)

The landuse types of the exterior space of this segment are relatively few, which basically include residential land and green space. The residential land covers about 53% of the total land area, and green space covers around 28%. This district does not have any commercial land, and the percentage of its public services land is relatively small. Unlike the previous segments, the greenway in this segment is located at the west bank of the Qinhuai River, with large areas of residential land surrounding both the east and west sides. Only a small area of public services land is centrally distributed on the edge of the northeastern part of the district.

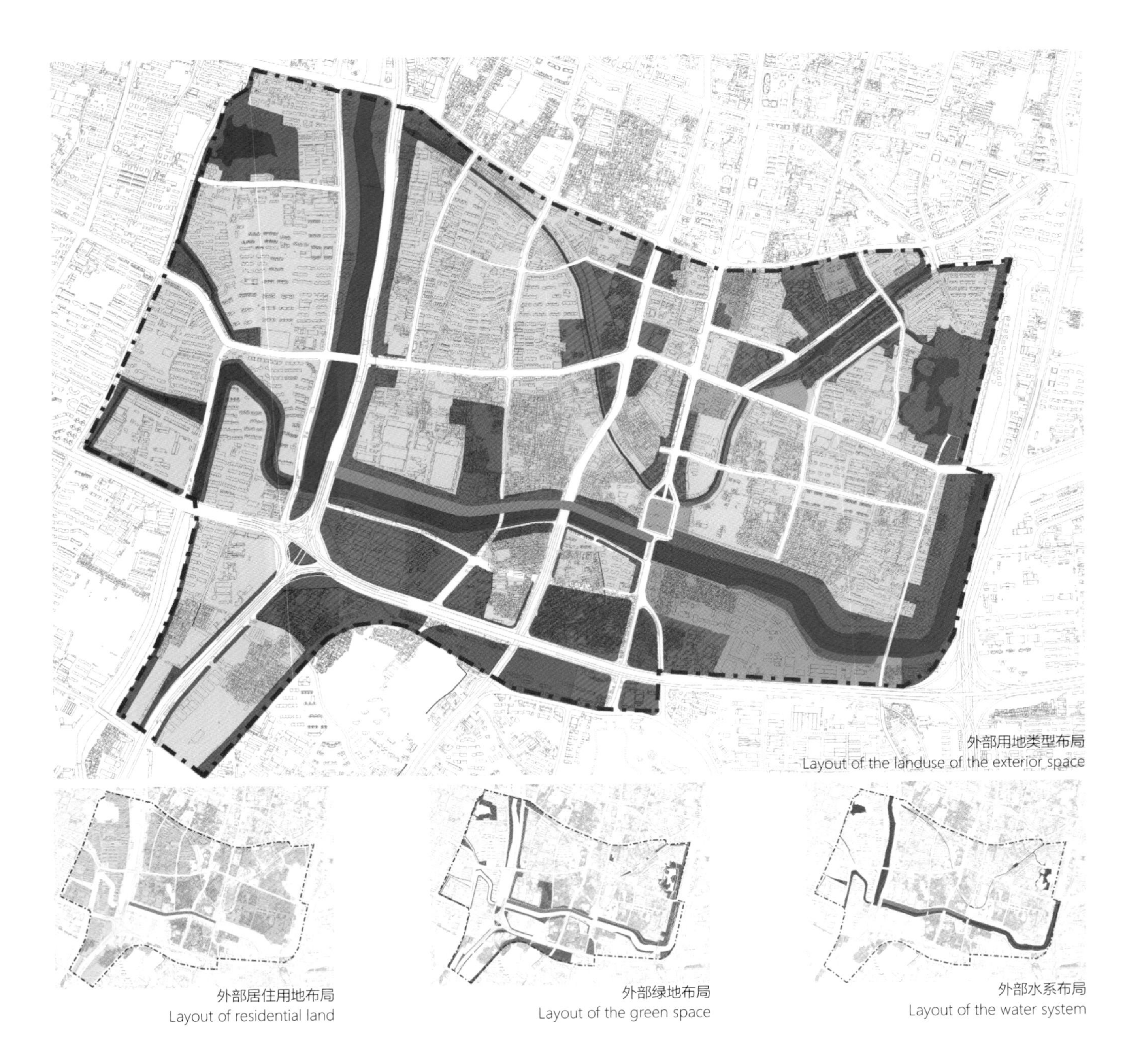

外部用地类型布局
Layout of the landuse of the exterior space

外部居住用地布局
Layout of residential land

外部绿地布局
Layout of the green space

外部水系布局
Layout of the water system

外部空间用地结构：区段 5（凤台路—中华门）

该区段所在地以居住用地为主，约占总用地面积 45%，商业用地比重相对也较高，约占 11%。绿地与公共服务用地占比均相对较小，合计共占约 13%。绿道西侧和北侧都是居住用地，东北则是商业和居住用地混合区域，南侧秦淮河对岸是大面积商业用地和公共服务用地。区域内还布有两块面积较大的公园绿地，分别是西北部莫愁湖公园和东北部白鹭洲公园。

Landuse structure of the exterior space of S5 (FTR - ZHG)

The district where this segment is located mainly consists of residential land, which covers about 45% of the total land area. The percentage of its commercial land is relatively big at 12%. The percentages of green space and public services land are relatively small, with a total of about 13%. Residential land is to the west and north of the greenway. The northeastern part is a mixed district of commercial and residential land. Across the Qinhuai River to the south are large areas of commercial land and public services land. This district is also equipped with two large parks, namely, Mochou Lake Park to the northwest and Bailuzhou Park to the northeast.

用地类型	比例	面积
其他类型的用地 Other landuse types	31.22%	186.36hm²
绿地 Green Space	10.93%	65.24hm²
商业用地 Commercial land	10.99%	65.67hm²
公共服务用地 Public Service land	2.43%	14.48hm²
居住用地 Residential land	44.43%	265.27hm²

外部居住用地布局
Layout of residential land

外部绿地布局
Layout of the green space

外部水系布局
Layout of the water system

外部用地类型布局
Layout of the landuse of the exterior space

外部空间用地结构：区段 6（中华门—雨花门）

区段 6 是区段 5（凤台路—中华门）向东延伸段，外部空间与上一区段大量重合，其中居住用地约占 38%，其次是商业用地和绿地，约占 11%，公共服务用地占比较小，仅不到 3%。从布局上看，绿道北部是大面积居住用地和商业用地（夫子庙商业街），南侧则是大报恩寺。区域内绿地面积较小，东侧的白鹭洲公园和武定门公园是该区域内最主要的公园绿地。

Landuse structure of the exterior space of S6 (ZHG - YHG)

This segment is the part where S5 FTR-ZHG spreads out to the east. Its exterior space overlaps with a great area of the previous segment. 38% of S5 FTR-ZHG is residential land; commercial land and green space covers about 11% of the segment's land area. The percentage of public services land is relatively small, at less than 3%. In terms of layout, to the north of the greenway are large areas of residential land and commercial land (Confucius Temple Business Street), while to the south is Dabao'en Temple. The green space of this district covers a small area. Bailuzhou Park and Wudingmen Park to the east are the main parks in this district.

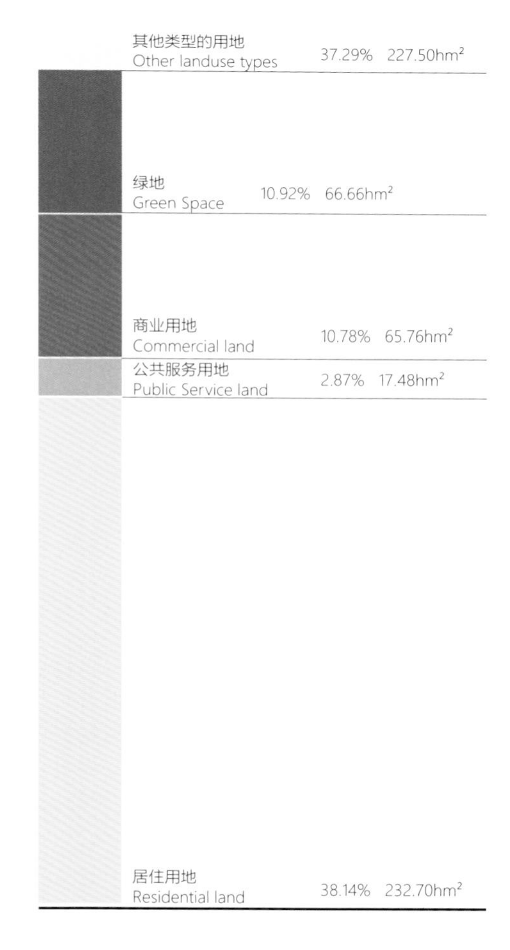

各用地类型占比
Percentages of different landuse types

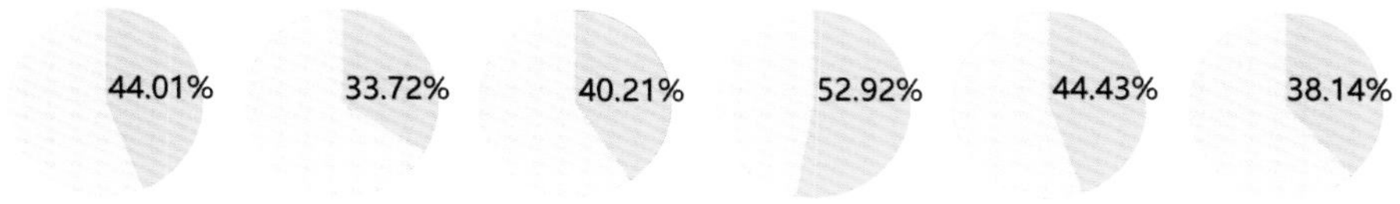

44.01% 33.72% 40.21% 52.92% 44.43% 38.14%

居住用地占比
Percentage of residential land

外部空间用地结构小结

6 个区段中：区段 4（水西门—集庆门）周边居住用地比重最大，是仅有居住面积比重超过一半的区段；区段 3（草场门—清凉门）比重最小，仅占总面积约三分之一。绿地面积比重是区段 4（水西门—集庆门）最高（27%），区段 1（挹江门—华严岗门）、区段 2（定淮门—草场门）、区段 3（草场门—清凉门）次之均为约 20%，区段 5（凤台路—中华门）、区段 6（中华门—雨花门）并列最低，均仅为 11%。公共服务用地占比区段 3（草场门—清凉门）、区段 2（定淮门—草场门）最高，均已接近 20%，区段 5（凤台路—中华门）、区段 6（中华门—雨花门）最低，均未超过 3%。商业用地比重上，区段 1（挹江门—华严岗门）、区段 5（凤台路—中华门）、区段 6（中华门—雨花门）较高，均为 12%，区段 2（定淮门—草场门）、区段 3（草场门—清凉门）次之，均为 6%，区段 4（水西门—集庆门）则没有商业用地。

Summary of landuse structure of the exterior space

Among the six segments, S4 SXG-JQG has the largest percentage of residential land, making it the only segment whose percentage of residential land covers more than half of the total land area. S3 CCG-QLG has the smallest percentage, which is only about one third of the total land area. In terms of the percentage of green space area, S4 SXG-JQG has the biggest percentage (about 27%). The ratio of green space of the districts located in S1 YJG-HYGG, S2 DHG-CCG, and S3 CCG-QLG is also high, each of which can reach about 20%. S5 FTR-ZHG and S6 ZHG-YHG rank fifth and sixth, each with only about 11% green space. In terms of the percentage of public services land, S3 CCG-QLG and S2 DHG-CCG are the top two, each with nearly 20%, while S5 FTR-ZHG and S6 ZHG-YHG are fifth and sixth, neither of which has more than 3%. S1 YJG-HYGG, S5 FTR-ZHG, and S6 ZHG-YHG have very large percentages of commercial land, each of which can reach more than 10%. S2 DHG-CCG and S3 CCG-QLG rank second, both with about 6%. S4 SXG-JQG does not have any commercial land.

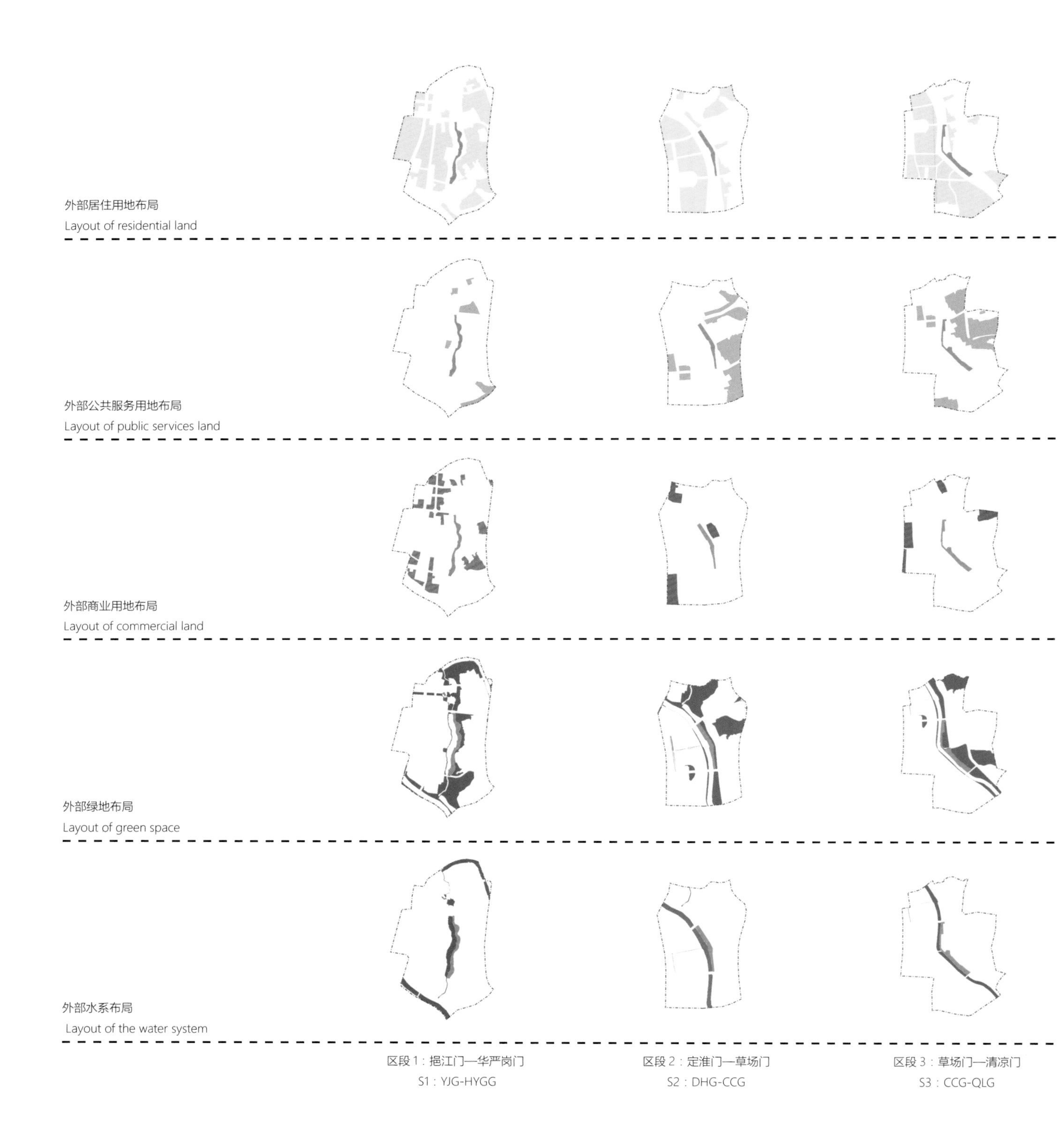
外部居住用地布局
Layout of residential land
外部公共服务用地布局
Layout of public services land
外部商业用地布局
Layout of commercial land
外部绿地布局
Layout of green space
外部水系布局
Layout of the water system
区段 1：挹江门—华严岗门
S1：YJG-HYGG
区段 2：定淮门—草场门
S2：DHG-CCG
区段 3：草场门—清凉门
S3：CCG-QLG

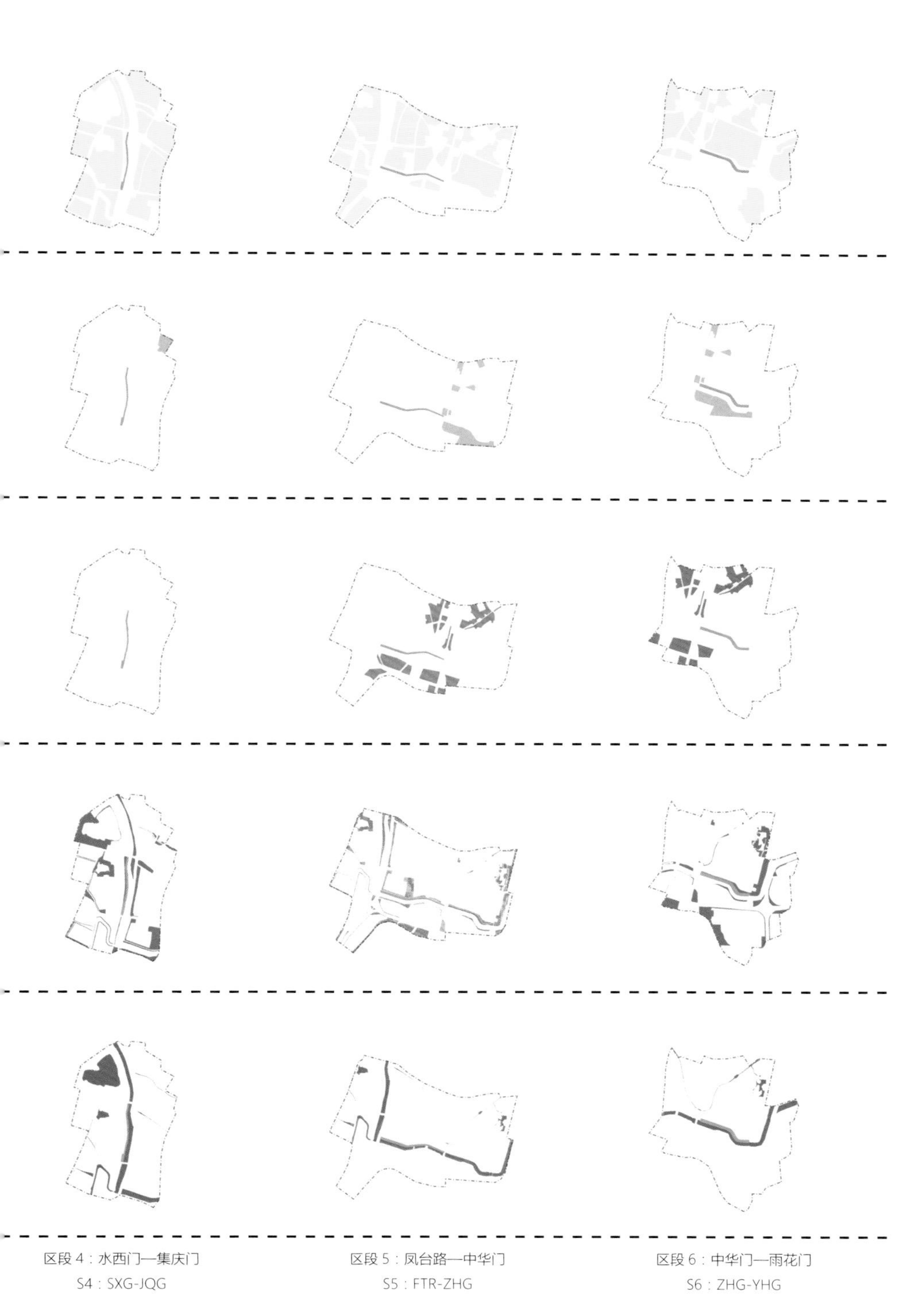

各用地类型布局
Layouts of different landuse types

外部空间用地布局

6个区段中，区段1（挹江门—华严岗门）和区段4（水西门—集庆门）居住用地集中在绿道区段两侧分布。而其他4个区段居住用地则主要集中在绿道一侧布局，其中区段5（凤台路—中华门）和区段6（中华门—雨花门）外部居住用地主要紧邻绿道区段分布，而区段2（定淮门—草场门）和区段3（草场门—清凉门）外部大部分居住用地则未直接与绿道相接，主要在河对岸分布。

区段2（定淮门—草场门）和区段3（草场门—清凉门）公共服务用地规模较大，并集中分布在绿道东侧，其他4个区段公共服务用地规模较小，主要在绿道周边零星分布。

区段1（挹江门—华严岗门）、区段5（凤台路—中华门）和区段6（中华门—雨花门）商业用地面积比重较高，并且单块用地面积较小，在绿道周边均匀分散分布。区段2（定淮门—草场门）和区段3（草场门—清凉门）商业用地面积比重较小，且单块用地面积较大，主要分布在研究范围边缘。

区段1（挹江门—华严岗门）、区段2（定淮门—草场门）、区段3（草场门—清凉门）、区段4（水西门—集庆门）绿地占比均较高，在空间上除了沿城墙和秦淮河布有带状绿地外，所在地域还有大型公园绿地。而区段5（凤台路—中华门）和区段6（中华门—雨花门）则以沿城墙和滨河带状绿地为主，尽管也有块状公园绿地，但单个公园绿地规模较前4个区段要小很多。

Layout of the landuse of the exterior space

Among the six segments, the external residential lands around S1 YJG-HYGG and S4 SXG-JQG are mainly distributed on both sides of the greenway segments, while the residential lands around the other four segments are mainly gathered around one side of the greenway. The external residential lands of S5 FTR-ZHG and S6 ZHG-YHG are right next to the greenway, while most of the residential lands surrounding S2 DHG-CCG and S3 CCG-QLG are not directly connected to the greenway and are mainly distributed across the Qinhuai River.

The external public services lands of S2 DHG-CCG and S3 CCG-QLG are large and are centrally distributed to the east of the greenway, while those of the other four segments are relatively small and are mainly dotted around the greenway's surroundings.

S1 YJG-HYGG, S5 FTR-ZHG, and S6 ZHG-YHG have very large percentages of external commercial land that is evenly distributed in the form of small land parcels surrounding the greenway. S2 DHG-CCG and S3 CCG-QLG have very small percentages of commercial land that is distributed in large blocks on the edge of the study district.

S1 YJG-HYGG, S2 DHG-CCG, S3 CCG-QLG, and S4 SXG-JQG have very large percentages of green space. Apart from ribbon-like green spaces distributed along the City Wall and river courses, large parks can also be found in the study district outside of the greenway. The green spaces in the districts located in S5 FTR-ZHG and S6 ZHG-YHG are mainly stripe-shaped along the City Wall and at the waterfront. Although there are patches of parks , the size of a single park is much smaller than that of the other four segments.

S1：YJG-HYGG
S4：SXG-JQG
两侧
On-both-sides
S2：DHG-CCG
S3：CCG-QLG
同侧
In-close-proximity
S5：FTR-ZHG
S6：ZHG-YHG
异侧
Across-the-river
S2：DHG-CCG
S3：CCG-QLG
集中大面积
Centralized
S1：YJG-HYGG
S4：SXG-JQG
S5：FTR-ZHG
S6：ZHG-YHG
分散小面积
Dispersed
S1：YJG-HYGG
S5：FTR-ZHG
S6：ZHG-YHG
分散、多
Dispersed
S2：DHG-CCG
S3：CCG-QLG
集中、少
Centralized
S4：SXG-JQG
无
None
S1：YJG-HYGG
S2：DHG-CCG
S3：CCG-QLG
S4：SXG-JQG
大
Large
S5：FTR-ZHG
S6：ZHG-YHG
小
Small

绿道外部空间结构简图 Study scope of the exterior space	绿道外部用地类型占比以及 6 区段排序 Percentages and rankings of land types of the exterior space		绿道外部用地布局模式 Pattern of land layout
区段 1：挹江门—华严岗门 S1：YJG-HYGG	R 44.01%	③	R
	A 4.66%	④	A
	B 12.07%	①	B
	G 19.72%	④	G
区段 2：定淮门—草场门 S3：DHG-CCG	R 33.72%	⑥	R
	A 17.66%	②	A
	B 6.44%	④	B
	G 21.28%	②	G
区段 3：草场门—清凉门 S3：CCG-QLG	R 40.21%	④	R
	A 19.19%	①	A
	B 5.60%	⑤	B
	G 21.77%	③	G
区段 4：水西门—集庆门 S4：SXG-JQG	R 52.92%		R
	A 8.92%	③	A
	B 0%	⑥	B
	G 27.54%	①	G
区段 5：凤台路—中华门 S5：FTR-ZHG	R 44.43%	②	R
	A 2.43%	⑥	A
	B 11.00%	②	B
	G 10.93%	⑤	G
区段 6：中华门—雨花门 S6：ZHG-YHG	R 38.14%	⑤	R
	A 2.87%	⑤	A
	B 10.78%	③	B
	G 10.93%	⑥	G

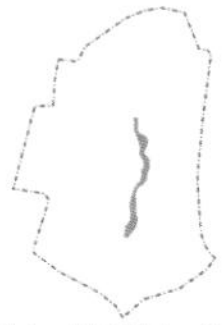

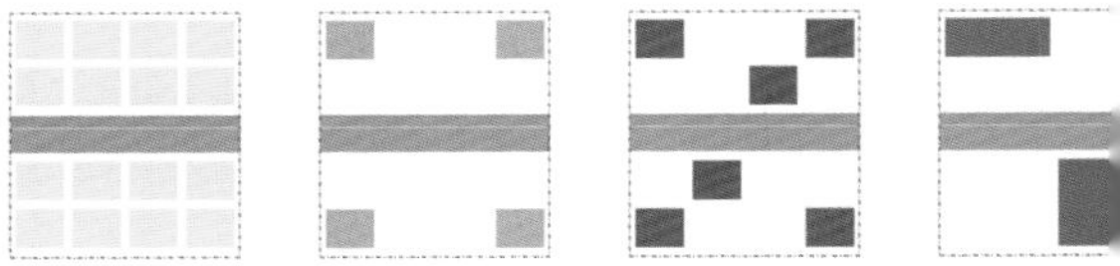

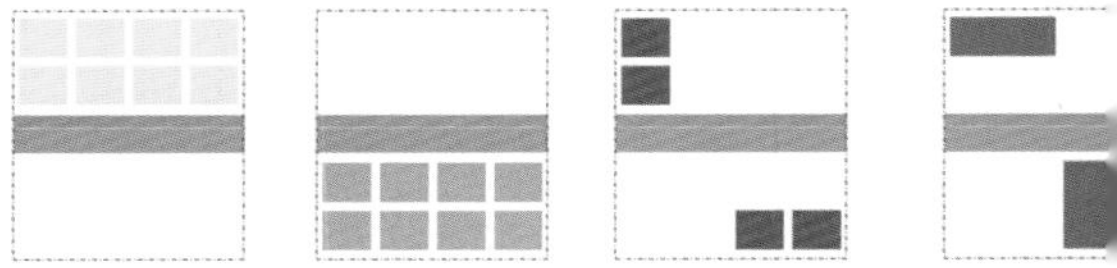

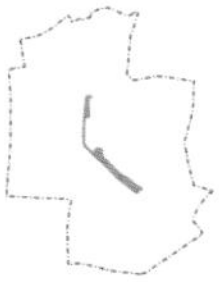

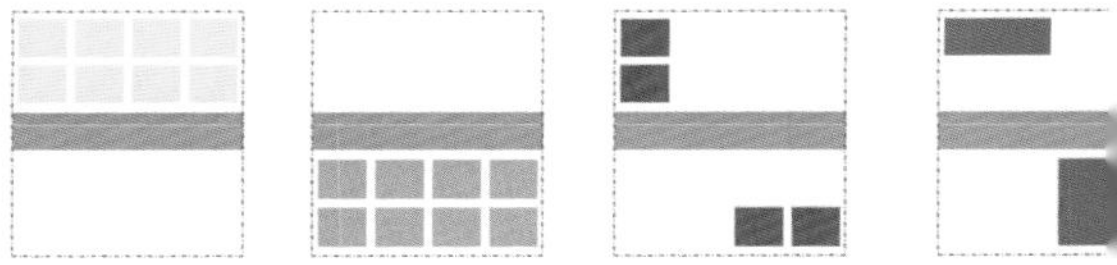

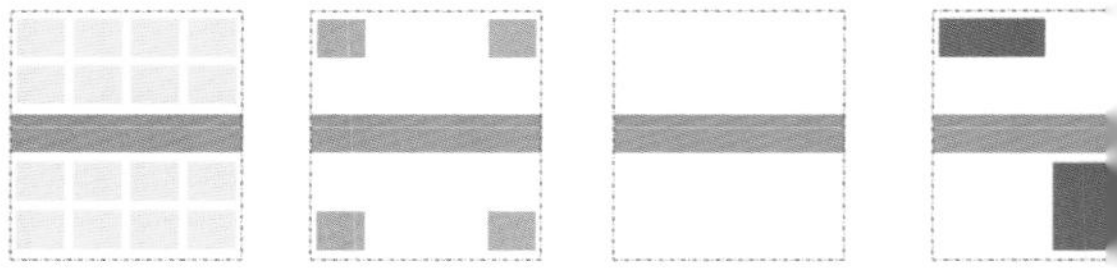

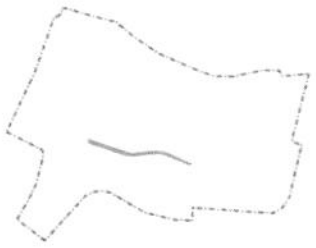

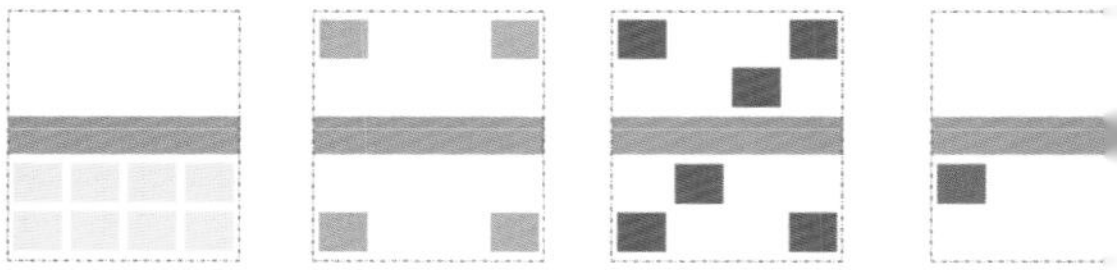

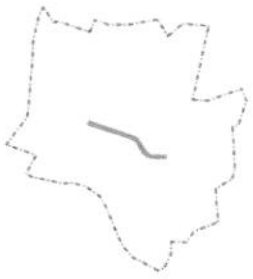

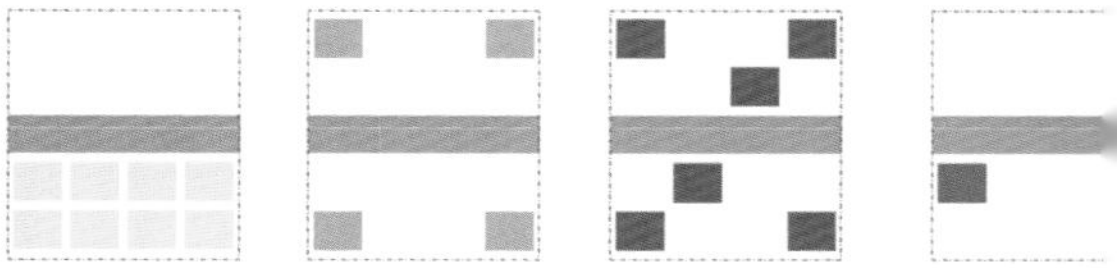

绿道使用人群规模以及 6 区段排序 Number of greenway users and ranking	绿道使用人群密度以及 6 区段排序 Density of greenway users and ranking	不同时间段绿道使用人群密度以及 6 区段排序 Densities of greenway users in two time periods and ranking
551 person ①	423 person/km ①	14:00-16:00 458 person/km ② 18:00-20:00 388 person/km ①
103 person ⑤	103 person/km ⑤	14:00-16:00 119 person/km ⑤ 18:00-20:00 87 person/km ⑤
499 person ②	332 person/km ②	14:00-16:00 487 person/km ① 18:00-20:00 178 person/km ④
78 person ⑥	64 person/km ⑥	14:00-16:00 53 person/km ⑥ 18:00-20:00 76 person/km ⑥
251 person ④	179 person/km ④	14:00-16:00 141 person/km ④ 18:00-20:00 217 person/km ③
347 person ③	315 person/km ③	14:00-16:00 289 person/km ③ 18:00-20:00 342 person/km ②

外部空间用地结构与服务绩效

指标对照可以发现，区段 1（挹江门—华严岗门）使用人群规模和各个使用人群密度指标均最高，其周边商业用地占比在各区段中排序第一，居住用地仅排序第三，其余用地居中。使用人群规模和密度相对较低的为区段 4（水西门—集庆门）以及区段 2（定淮门—草场门）。其中区段 4（水西门—集庆门）除了日常傍晚使用人群密度指标排序第四外，使用人群规模和其他密度指标均排序垫底，但其周边地区用地结构中，居住用地和绿地占比却排序第一，同时也是唯一周边没有商业用地的区段。区段 2（定淮门—草场门）则是除日常傍晚使用人群密度指标排序垫底外，使用人群规模和其他密度指标均排序倒数第二，在周边其用地结构中，居住用地占比为 6 区段最低，绿地和公共服务用地占比排序第二，商业用地占比居中。

Land structure and service performance of the exterior space

Through the comparison of indicators, we found that S1 YJG-HYGG ranks first in terms of number of users, the average user density in two time periods, and the user density every day at nightfall. The percentage of its external commercial land ranks first among all the segments, that of its external residential land ranks third, and that of the other types of land rank in the middle. S4 SXG-JQG and S2 DHG-CCG have a small number of users and relatively low user density. As for S4 SXG-JQG, except for the user density, which ranks fourth, its number of users and the other density indicators all rank last. However, in terms of its external land structure, the percentages of its residential land and green space both rank first but it is the only segment without any surrounding commercial land. The user density of S2 DHG-CCG every day at nightfall ranks last, and its number of users and the other density indicators all rank fifth. In terms of external land structure, the percentage of its residential land is the smallest among the six segments, those of green space and public services rank second, and that of commercial land ranks in the middle.

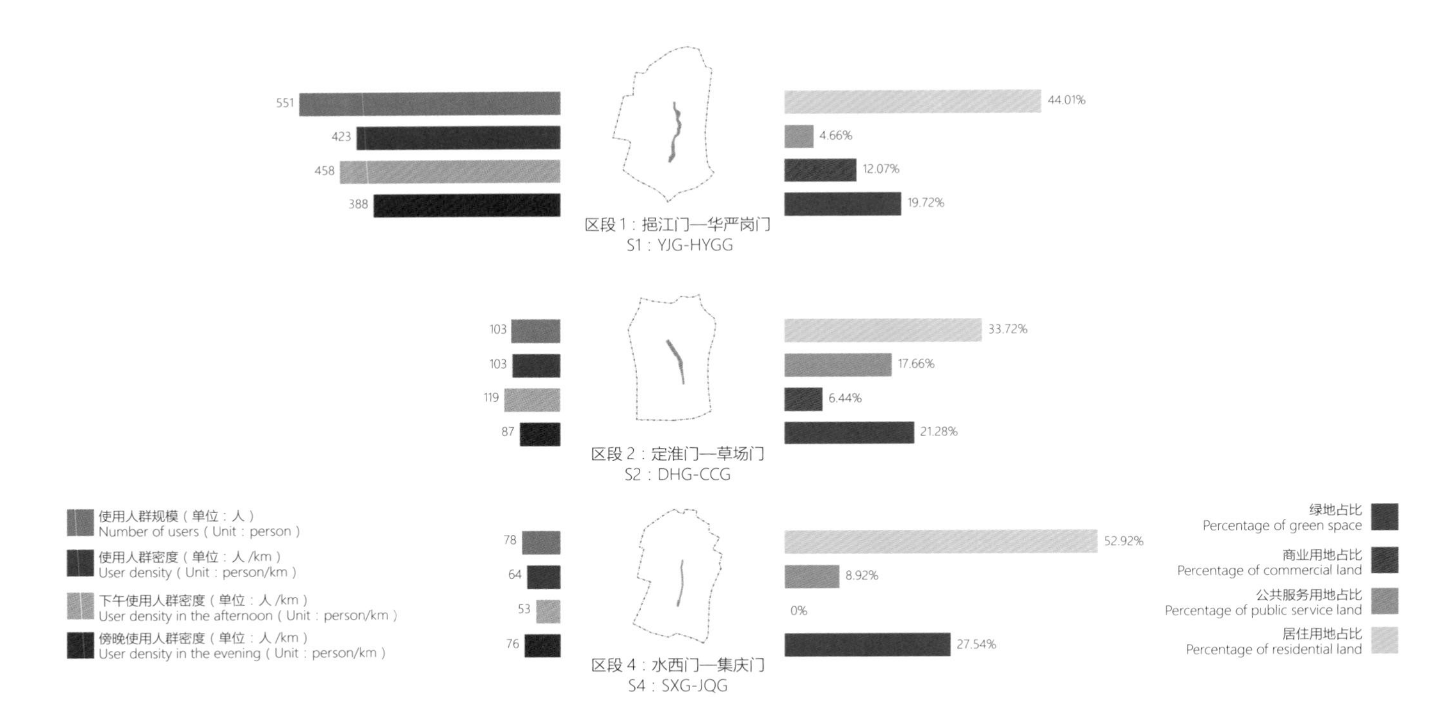

在用地布局上，我们主要围绕与绿道使用关联最紧密的居住用地布局形式展开讨论。在南京明城墙绿道中，由于秦淮河这个重要空间要素始终存在，因而居住用地布局与绿道的相对位置主要可分为两侧布局、对岸布局和紧邻布局三个类型。从理论上看，秦淮河既是绿道的景观游憩资源，同时也是绿道的空间阻隔要素，即对河对面居民日常使用绿道造成空间阻碍。6 个区段中，采用两侧布局的有区段 1（挹江门—华严岗门）和区段 4（水西门—集庆门），对岸布局的有区段 2（定淮门—草场门）和区段 3（草场门—清凉门），紧邻布局的有区段 5（凤台路—中华门）和区段 6（中华门—雨花门）。除区段 4（水西门 - 集庆门）指标比较特殊外，总体使用人群密度指标的变化情况与理论假定情况吻合，即两侧布局使用人群密度最高，紧邻布局次之，对岸布局最低。

To keep in line with the goal of this study, we mainly focus our discussion on the layout of residential land that is most closely linked to the use of the greenway. For the Nanjing Ming Dynasty City Wall Greenway, since the moat is a very important spatial element, the layout of its external residential land and the relative land position to the greenway can mainly be divided into three patterns: being on both sides of the greenway, being across the river, and being in close proximity to the greenway. Among the six segments, S1 YJG-HYGG and S4 SXG-JQG are laid out on both sides of the greenway. S2 DHG-CCG and S3 CCG-QLG are laid out across the other side of the river. S5 FTR-ZHG and S6 ZHG-YHG are laid out in close proximity. Except for S4 SXG-JQG, which has abnormal indicators, the user densities of the segments that are laid out on both sides are the highest. The exterior space laid out in close proximity to the greenway ranks second, and the exterior space of that is laid out across the other side of the river is the lowest.

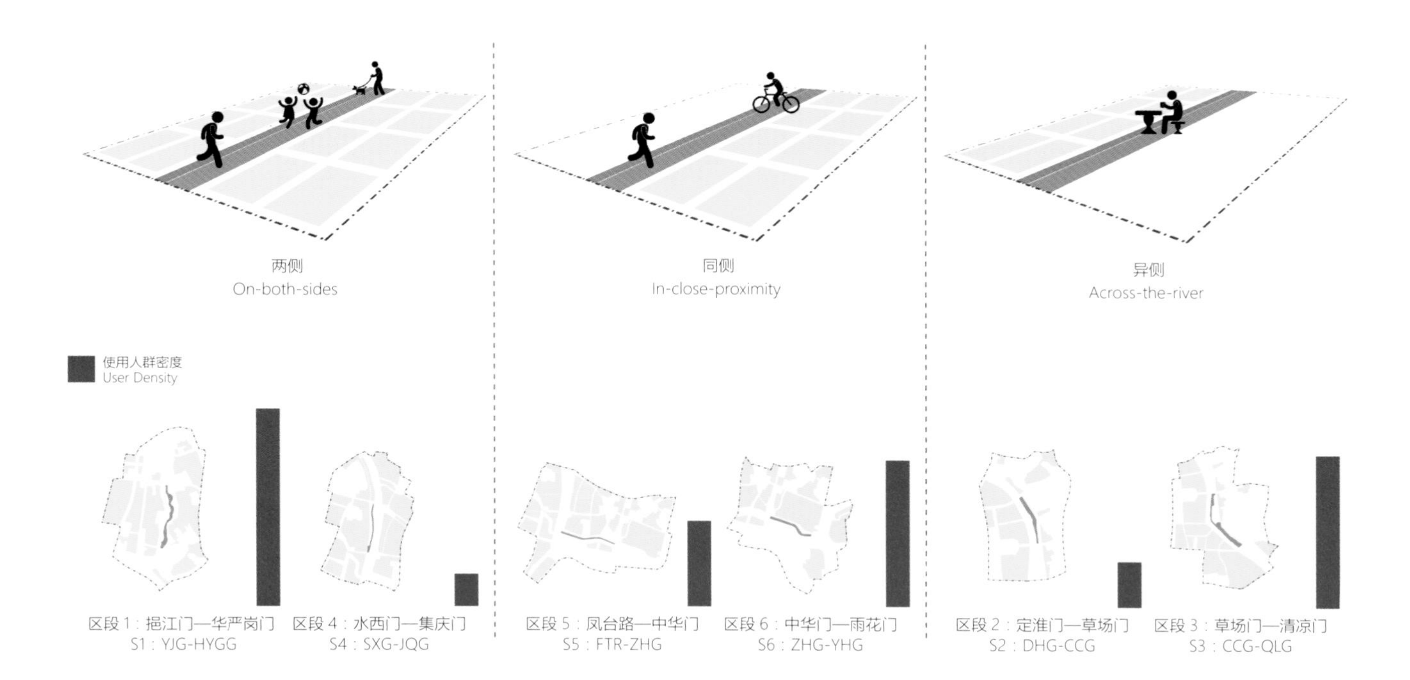

外部楼房高度
Number of building stories of the exterior space

外部空间开发强度：区段 1（挹江门—华严岗门）

该区段所在研究区域总体容积率为 1.41，建筑密度约 31.41%。区域内建筑以低层（1-3 层）为主，约占总量60%，多层（4-6 层）建筑次之，占 28%，中高层（7-9 层）建筑约占 9%，高层（10 层以上）建筑占比最小，约占 3%。布局上，绿道东侧以多层建筑和中高层建筑为主，西侧是大面积的低层建筑，并有少数高层建筑散布。

Development intensity of the exterior space of S1 (YYG - HYGG)

The total plot ratio of the district in study where this segment is located is 1.41, and building density about 31.41%. Within the district are mainly low-rise(floors 1-3) buildings, covering about 60% of the total. Multistoried (floors 4-6) buildings have the next greatest, covering 28%. Mid-rise (floors 7-9) buildings cover about 9%. The percentage of high-rise (floors 10+) buildings is the smallest, at about 3%. In terms of layout, to the east of the greenway are mainly multistoried buildings and mid-rise buildings, while to the west are large areas of low-rise buildings, with a small dispersed amount of high-rise buildings.

10+ 层 Floors 10+

7-9 层 Floors 7-9

4-6 层 Floors 4-6

1-3 层 Floors 1-3

数据统计表 Statistics of development intensity		
总建筑面积 Overall floorage		6858200m²
区域容积率 District gross plot ratio		1.41
建筑密度 Building density		31.41%
各高度建筑占地面积占比 (%) Proportion of floor space of buildings with different stories	Floors 1-3	59.89%
	Floors 4-6	27.96%
	Floors 7-9	8.99%
	Floors 10+	3.16%

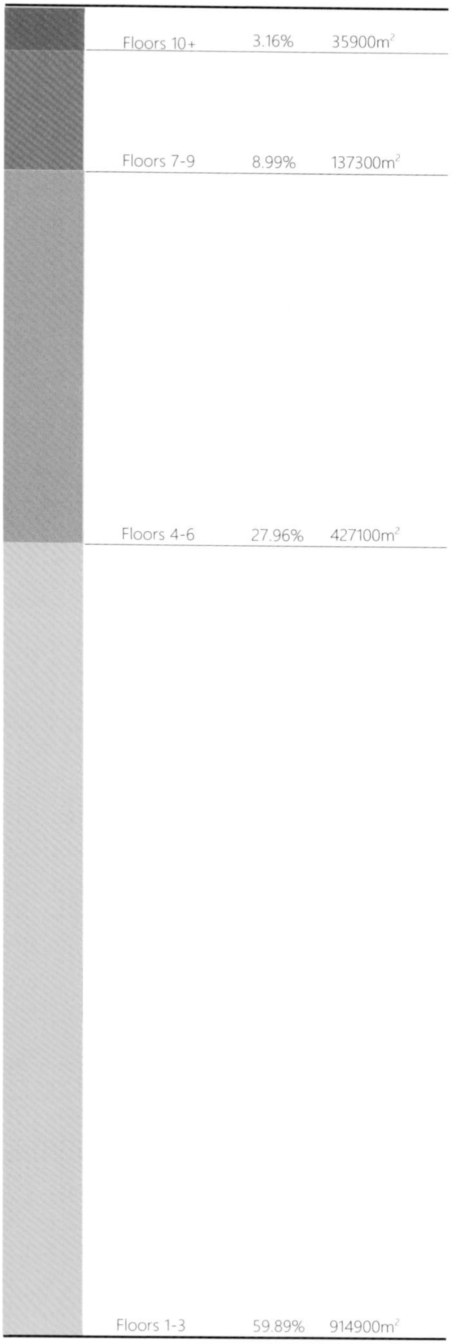

Floors 10+	3.16%	35900m²
Floors 7-9	8.99%	137300m²
Floors 4-6	27.96%	427100m²
Floors 1-3	59.89%	914900m²

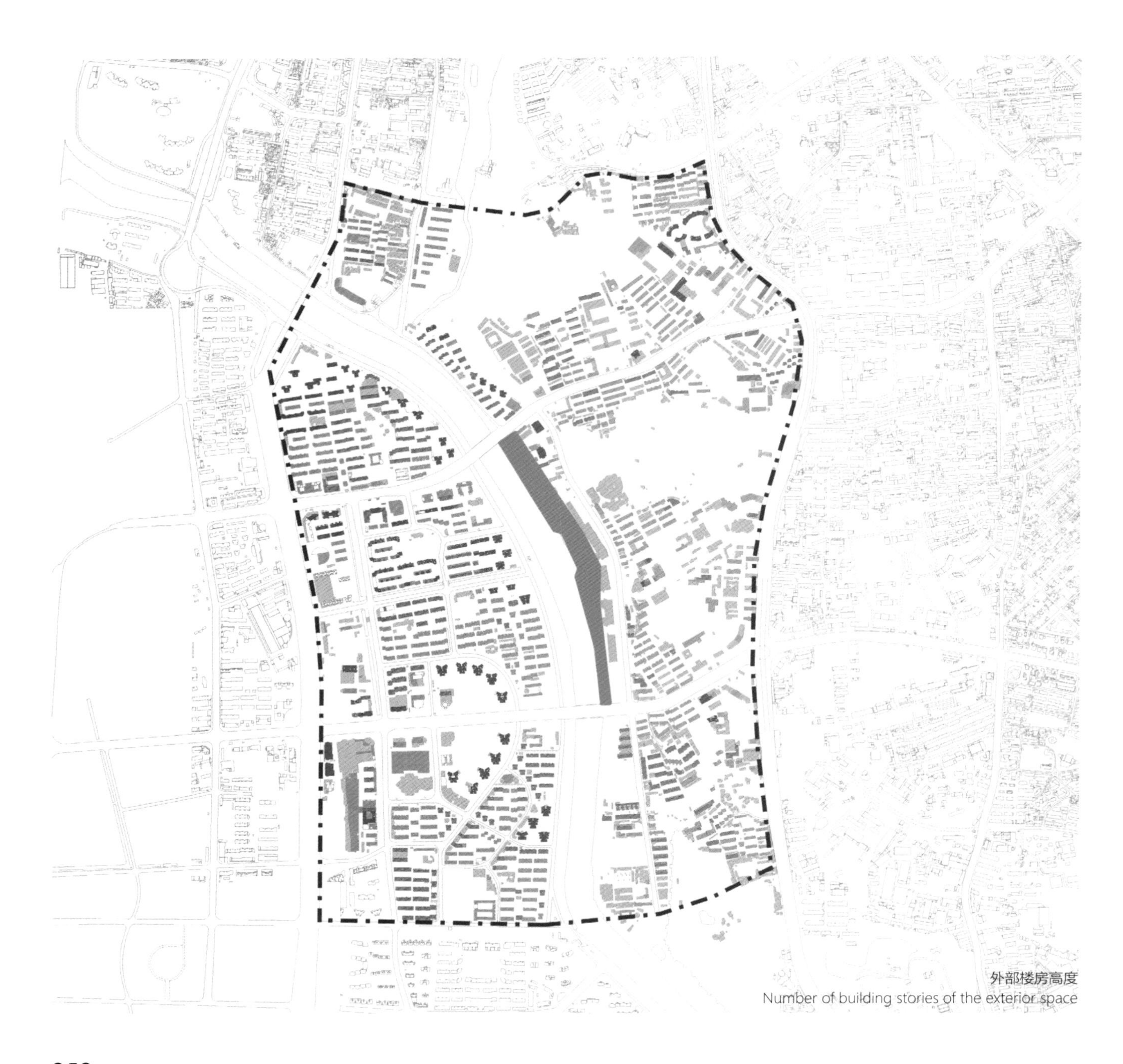

外部楼房高度
Number of building stories of the exterior space

外部空间开发强度：区段 2（定淮门—草场门）

该区段所在研究区域总体容积率为 0.97，建筑密度约 23.22%，为 6 个区段最低。低层建筑占总量的约 41%，多层建筑占总量约三分之一，中高层建筑占比相对较高，达到约 16%，高层建筑则占比约 9%。从整体上看，区域内开发强度适中，其中东北部虎踞北路与西南的草场门大街沿线开发强度较高，区段内大部分高层建筑均位于该地段。

Development intensity of the exterior space of S2 (DHG - CCG)

The total plot ratio of the district in study where this segment is located is 0.97, and building density about 23.22%, making it the lowest out of all six segments. Low-rise buildings cover about 41% of the total; multistoried buildings cover about one third of the total; the percentage of mid-rise buildings is relatively large at around 16%; and the percentage of high-rise buildings is about 9%. Overall, the development intensity of this district is moderate. The development intensity of North Huju Road to the northeast and the area along Caochangmen Avenue to the southwest is relatively strong, and most of the high-rise buildings in this segment are located in this district.

10+ 层 Floors 10+

7-9 层 Floors 7-9

4-6 层 Floors 4-6

1-3 层 Floors 1-3

数据统计表 Statistics of development intensity		
总建筑面积 Overall floorage		3947000m²
区域容积率 District gross plot ratio		0.97
建筑密度 Building density		23.22%
各高度建筑占地面积占比 (%) Proportion of floor space of buildings with different stories	Floors 1-3	41.39%
	Floors 4-6	33.99%
	Floors 7-9	15.81%
	Floors 10+	8.81%

Floors	Percentage	Floorage
Floors 10+	8.81%	79500m²
Floors 7-9	15.81%	149400m²
Floors 4-6	33.99%	321100m²
Floors 1-3	41.39%	391000m²

外部楼房高度
Number of building stories of the exterior space

外部空间开发强度：区段 3（草场门—清凉门）

该区段所在研究区域总体容积率 1.50，为 6 个区段最高，建筑密度约 27.28%，排序倒数第二。绿道周边中低层和多层建筑居多，其中低层建筑占比约 54%，多层建筑占比约 37%，合计超过总量九成。中高层和高层建筑合计占比不到一成。在整体布局上，绿道东侧以低层建筑为主，南侧也基本是低层和多层建筑，而中高层和高层建筑主要分布在绿道西侧，秦淮河对岸区域。

Development intensity of the exterior space of S3 (CCG - QLG)

The total plot ratio of the district in study where this segment is located is 1.50, making it the highest among the six segments, with a building density of about 27.28%, which ranks fifth. The surrounding area of the greenway is mostly equipped with low-rise and multistoried buildings; the percentage of low-rise buildings is about 54%, while that of multistoried buildings is about 37%, so together they represent over 90% of the buildings around this segment. The percentages of mid-rise buildings and high-rise buildings are smaller than 10%. In terms of overall layout, to the east of the greenway are mainly low-rise buildings. The south side is also basically set with low-rise buildings and multistoried buildings. Mid-rise buildings and high-rise buildings are mainly distributed to the west of the greenway in the district across the Qinhuai River.

数据统计表 Statistics of development intensity		
总建筑面积 Overall floorage		8959500m²
区域容积率 District gross plot ratio		1.50
建筑密度 Building density		27.28%
各高度建筑占地面积占比 (%) Proportion of floor space of buildings with different stories	Floors 1-3	53.57%
	Floors 4-6	36.69%
	Floors 7-9	4.01%
	Floors 10+	5.73%

Floors	Proportion	Area
Floors 10+	5.73%	89300m²
Floors 7-9	4.01%	65300m²
Floors 4-6	36.69%	597700m²
Floors 1-3	53.57%	872700m²

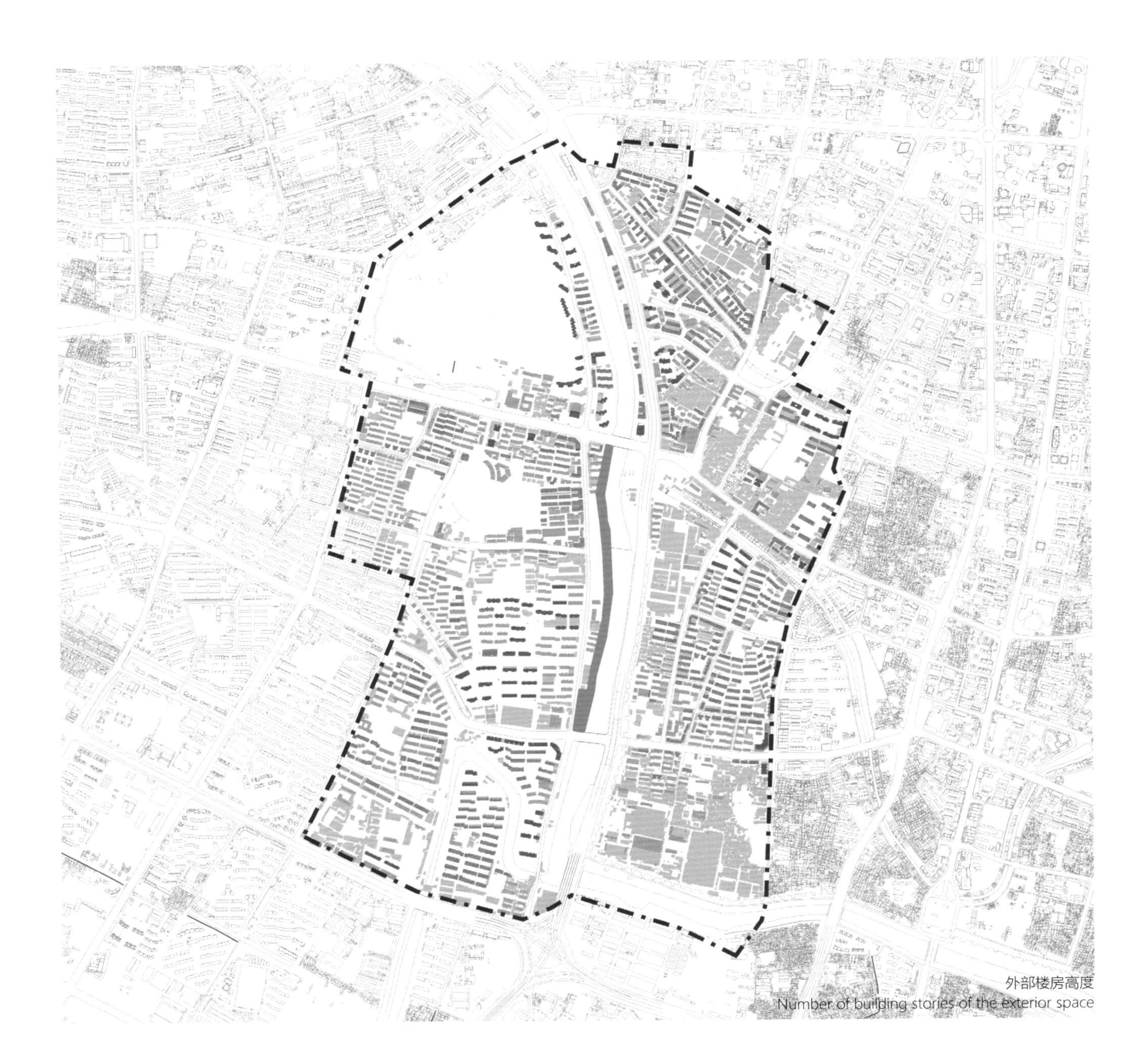
外部楼房高度
Number of building stories of the exterior space

外部空间开发强度：区段 4（水西门—集庆门）

该区段所在研究区域总体容积率 1.25，建筑密度约 33.37%。区段周边低层建筑约占总量一半，多层占到约四分之一，剩下四分之一为中高层和高层建筑。在分布上，低层建筑分布于集庆路东段南侧以及秦淮河对岸区域，绿道西侧以及东北部是中高层和高层建筑密集区。

Development intensity of the exterior space of S4 (SXG - JQG)

The total plot ratio of the district in study where this segment is located is 1.25, and building density about 33.37%. Low-rise buildings surrounding the segment cover about half of the total amount. Multistoried buildings cover about one fourth. The remaining fourth consists of mid-rise and high-rise buildings. In terms of layout, low-rise buildings are distributed on the south side of the eastern part of Jiqing Road and in the district across the Qinhuai River, while clusters of mid-rise and high-rise buildings are distributed to the west and northeast of the greenway.

10+ 层 Floors 10+

7-9 层 Floors 7-9

4-6 层 Floors 4-6

1-3 层 Floors 1-3

数据统计表 Statistics of development intensity		
总建筑面积 Overall floorage		6666300m²
区域容积率 District gross plot ratio		1.25
建筑密度 Building density		33.37%
各高度建筑占地面积占比（%） Proportion of floor space of buildings with different stories	Floors 1-3	52.05%
	Floors 4-6	25.23%
	Floors 7-9	17.49%
	Floors 10+	5.23%

Floors 10+	5.23%	89700m²
Floors 7-9	17.49%	311200m²
Floors 4-6	25.23%	449000m²
Floors 1-3	52.05%	926200m²

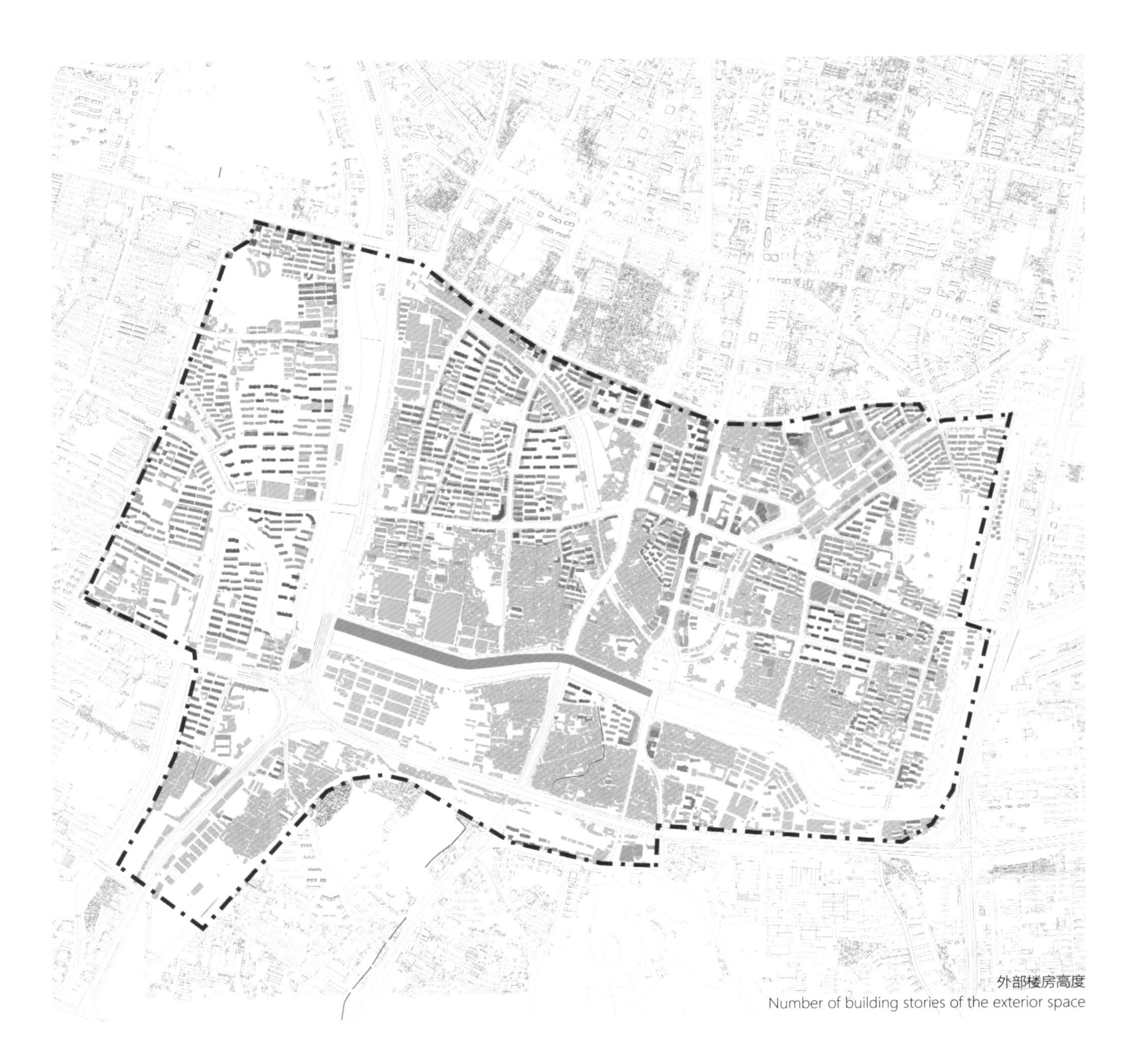

外部楼房高度
Number of building stories of the exterior space

Floors 10+	4.77%	$102400m^2$
Floors 7-9	10.69%	$251700m^2$
Floors 4-6	20.33%	$478700m^2$
Floors 1-3	64.21%	$1511900m^2$

外部空间开发强度：区段 5（凤台路—中华门）

该区段所在研究区域总体容积率最低，为 0.85，建筑密度约 39.44%，为 6 个区段最高。绿道区段周边有大量历史文化建筑和街区，因而建筑整体高度较低，低层建筑比重最大，约占 64%，多层与中高层建筑合占约 31%，高层建筑占比仅为 5%。区段北侧和南侧都被低层建筑围绕，在区域东、西、北部边缘则有较多多层和中高层建筑分布。

Development intensity of the exterior space of S5（FTR - ZHG）

The total plot ratio of the district in study where this segment is located is the lowest, at 0.85, and its building density is about 39.44%, making it the highest among the six segments. There are a lot of historic and cultural buildings and street blocks surrounding the greenway, so the overall height of the buildings is relatively low, with the biggest percentage of low-rise buildings, which is about 64%. The sum of the percentages of multistoried and mid-rise buildings is about 31%, and the percentage of high-rise buildings is only 5%. Low-rise buildings surround both the north and south sides of the segment. A lot of multistoried and mid-rise buildings are distributed on the edges of the eastern, western, and northern parts of the district.

10+ 层 Floors 10+

7-9 层 Floors 7-9

4-6 层 Floors 4-6

1-3 层 Floors 1-3

数据统计表 Statistics of development intensity		
总建筑面积 Overall floorage		$5074700m^2$
区域容积率 District gross plot ratio		0.85
建筑密度 Building density		39.44%
各高度建筑占地面积占比（%）Proportion of floor space of buildings with different stories	Floors 1-3	64.21%
	Floors 4-6	20.33%
	Floors 7-9	10.69%
	Floors 10+	4.77%

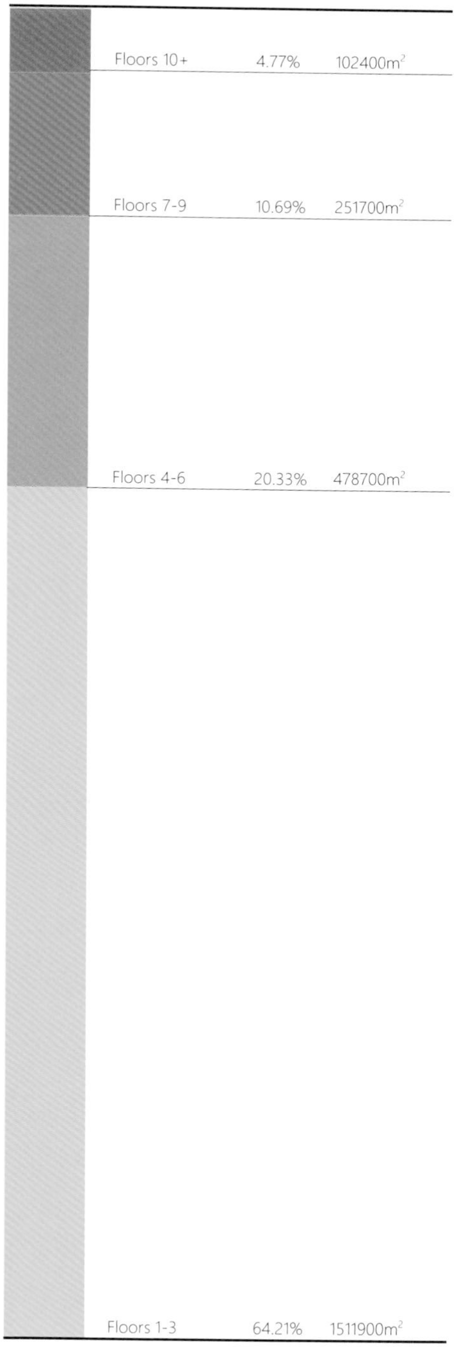

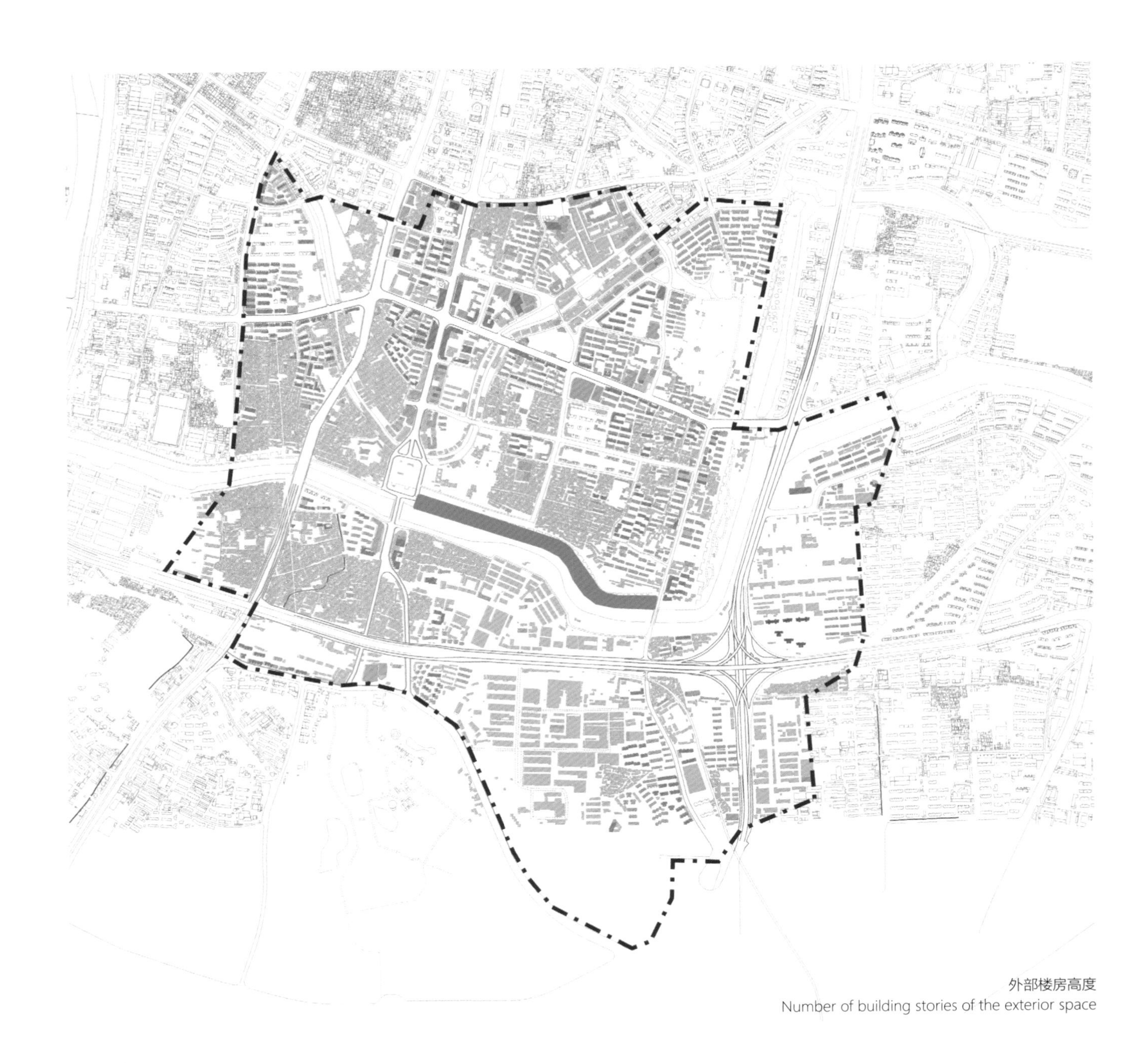

外部楼房高度
Number of building stories of the exterior space

外部空间开发强度：区段 6（中华门—雨花门）

该区段所在研究区域总体容积率 1.22，建筑密度约 34.40%。与上一区段情况类似，外部也以低层建筑为主，占比达到约 82%，多层建筑占比约 12%，而中高层与高层建筑合计占比不到 6%。在布局上看，低层建筑密度较高，并形成了区域主导肌理，而其他高度建筑主要分布在区段北侧三山街街区以及东侧江宁路沿线。

Development intensity of the exterior space of S6 (ZHG - YHG)

The total plot ratio of the district in study where this segment is located is 1.22, and building density about 34.40%. Similar to the previous segment, the exterior of the segment is mainly set with low-rise buildings that cover about 82%. The percentage of multistoried buildings is about 12%, while the sum of the percentages of mid-rise and high-rise buildings is smaller than 6%. In terms of layout, the building density of low-rise buildings is relatively high. The other kinds of buildings are mainly distributed on Sanshanjie Street Block on the northern side of the segment and along Jiangning Road to the east.

10+ 层 Floors 10+

7-9 层 Floors 7-9

4-6 层 Floors 4-6

1-3 层 Floors 1-3

数据统计表 Statistics of development intensity		
总建筑面积 Overall floorage		7443200m²
区域容积率 District gross plot ratio		1.22
建筑密度 Building density		34.40
各高度建筑占地面积占比 (%) Proportion of floor space of buildings with different stories	Floors 1-3	81.56%
	Floors 4-6	12.06%
	Floors 7-9	5.17%
	Floors 10+	1.21%

Floors 10+	1.21%	21200m²
Floors 7-9	5.17%	108600m²
Floors 4-6	12.06%	253200m²
Floors 1-3	81.56%	1712000m²

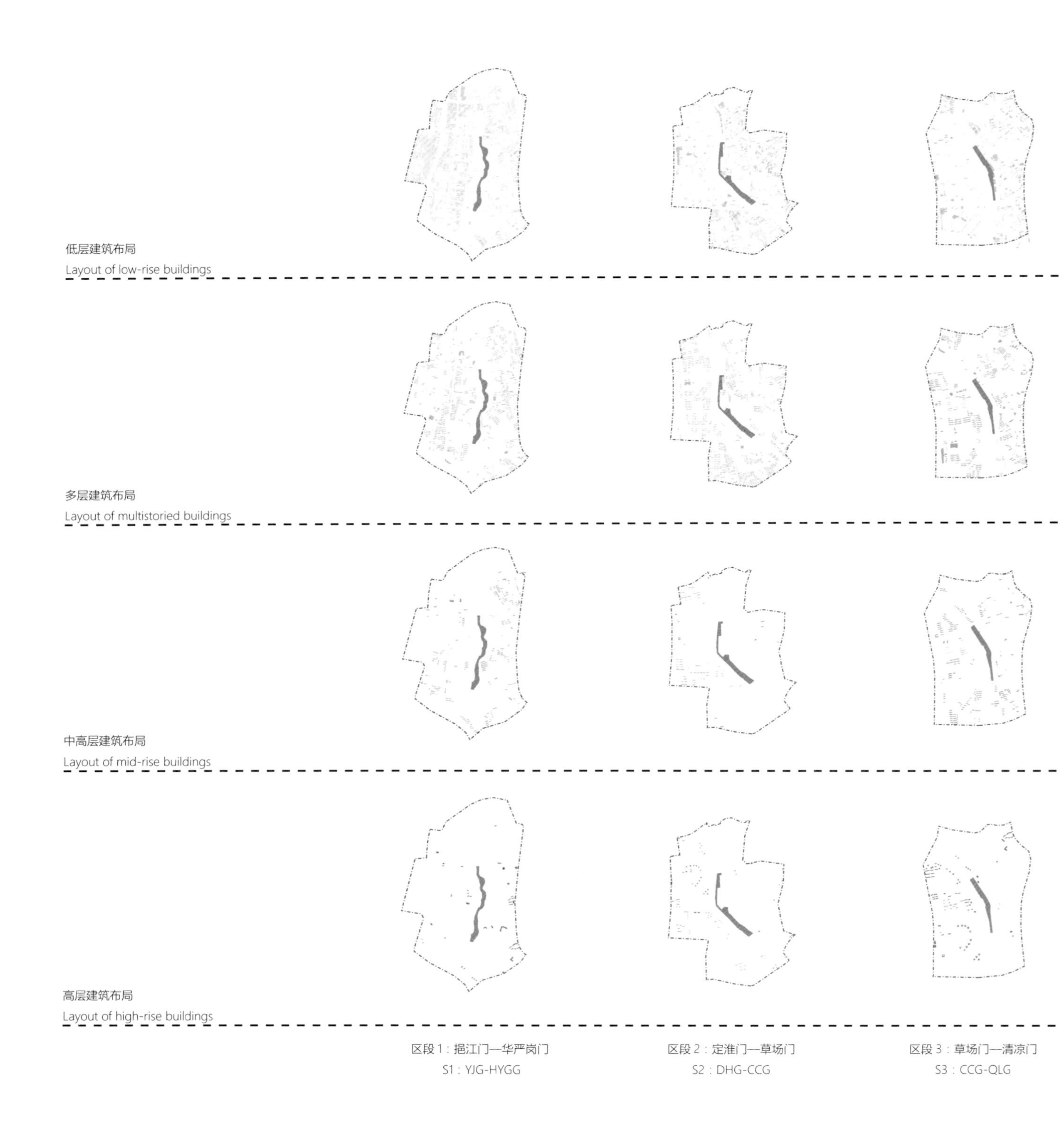
低层建筑布局
Layout of low-rise buildings
多层建筑布局
Layout of multistoried buildings
中高层建筑布局
Layout of mid-rise buildings
高层建筑布局
Layout of high-rise buildings
区段 1：挹江门—华严岗门
S1：YJG-HYGG
区段 2：定淮门—草场门
S2：DHG-CCG
区段 3：草场门—清凉门
S3：CCG-QLG

区段 4：水西门—集庆门
S4：SXG-JQG

区段 5：凤台路—中华门
S5：FTR-ZHG

区段 6：中华门—雨花门
S6：ZHG-YHG

各楼层高度布局
Layout of buildings with different stories

外部空间开发强度小结

由于绿道位于老城，因此周边地段大多均属于高建筑密度、中低容积率的开发模式。6 个区段所在区域的总体容积率均不超过 2，其中容积率在 1 以下的有区段 2（定淮门—草场门）和区段 5（风台路—中华门），其余 4 个区段相对较接近，容积率在 1.2 至 1.5 之间。而除了区段 2（定淮门—草场门）和区段 3（草场门—清凉门）外，其他四个区段所在研究区域建筑密度均在 30% 以上，区段 5（凤台路—中华门）周边建筑密度更是接近 40%。

绿道周边大多为低层建筑，除区段 2（定淮门—草场门）外，其他区段低层建筑占比均超过一半，区段 6（中华门—雨花门）周边有大量的传统街区，低层建筑占比甚至超过了 80%。中高层和高层建筑合计占比最高的为区段 2（定淮门—草场门）以及区段 4（水西门—集庆门），合计占比均超过了 20%，而区段 6（中华门—雨花门）最低，合计占比仅为 6.37%。

Summary of development intensity of the exterior space

Since the greenway is located in the old city of Nanjing, the surrounding area is mostly developed such that building density is high and plot ratio is at a moderate or low level due to limitation of most heights of the buildings. The total plot ratio of the districts where the six segments are located does not exceed 2. The plot ratio of the districts located on S2 DHG-CCG and S5 FTR-ZHG are lower than 1. The plot ratios of the district where the other four segments are located are relatively similar to one another, with the figure ranging from 1.2 to 1.5. Except in the cases of S2 DHG-CCG and S3 CCG-QLG, the building densities of the districts are all above 30%. The building density surrounding S5 FTR-ZHG is close to 40%.

In the study, the greenway is mostly surrounded by low-rise buildings. Except for in S2 DHG-CCG, the percentages of the low-rise buildings in the other segments all exceed 50%. A lot of traditional street blocks surround S6 ZHG-YHG, with the percentage of low-rise buildings even exceeding 80%. S2 DHG-CCG and S4 SXG-JQG have the highest sums of the combined percentages of mid-rise and high-rise buildings, which exceed 20% in total, while the combined mid- and high-rise percentage of 6.37% in S6 ZHG-YHG is the lowest.

外部区域总面积
Total area of the exterior space
外部建筑总面积
Overall floorage of the exterior space

486.40hm²
685.82hm²
406.91hm²
394.70hm²
587.30hm²
895.95hm²
533.30hm²
666.63hm²
597.02hm²
507.47hm²
610.10hm²
744.32hm²

S1 YJG-HYGG
S2 DHG-CCG
S3 CCG-QLG
S4 SXG-JQG
S5 FTR-ZHG
S6 ZHG-YHG

各区段外部区域总面积与建筑总面积

Total area and overall floorage of the exterior space

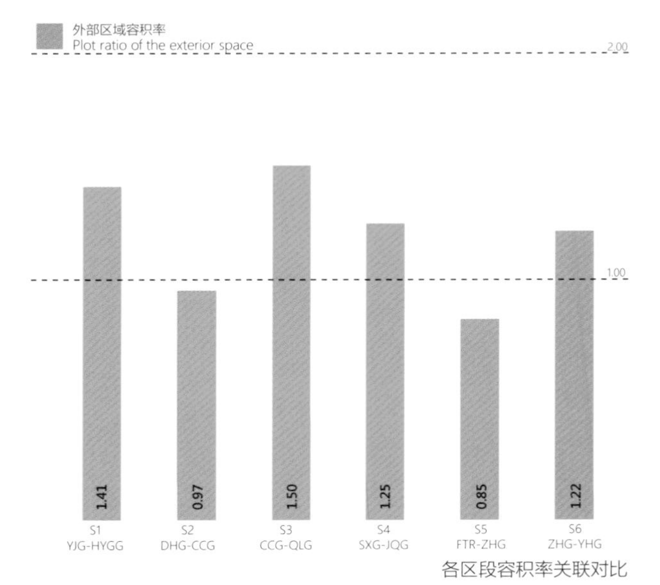

各区段容积率关联对比

Correlation and comparison of plot ratio

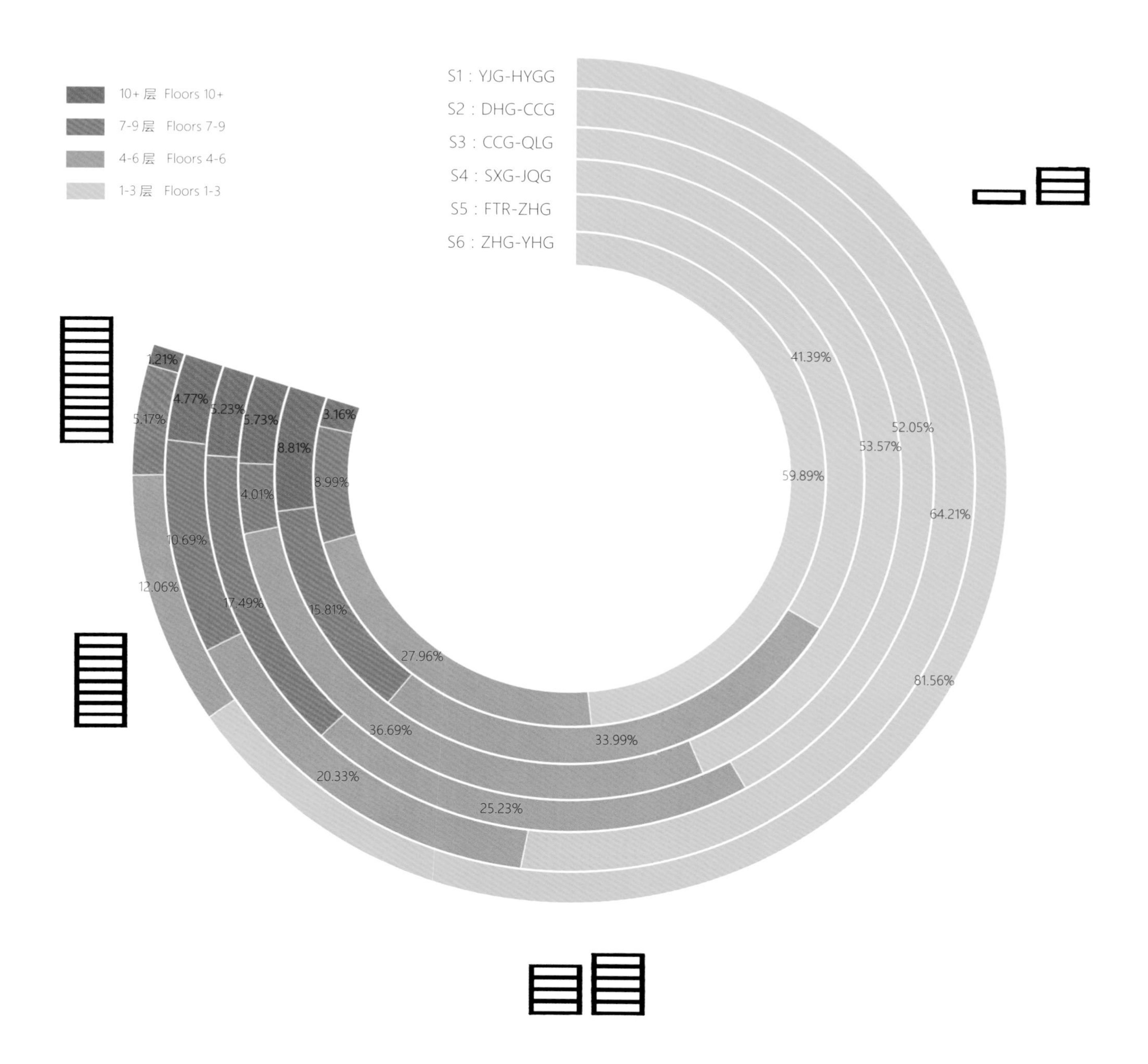
10+ 层 Floors 10+
7-9 层 Floors 7-9
4-6 层 Floors 4-6
1-3 层 Floors 1-3
S1：YJG-HYGG
S2：DHG-CCG
S3：CCG-QLG
S4：SXG-JQG
S5：FTR-ZHG
S6：ZHG-YHG
41.39%
52.05%
53.57%
59.89%
64.21%
81.56%
1.21%
4.77%
5.23%
5.73%
3.16%
5.17%
8.81%
8.99%
4.01%
10.69%
12.06%
17.49%
15.81%
27.96%
36.69%
33.99%
20.33%
25.23%

绿道外部空间结构简图 Study scope of the exterior space	绿道外部不同层高建筑占比 Percentages of floor space of buildings with different stories and rankings	绿道外部区域容积率以及 6 区段排序 Plot ratio of the study district and ranking	绿道外部区域建筑密度以及 6 区段排序 Density of buildings in the study district and ranking	绿道外部居住人口数量与密度及 6 区段排序 Number and density of residents in the study district and ranking
区段 1：挹江门—华严岗门 S1：YJG-HYGG	10+ 3.16% ⑤ 7-9 8.99% ④ 4-6 27.96% ③ 1-3 59.89% ③	1.41 ②	31.41%	8268 person 170.00 person/h
区段 2：定淮门—草场门 S2：DHG-CCG	10+ 8.81% ① 7-9 15.81% ② 4-6 33.99% ② 1-3 41.39% ⑥	0.97 ⑤	23.22%	36459 person 89.60 person/hm
区段 3：草场门—清凉门 S3：CCG-QLG	10+ 5.73% ③ 7-9 4.01% ⑥ 4-6 36.69% ① 1-3 53.57% ④	1.50 ①	27.28%	98704 person 165.25 person/h
区段 4：水西门—集庆门 S4：SXG-JQG	10+ 5.23% ④ 7-9 17.49% ① 4-6 25.23% ④ 1-3 52.05% ⑤	1.25 ③	33.37%	96654 person 181.24 person/h
区段 5：凤台路—中华门 S5：FTR-ZHG	10+ 4.77% ④ 7-9 10.69% ③ 4-6 20.33% ⑤ 1-3 64.21% ②	0.85 ⑥	39.44%	61775 person 103.47 person/h
区段 6：中华门—雨花门 S6：ZHG-YHG	10+ 1.21% ⑥ 7-9 5.17% ⑤ 4-6 12.06% ⑥ 1-3 81.56% ①	1.22 ④	34.40%	77779 person 127.49 person/

注：图表中研究区域居民规模指标是将居住用地上住宅总建筑面积除以 2015 年南京城镇常住居民人均住房建筑面积（36.5m²）获得（数据来源于《江苏统计年鉴 2016》）。

	绿道使用人群规模以及 6 区段排序 Number of greenway users and ranking	绿道使用人群密度以及 6 区段排序 Density of greenway users and ranking	不同时间段绿道使用人群密度以及 6 区段排序 Densities of greenway users in two different time periods and rankings
③ ②	551 person ①	423 person/km ①	14:00-16:00 458 person/km ② 18:00-20:00 388 person/km ①
⑥ ⑥	103 person ⑤	103 person/km ⑤	14:00-16:00 119 person/km ⑤ 18:00-20:00 87 person/km ⑤
① ③	499 person ②	332 person/km ②	14:00-16:00 487 person/km ① 18:00-20:00 178 person/km ④
② ①	78 person ⑥	64 person/km ⑥	14:00-16:00 53 person/km ⑥ 18:00-20:00 76 person/km ⑥
⑤ ⑤	251 person ④	179 person/km ④	14:00-16:00 141 person/km ④ 18:00-20:00 217 person/km ③
④ ④	347 person ③	315 person/km ③	14:00-16:00 289 person/km ③ 18:00-20:00 342 person/km ②

Note: The number of residential population index of the district in study on the diagram is calculated by dividing the overall floorage of residential buildings on residential land by per capita housing space of the permanent residents of the city of Nanjing in 2015 (36.5m^2) (data source: 2016 Jiangsu Statistical Yearbook).

=1hm²   =10 人 person

区段 4：水西门—集庆门 S1：SXG-JQG

147.00 person/hm²

区段 1：挹江门—华严岗门 S1：YJG-HYGG

137.89 person/hm²

区段 3：草场门—清凉门 S3：CCG-QLG

134.04 person/hm²

区段 6：中华门—雨花门 S6：ZHG-YHG

103.4 person/hm²

区段 5：凤台路—中华门 S5：FTR-ZHG

83.93 person/hm²

区段 2：定淮门—草场门 S2：DHG-CCG

72.67 person/hm²

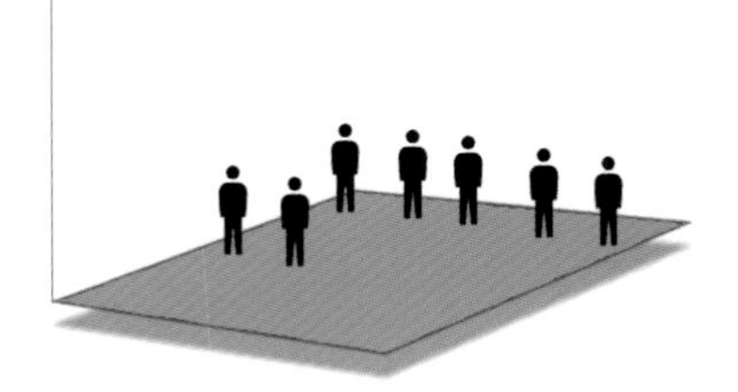

	绿道外部居住人口密度 Density of residents in the exterior space of the greenway	绿道使用人群密度（均值） Density of greenway users (Average)
区段 1：挹江门—华严岗门 S1：YJG-HYGG	170.00person/hm²	98person/hm² 423person/km
区段 2：定淮门—草场门 S2：DHG-CCG	89.60person/hm²	29person/hm² 103person/km
区段 3：草场门—清凉门 S3：CCG-QLG	165.25person/hm²	51person/hm² 332person/km
区段 4：水西门—集庆门 S4：SXG-JQG	181.24person/hm²	29person/hm² 64person/km
区段 5：凤台路—中华门 S5：FTR-ZHG	103.47person/hm²	61person/hm² 179person/km
区段 6：中华门—雨花门 S6：ZHG-YHG	127.49person/hm² 南京平均 Nanjing average 123 985 人 /hm²	79person/hm² 315person/km

外部空间开发强度与服务绩效

在理论上，绿道周边区域居民规模和密度越高，意味着绿道的潜在使用人群规模和强度越高。6个区段中，除区段4（水西门—集庆门）前后数据排位反差较大外，其他区段总体上仍体现了这一特点，但在少部分使用人群规模和密度指标较接近的相邻排位区段仍会出现与其周边居民规模和密度排位不完全一致的情况，例如，周边居民规模最高的区段3（草场门—清凉门）使用人群规模排位第二，而使用人群规模排位最高的区段1（挹江门—华严岗门）周边居民规模排位第三（除区段4则排位第二）。又如使用人群密度排位第二的区段6（中华门—雨花门）周边居民密度仅排位第四（除区段4则排位第三）。这种排位前后错位的现象实际上也反映出绿道使用人群规模与密度除了受周边居民规模、密度影响外，还很可能会受到其他诸如用地布局、交通条件、绿道自身空间形态、游憩资源吸引力等多方面因素的影响。

6个区段中，唯有区段4（水西门—集庆门）是较特殊的特例。该区段所在研究区域居民密度高居首位，居民规模也排序第二，但人群使用规模和使用密度指标均排位最后，前后指标出现强烈反差，反映出该区段绿道的使用受到了周边居民规模和密度以外其他因素的强烈影响。

Development intensity and service performance of the exterior space

In theory, a larger number of residents and higher density of residential population in the surroundings of the greenway can potentially lead to a larger number of potential users and higher intensity of the users of the greenway. Among the six segments, with the exception of S4 SXG-JQG, which has a sharp contrast between the previous and later data rankings, the other segments provide results consistent with the aforementioned theory. However, in some of the segments that have a similar number of users and user density, there is still inconsistency with the number of surrounding residents and the density of residential population. For example, S3 CCG-QLG has the biggest number of the surrounding residents, but the number of its users ranks second. However, S1 YJG-HYGG ranks first in terms of the number of users, while the number of its surrounding residents ranks third (if we ignore S4 SXG-JQG for discussion and individual analysis, it ranks the second). Another example is S6 ZHG-YHG. The user density ranks second, but the density of its surrounding residents ranks fourth (if one excludes S4 SXG-JQG, it ranks the third). The phenomenon of this unmatched ranking suggests that the number and density of greenway users is probably influenced by many other factors including the layout of land use, traffic conditions, the spatial structure of the greenway, the attraction of recreational resources and so on, in addition to the number and density of surrounding residents.

Among the six segments, S4 SXG-JQG is the only special case. The density of the residential population of the district where the segment is located ranks first, and the number of its residents ranks second, but the number of its users and the intensity of its usage both rank last. The sharp contrast between the previous and later indicators reflects that the use of the greenway on this segment is strongly influenced by factors beyond the number and density of surrounding residents.

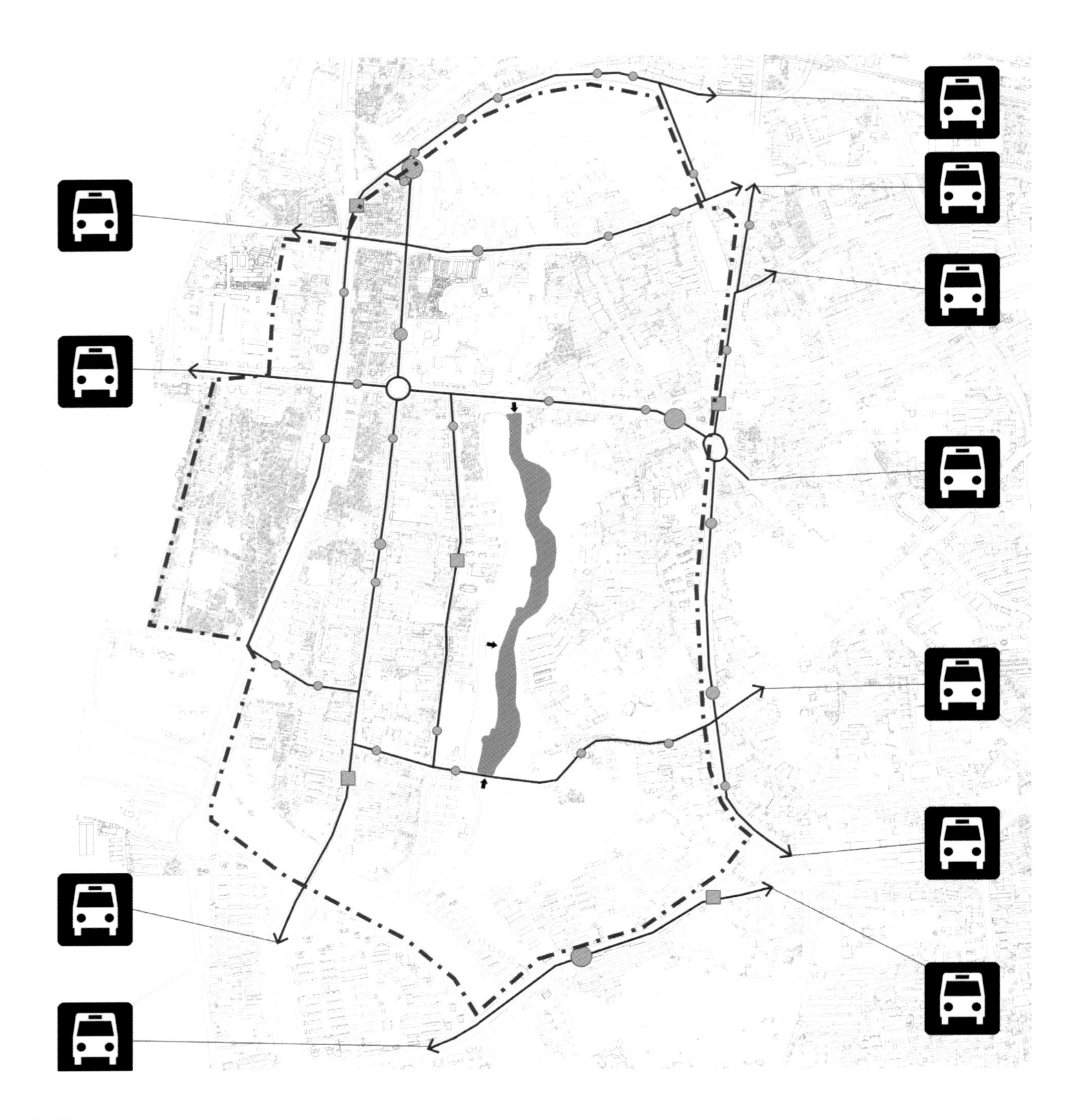

外部空间交通条件：区段1（挹江门—华严岗门）

该区段外部支路网发达，占据道路网总长度一半以上，快速路较少，仅有一小段位于绿道对岸以西约800m处，对绿道使用人群干扰有限。该区段两个出入口位于绿道南北两端与察哈尔路、中山北路两条城市干路相连处，中部鹊桥是该区段第三个出入口，与护城河中岛屿及对岸连通。区域内的干路上均布有公交站点，密度最高，达到8.22个/km，但整个区域尚无地铁站点。

Traffic conditions of the exterior space of S1（YYG - HYGG）

The external branch road network of this segment is fully developed, which covers more than half of the total length of the road network. It does not have many expressways. Only a small fraction of it is located about 800m on the west of the bank across the greenway, so it does not interfere too much with the accessibility of the greenway. The two entrances in this segment are located on the north and south ends and are connected to Chahar Road and North Zhongshan Road, two primary roads of the city. The Magpie Bridge in the middle is the third entrance of this segment and connected to the isles in the moat and the opposite side of the moat. The primary road of this district is dotted with bus stops, the density of which is the highest in six segments and reaches up to 8.22 per kilometer, but there are no metro stops in the entire district.

- - - - - - 地铁路线 Metro lines
——— 公共汽车路线 Bus lines
○ 公共汽车经停站 Bus stops
□ 公共汽车起终点站 Start and end of bus lines
⬭ 地铁站 Metro stations

数据统计表 Statistics of traffic conditions	
公共汽车站点密度 Density of bus stops	8.22 piece/km²
地铁站点密度 Density of metro stations	0 piece/km²
快速路密度 Density of urban expressways	0.31km/km²
次干道路密度 Density of primary roads	1.86km/km²
支路密度 Density of branch roads	2.63km/km²
总路网密度 Density of overall road network	4.80km/km²
绿道出入口密度 Density of entrances of the greenway	2.31 piece/km

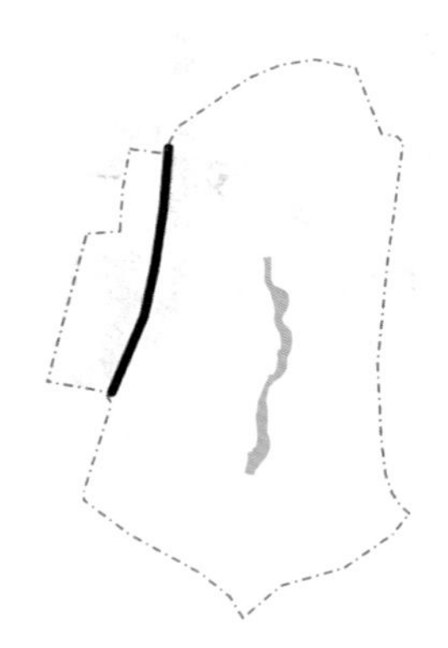

城市快速路
Expressways

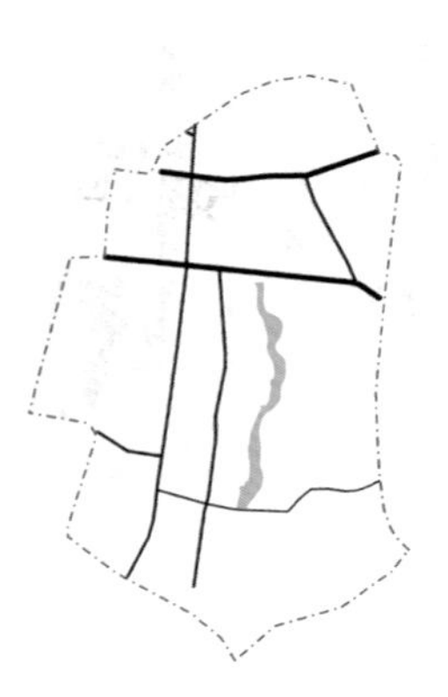

城市主次干路
Primary roads

城市支路
Branch roads

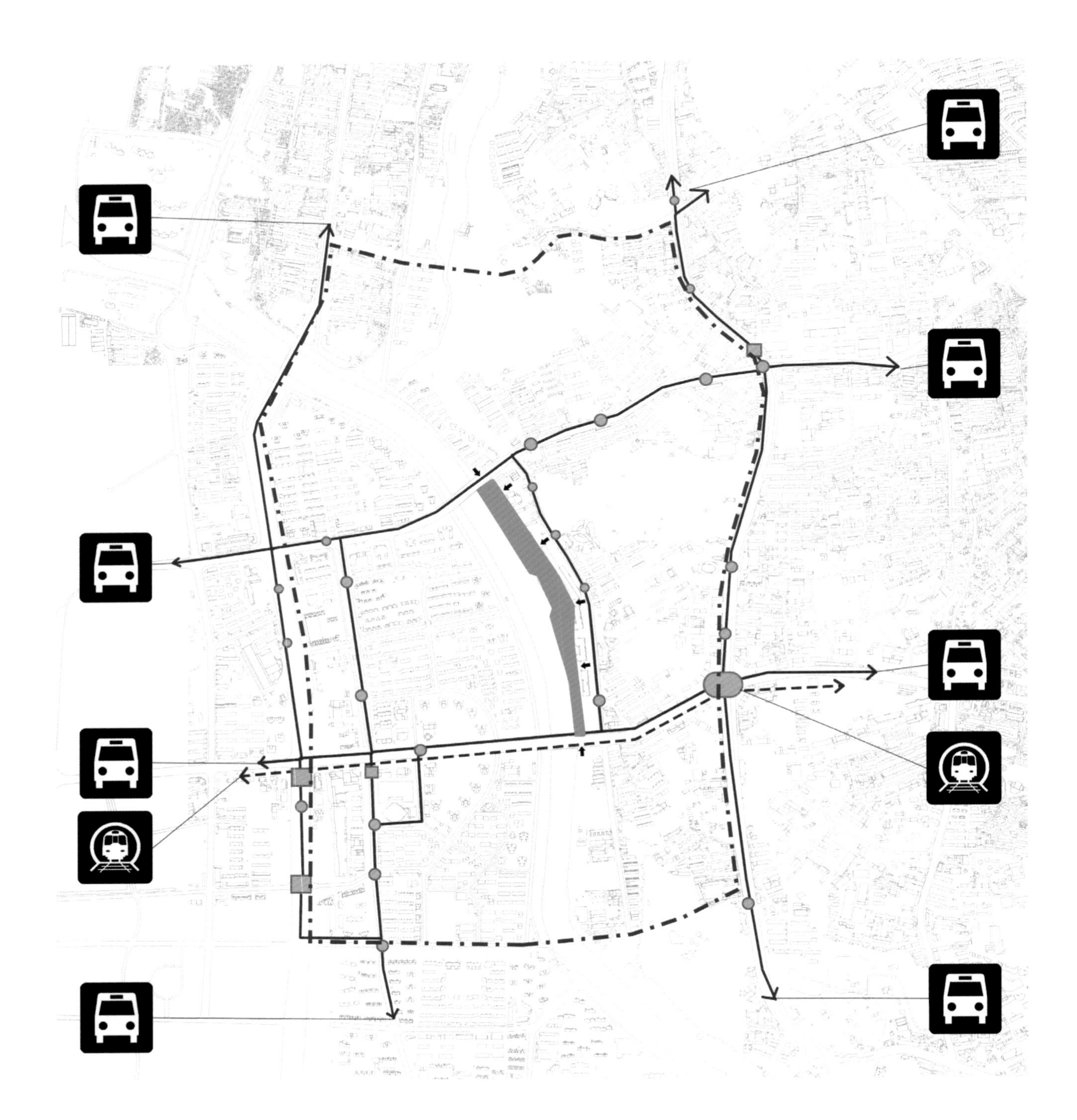

外部空间交通条件：区段 2（定淮门—草场门）

该区段外部干路较发达，接近整个道路网总长一半。研究区域东侧边界是城市快速路虎踞北路。区段南北分别与城市主干路草场门大街和定淮门大街相交，东面石头城路方向均匀分布 4 个出入口，但由于东面古林公园和南京艺术学院地块较大，因此并没有东西向路网与该路相连，该路上的人群主要仍是从南北方向导入，从一定程度上弱化了 4 个出入口的作用，在区段中部有一处码头，但平常较少有游船在此停靠。该区段周边 1 个地铁站。

Traffic conditions of the exterior space of S2 (DHG - CCG)

The external branch road network of this segment is developed, covering nearly half of the entire road network. The edge of the east side of the district in study is North Huju Road, an urban expressway. The south and east of the segment intersect, respectively, with Caochangmen Avenue and Dinghuaimen Avenue, two primary roads of the city. Four entrances are evenly distributed along Stone City Road, which is to the east. However, due to the superblock of Gulin Park and Nanjing University of the Arts to the east, there is no road network from east to west connected to this road. The crowds on this road are mainly led in from north and south, which significantly weakens the role of the four entrances along Stone City Road. There is a wharf in the middle of this segment, but few recreational boats stop here. The primary roads surrounding this segment are equipped with one metro station.

- - - - - - 地铁路线 Metro lines
—— 公共汽车路线 Bus lines
● 公共汽车经停站 Bus stops
■ 公共汽车起终点站 Start and end of bus lines
⬭ 地铁站 Metro stations

数据统计表 Statistics of traffic conditions	
公共汽车站点密度 Density of bus stops	6.64 piece/km^2
地铁站点密度 Density of metro stations	0.25 piece/km^2
快速路密度 Density of urban expressways	0.63km/km^2
次干道路密度 Density of primary roads	2.23km/km^2
支路密度 Density of branch roads	1.84km/km^2
总路网密度 Density of overall road network	4.70km/km^2
绿道出入口密度 Density of entrances of the greenway	6.00 piece/km

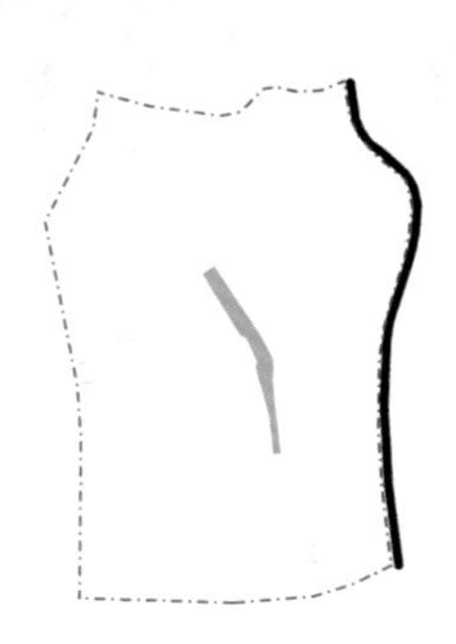

城市快速路
Expressways

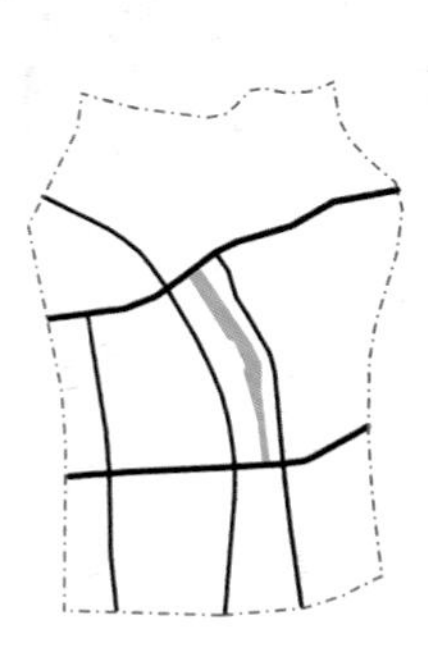

城市主次干路
Primary roads

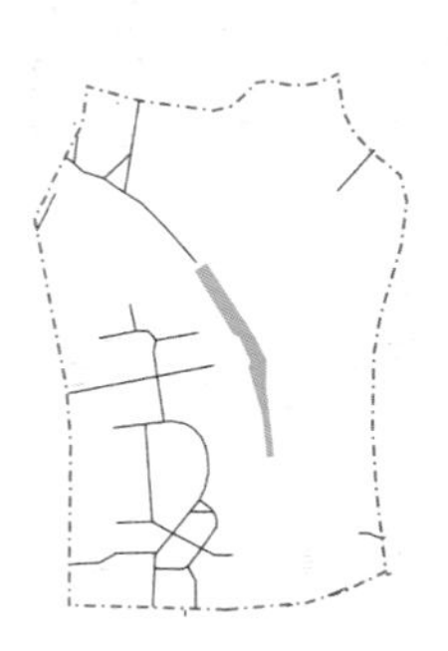

城市支路
Branch roads

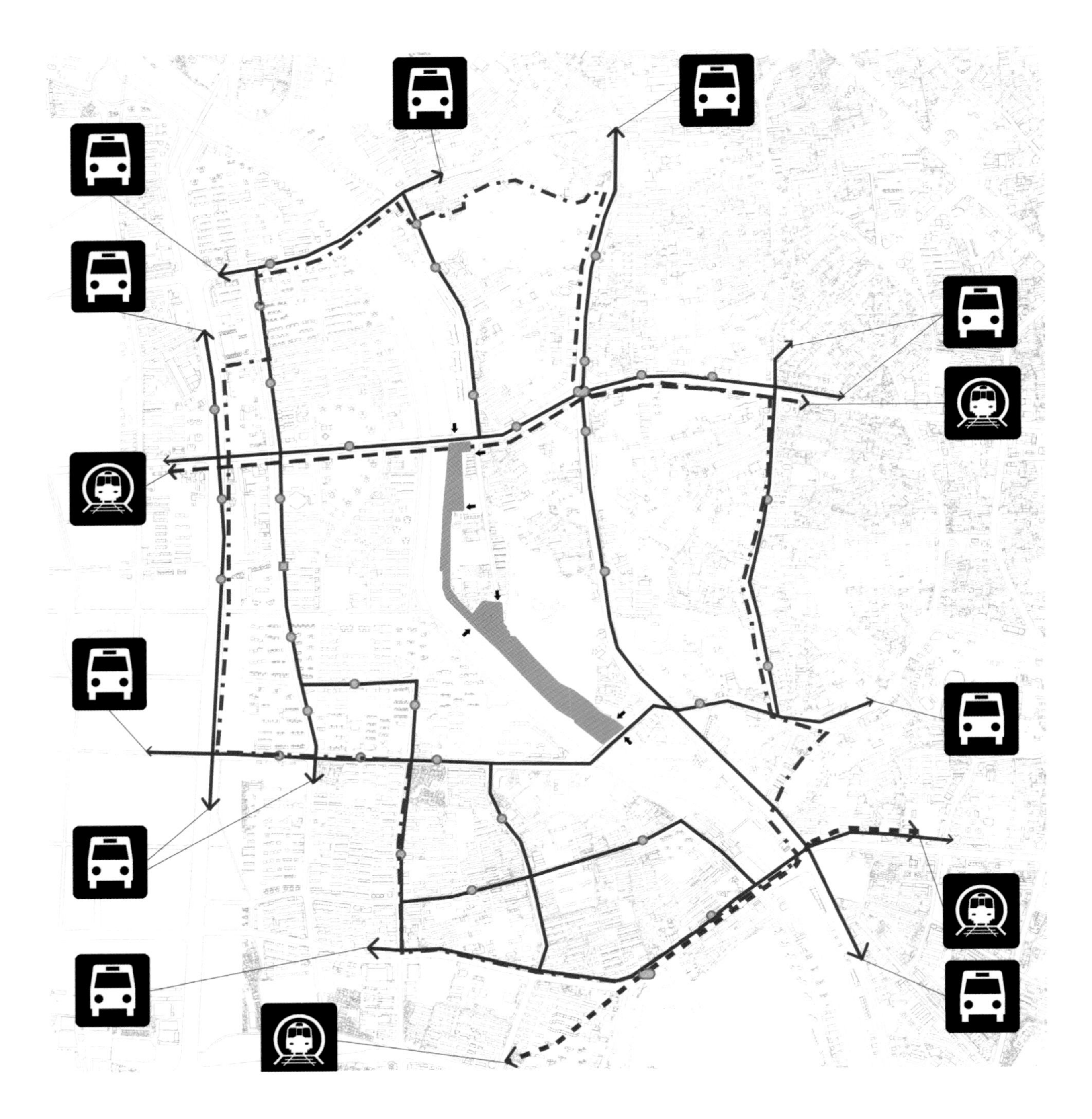

外部空间交通条件：区段 3（草场门—清凉门）

该区段外部干路比重最高，超过路网总长一半，其东侧虎踞路距绿道较近，并对外侧使用人群的进入形成一定阻隔。区段南北分别与城市主干路广州路和草场门大街相接，沿石头城路方向还有 2 个出入口。但与区段 2（定淮门—草场门）类似，由于没有东西向道路相连，因此人流仍主要从南北方向导入。区段中部和南部各有 1 个码头，并会有一些游船在此停靠。该区段周边有 2 个地铁站，公共交通设施条件较好。

Traffic conditions of the exterior space of S3 (CCG - QLG)

The percentage of the external primary roads on this segment is the largest, with the length of those roads exceeding half of the total length of the road network. Huju Road on to the east is close to the greenway and considerably obstructs the entrance for greenway users. The north and south of this segment are respectively connected to Guangzhou Road and Caochangmen Avenue, two primary roads of the city. Two entrances are located along Shitoucheng Road. However, similar to S2 DHG-CCG, since this segment is not connected to roads from east to west, streams of people are mainly led in from the north and south. Both the middle and southern parts of the segment are equipped with a wharf, with some recreational boats docking there. The surrounding area of this segment has two metro stations.

地铁路线 Metro lines

公共汽车路线 Bus lines

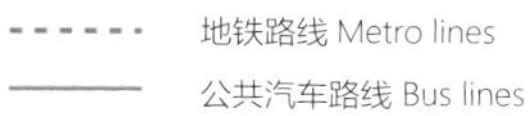

公共汽车经停站 Bus stops

公共汽车起终点站 Start and end of bus lines

地铁站 Metro stations

数据统计表 Statistics of traffic conditions	
公共汽车站点密度 Density of bus stops	5.69 piece/km^2
地铁站点密度 Density of metro stations	0.17 piece/km^2
快速路密度 Density of urban expressways	0.35km/km^2
次干道路密度 Density of primary roads	2.12km/km^2
支路密度 Density of branch roads	1.46km/km^2
总路网密度 Density of overall road network	3.93km/km^2
绿道出入口密度 Density of entrances of the greenway	4.67 piece/km

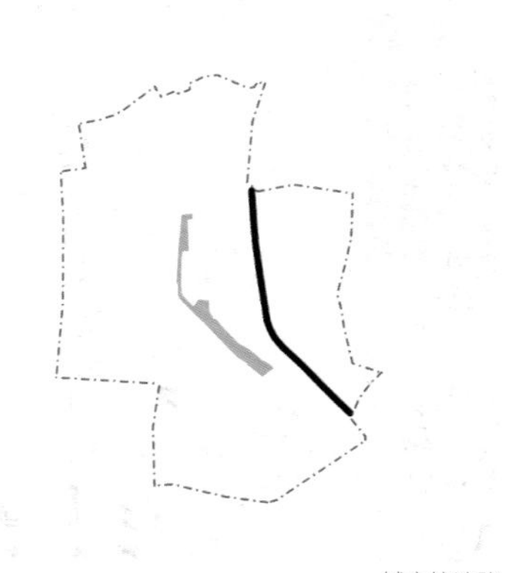

城市快速路
Expressways

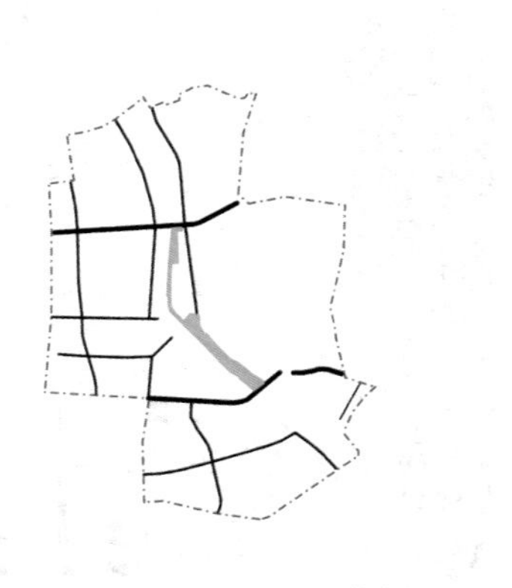

城市主次干路
Primary roads

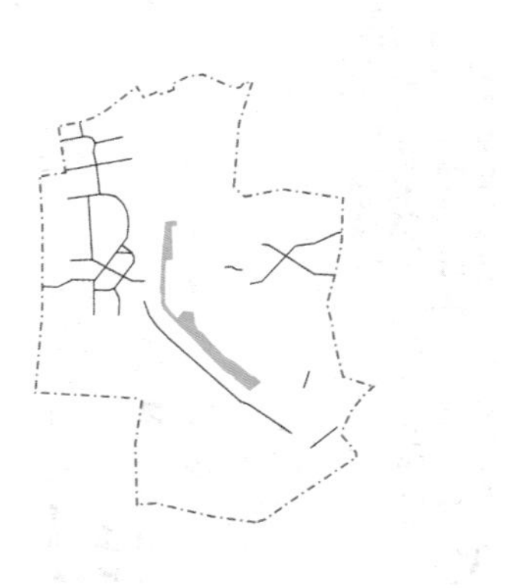

城市支路
Branch roads

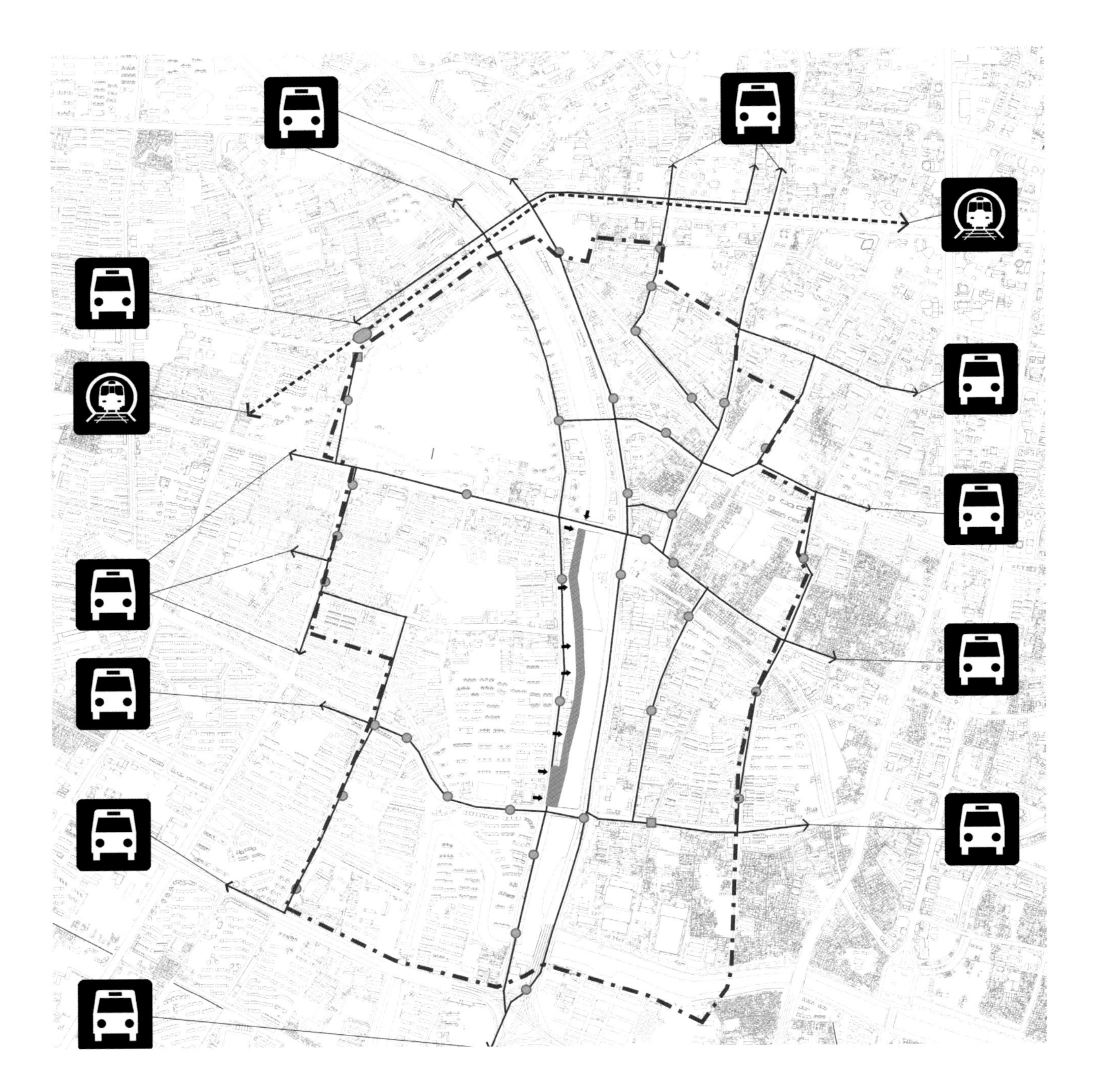

外部空间交通条件：区段 4(水西门—集庆门)

该区段外部交通网络中干路和支路网络密度接近，均在 40% 以上。区段东面除秦淮河外，河东岸滨河地段的城市快速路凤台路也会对东面使用人群的进入造成一定阻隔。区段南北分别与水西门大街和集庆路两条城市主干路相交，西侧沿长虹路另有 5 个出入口。长虹路中段有东西向的南湖中路相连，能从西面引入使用人群，因而该侧可达性相对较好。该区段周边西北有 1 个地铁站。

Traffic conditions of the exterior space of S4 (SXG - JQG)

In the external traffic network of this segment, the density of primary roads is close to that of branch roads, both of which are more than 40%. To the east of this segment, Fengtai Road, an urban expressway at the waterfront of the east of the river, also considerably obstructs the access of the users from the east side to this segment. The north and south of the segment respectively intersect with Shuiximen Avenue and Jiqing Road, two primary roads of the city. The west side of the segment has five entrances along Changhong Road. The middle of Changhong Road is connected to Mid Nanhu Road, which spreads from east to west, so it can lead in users from the west side, enhancing the accessibility of this side to a large extent. The surrounding area of this segment has one metro station in the northwest.

- - - - - - 地铁路线 Metro lines
——— 公共汽车路线 Bus lines
● 公共汽车经停站 Bus stops
■ 公共汽车起终点站 Start and end of bus lines
⬭ 地铁站 Metro stations

数据统计表 Statistics of traffic conditions	
公共汽车站点密度 Density of bus stops	7.31 piece/km²
地铁站点密度 Density ofmetro stations	0.19 piece/km²
快速路密度 Density of urban expressways	0.59km/km²
次干道路密度 Density of primary roads	2.26km/km²
支路密度 Density of branch roads	1.94km/km²
总路网密度 Density of overall road network	4.79km/km²
绿道出入口密度 Density of entrances of the greenway	6.67 piece/km

城市快速路
Expressways

城市主次干路
Primary roads

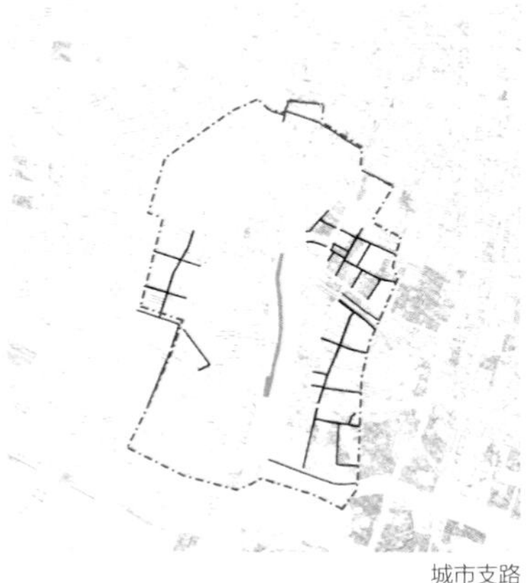
城市支路
Branch roads

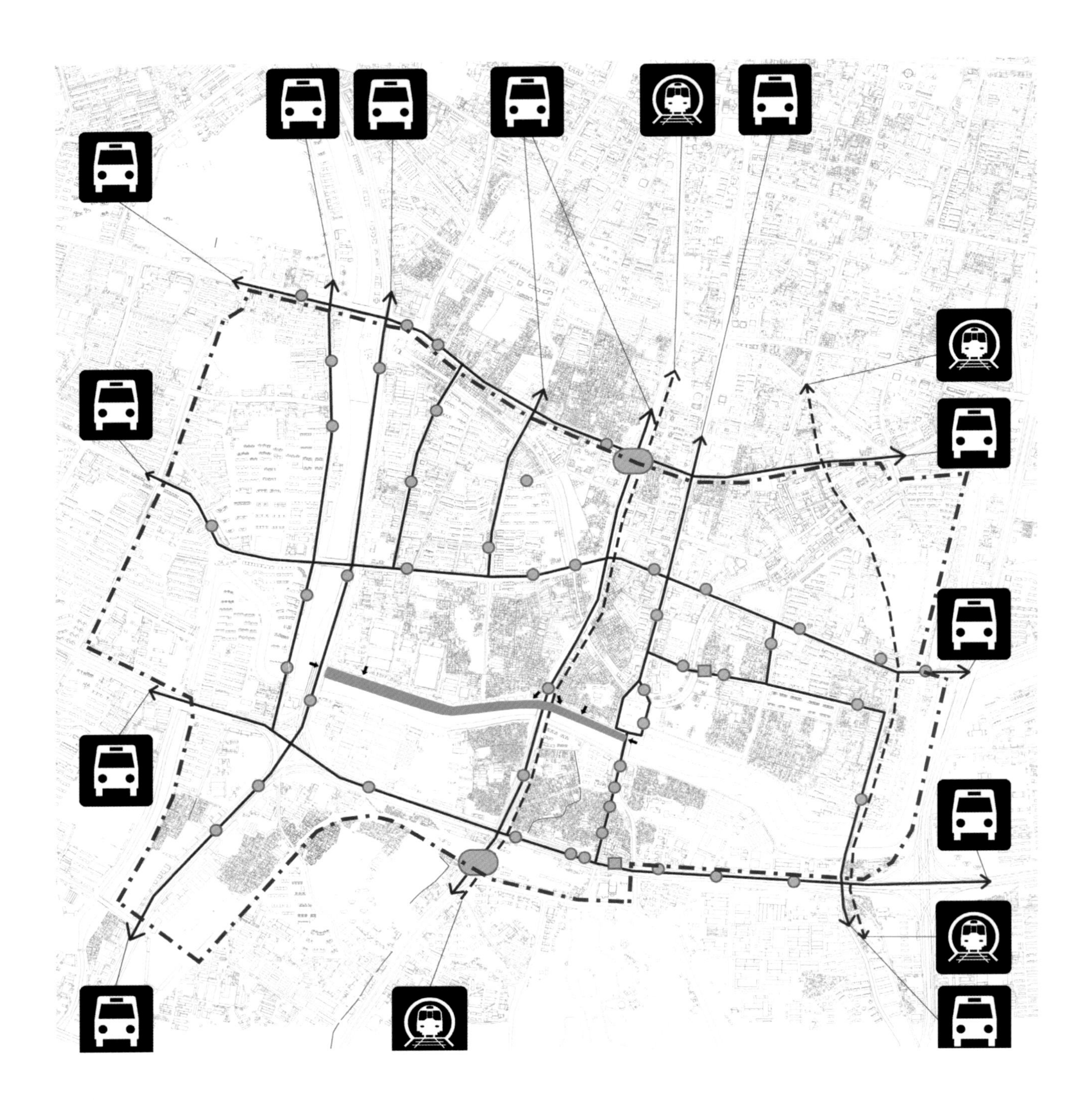

外部空间交通条件：区段 5（凤台路—中华门）

该区段外部空间支路占到道路网络总长近一半，区域内快速路为同属内环线的凤台路（西侧）和秦淮河对岸的应天大街（南侧）。区段东西分别与快速路凤台路和主干路中华路相交，沿西干长巷区段北面共有 4 个出入口，其中与西干长巷相交的中山南路能从南北方向引入城墙内部使用人群。区段周边布有 2 个地铁站。

Traffic conditions of the exterior space of S5 (FTR - ZHG)

The branch roads of the exterior space of this segment cover nearly half of the total length of the road network. The expressways of this segment are Fengtai Road (on the west side) and Yingtian Avenue (on the south side) across the Qinhuai River, both of which are part of inner-ring roads of the city. The east and west of the segment respectively intersect with Fengtai Road, an expressway, and Zhonghua Road, a primary road. The north side of the segment along Xiganchangxiang has four entrances. South Zhongshan Road, which intersects with Xiganchangxiang Road, can lead in users inside the City Wall from the north to south. The surrounding area of the segment locates the second densest bus stops among 6 segments and two metro stations.

地铁路线 Metro lines
公共汽车路线 Bus lines
公共汽车经停站 Bus stops
公共汽车起终点站 Start and end of bus lines
地铁站 Metro stations

数据统计表 Statistics of traffic conditions	
公共汽车站点密度 Density of bus stops	8.21 piece/km^2
地铁站点密度 Density of metro stations	0.33 piece/km^2
快速路密度 Density of urban expressways	0.90km/km^2
次干道路密度 Density of primary roads	3.34km/km^2
支路密度 Density of branch roads	4.22km/km^2
总路网密度 Density of overall road network	8.46km/km^2
绿道出入口密度 Density of entrances of the greenway	4.29 piece/km

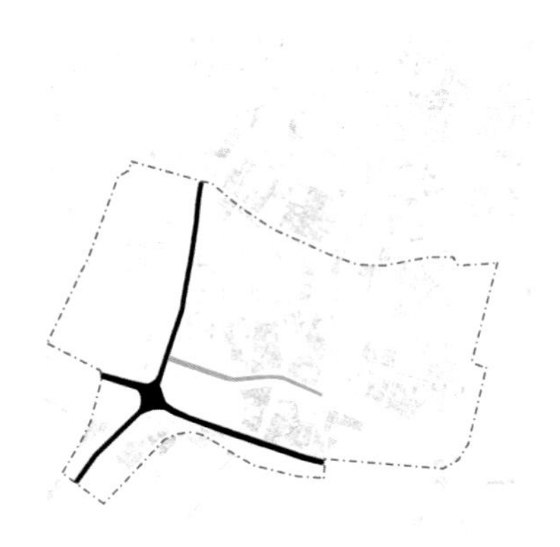
城市快速路
Expressways

城市主次干路
Primary roads

城市支路
Branch roads

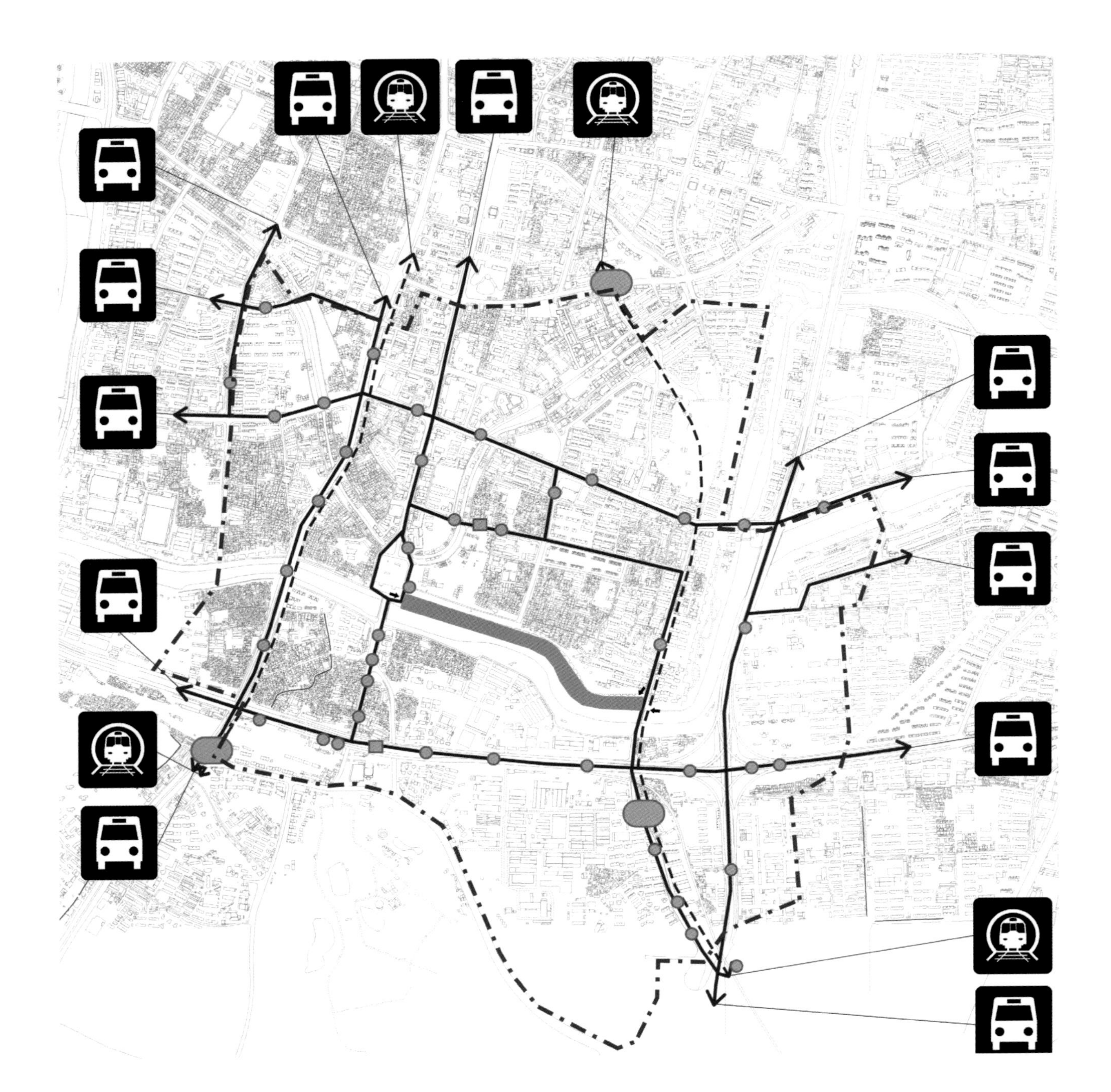

外部空间交通条件：区段 6（中华门—雨花门）

与区段 5（凤台路—中华门）相似，该区段所在区域支路密布，也超过总长一半。研究区域内快速路为南部应天大街和东部龙蟠南路。区段东西分别与中华路和江宁路两条城市干路相接，该区段北面紧靠城墙，没有出入口，仅在东面有一处出入口与城市道路江宁路相连，因此区段人群主要从东西方向引入。该区段周边有多达 3 个地铁站。

城市快速路
Expressways

Traffic conditions of the exterior space of S6 (ZHG - YHG)

Similar to S5 FTR-ZHG, the district where this segment is located is densely equipped with branch roads that cover more than half of the total length. The expressways in the district are Yingtian Avenue to the south and South Longpan Road to the east. The east and west of the segment are respectively connected to Zhonghua Road and Jiangning Road, two primary roads of the city. The north side of this segment adjoins to the City Wall without any entrance. There is only one entrance on its east side connected to Jiangning Road, an urban road. Therefore, the crowds in this segment are mainly led in from east and west. The surrounding area of this segment has three metro stations.

城市主次干路
Primary roads

- - - - - - 地铁路线 Metro lines
——— 公共汽车路线 Bus lines
● 公共汽车经停站 Bus stops
■ 公共汽车起终点站 Start and end of bus lines
⬭ 地铁站 Metro stations

数据统计表 Statistics of traffic conditions	
公共汽车站点密度 Density of bus stops	7.05 piece/km^2
地铁站点密度 Density of metro stations	0.49 piece/km^2
快速路密度 Density of urban expressways	0.69km/km^2
次干道路密度 Density of primary roads	2.30km/km^2
支路密度 Density of branch roads	3.39km/km^2
总路网密度 Density of overall road network	6.38km/km^2
绿道出入口密度 Density of entrances of the greenway	2.73 piece/km

城市支路
Branch roads

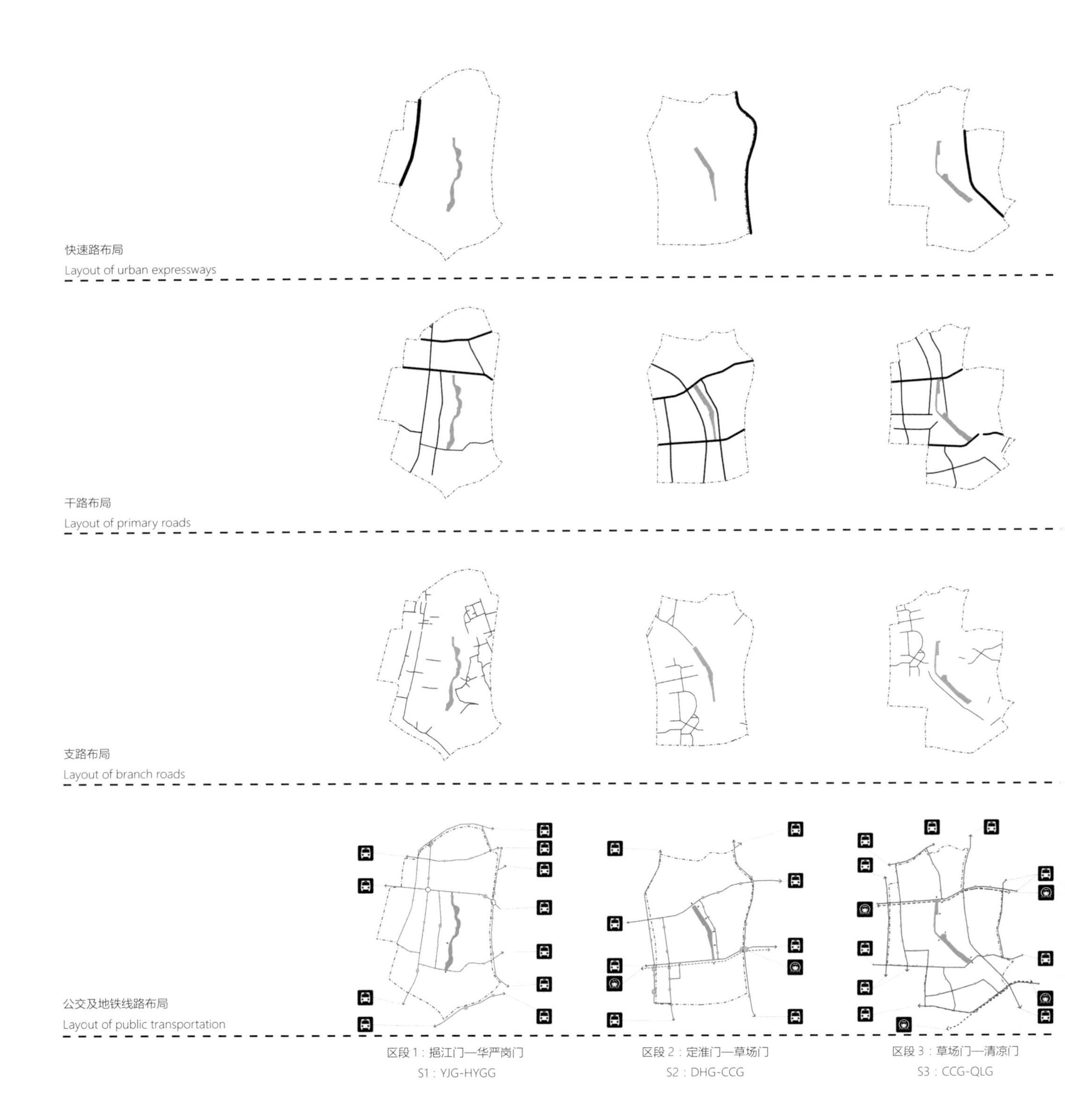
快速路布局
Layout of urban expressways
干路布局
Layout of primary roads
支路布局
Layout of branch roads
公交及地铁线路布局
Layout of public transportation
区段 1：挹江门—华严岗门
S1：YJG-HYGG
区段 2：定淮门—草场门
S2：DHG-CCG
区段 3：草场门—清凉门
S3：CCG-QLG

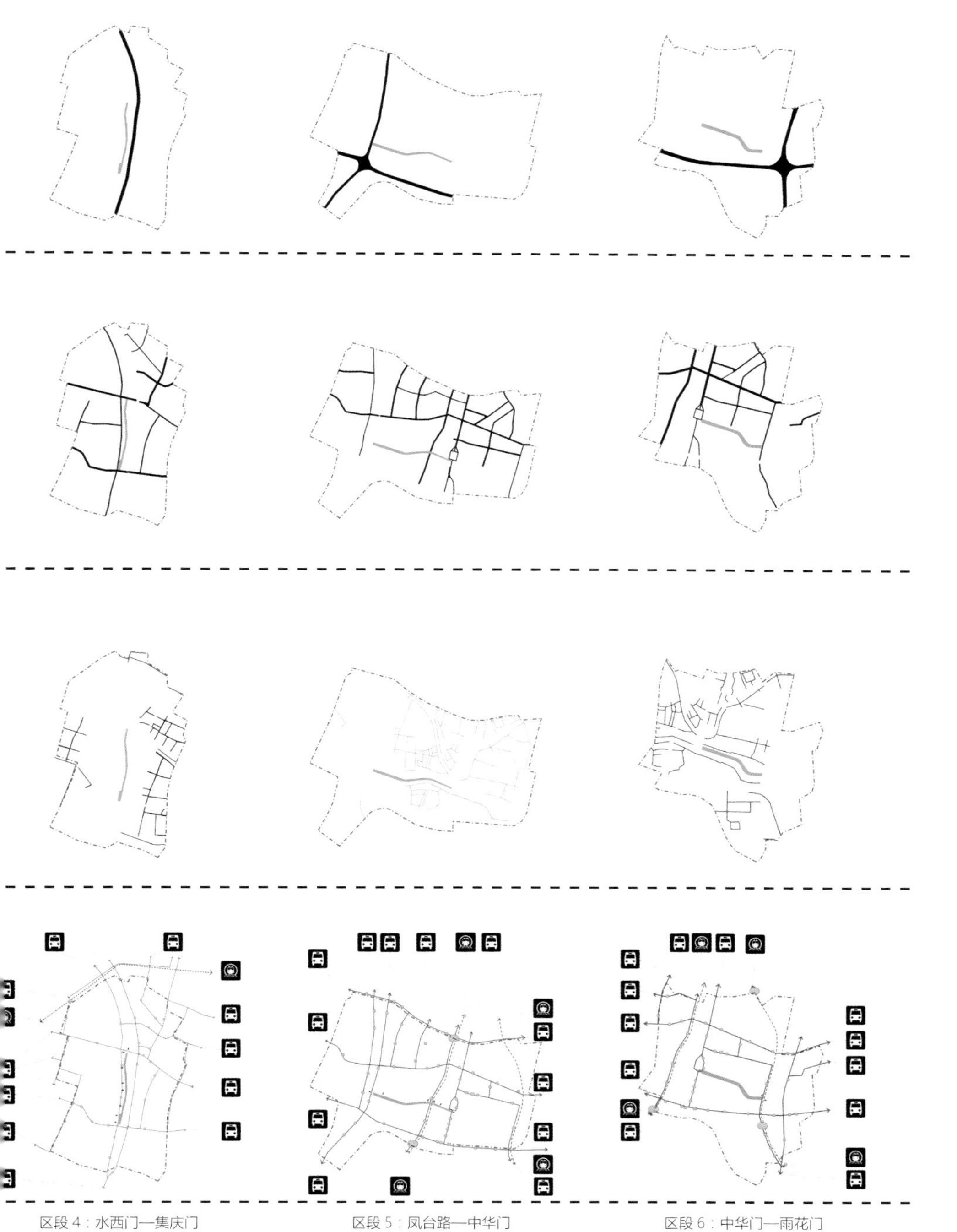

各级路网及公交与地铁路线布局
Layout of urban roads and public transportation

0.31
1.86
2.63
0
8.22
2.31
区段 1：挹江门—华严岗门
S1：YJG-HYGG
0.63
2.23
1.84
0.25
6.64
6.00
区段 2：定淮门—草场门
S2：DHG-CCG
0.35
2.12
1.46
0.17
5.69
4.67
区段 3：草场门—清凉门
S3：CCG-QLG
0.59
2.26
1.94
0.19
7.31
6.69
区段 4：水西门—集庆门
S4：SXG-JQG
0.90
3.34
4.22
0.33
8.21
4.29
区段 5：凤台路—中华门
S5：FTR-ZHG
0.69
2.30
3.39
0.49
7.05
2.73
区段 6：中华门—雨花门
S6：ZHG-YHG
快速路密度（单位：km/km²）
Density of urban expressways(Unit：km/km²）
干路密度（单位：km/km²）
Density of primary roads （Unit：km/km²）
支路密度（单位：km/km²）
Density of branch roads （Unit：km/km²）
地铁站点密度（单位：个 /km²）
Density ofmetro stations(Unit：piece/km²）
公交汽车站点密度（单位：个 /km²）
Density of bus stops（Unit：piece/km²）
绿道出入口密度（单位：个 /km）
Density of entrances of the greenway（Unit：piece/km）

外部空间交通条件小结

6个区段所在区域，区段6（中华门—雨花门）和区段5（凤台路—中华门）周边路网密度较高，均超过6km/km²，其中区段5（凤台路—中华路）周边路网密度最高达8.46km/km²，其他4个区段周边路网密度值均在3.9-4.8km/km²范围内。路网结构中，支路占比接近或超过一半的有区段1（挹江门—华严岗门）、区段6（中华门—雨花门）和区段5（凤台路—中华门）3个区段所在区域，该3个区段周边支路密度也均超过2.5km/km²，区段5（凤台路—中华门）周边支路密度高达4.22km/km²，其余3个区段周边干路占比均在支路占比之上。另外，除区段1（挹江门—华严岗门）和区段2（定淮门—草场门）所在区域，其他4个区段外部空间快速路占比均在10%以上。

公交站密度高于8个/km²的有区段1（挹江门—华严岗门）和区段5（凤台路—中华门）2个区段所在区域，区段3（草场门—清凉门）周边是唯一公交站密度低于6个/km²的区段。6个区段，除区段1（挹江门—华严岗门）周边无地铁站分布外，其他5个区段周边均布有地铁站，其中区段6（中华门—雨花门）周边地铁站密度最高，达到0.49个/km²。

绿道区段的出入口密度差距较大，其中密度最高的区段2（定淮门—草场门）和区段4（水西门—集庆门）密度均在6个/km以上，而密度较低的区段1（挹江门—华严岗门）和区段6（中华门—雨花门）均低于3个/km。

Summary of traffic conditions of the exterior space

Among the districts where the six segments are located, the densities of surrounding road networks of S6 ZHG-YHG and S5 FTR-ZHG are high, exceeding 6km/km². The density of the road network surrounding S5 FTR-ZHG is the highest, which reaches up to 8.46km/km², while the densities of road networks surrounding the other four segments range from 3.9-4.8km/km². In the structure of the road networks, the percentages of branch roads of the districts where S1 YJG-HYGG, S6 ZHG-YHG, and S5 FTR-ZHG are located near or exceed half of the total network, with the branch road density exceeding 2.5km/km². Among them, the density of the external branch road of S5 FTR-ZHG is the highest and reaches up to 4.22km/km². For the remaining other segments, the percentages of surrounding primary roads are all higher than those of surrounding branch roads. Aside from the districts where S1 YJG-HYGG and S2 DHG-CCG are located, the percentages of expressways of the exterior space the segments are all higher than 10%.

The densities of bus stops in the districts where S1 YJG-HYGG and S5 FTR-ZHG are located exceed eight per square kilometer. S3 CCG-QLG is the only segment with a bus stop density below 6 per square kilometer. Among the six segments, except for the exterior space of S1 YJG-HYGG, which does not have a metro station, the density of surrounding metro stations of S6 ZHG-YHG is the highest, at 0.49 per square kilometer.

The segments of the greenway varies a lot in terms of entrance density. S2 DHG-CCG and S4 SXG-JQG have the highest densities, both of which are fewer than six per kilometer, while S1 YJG-HYGG and S6 ZHG-YHG have low densities, both of which are no more than three per kilometer.

绿道外部空间结构简图 Study scope of the exterior space	区域各级路网密度以及 6 段排序 Densities of different levels of roads and rankings in the study district	区域各级路网比例以及 6 段排序 Percentages of different levels of roads and rankings in the study district	区域公共交通网点密度、绿道出入口密度以及 6 段排序 Density of public transportation network ; Density of entrance-exits of the greenway and ranking
区段 1：挹江门—华严岗门 S1 : YJG-HYGG	快速路 Expressways 0.31km/km^2 ⑥ 干路 Primary roads 1.86km/km^2 ⑥ 支路 Branch roads 2.63km/km^2 ③	6.47% ⑥ 38.79% ⑤ 54.74% ①	8.22 piece/km^2 0 piece/km^2 0 piece/km^2 2.31 piece/km
区段 2：定淮门—草场门 S2 : DHG-CCG	快速路 Expressways 0.63km/km^2 ③ 干路 Primary roads 2.23km/km^2 ④ 支路 Branch roads 1.84km/km^2 ⑤	13.47% ① 47.29% ② 39.24% ⑤	6.64 piece/km^2 0.25 piece/km^2 0.12 piece/km^2 6.00 piece/km
区段 3：草场门—清凉门 S3 : CCG-QLG	快速路 Expressways 0.35km/km^2 ⑤ 干路 Primary roads 2.12km/km^2 ⑤ 支路 Branch roads 1.46km/km^2 ⑥	8.89% ⑤ 53.97% ① 37.14% ⑥	5.69 piece/km^2 0.17 piece/km^2 017 piece/km^2 4.67 piece/km
区段 4：水西门—集庆门 S4 : SXG-JQG	快速路 Expressways 0.59km/km^2 ④ 干路 Primary roads 2.26km/km^2 ③ 支路 Branch roads 1.94km/km^2 ④	12.30% ② 47.12% ③ 40.58% ④	7.31 piece/km^2 0.19 piece/km^2 0 piece/km^2 6.67 piece/km
区段 5：凤台路—中华门 S5 : FTR-ZHG	快速路 Expressways 0.90km/km^2 ① 干路 Primary roads 3.34km/km^2 ① 支路 Branch roads 4.22km/km^2 ①	10.68% ④ 39.46% ④ 49.87% ③	8.21 piece/km^2 0.33 piece/km^2 0 piece/km^2 4.29 piece/km
区段 6：中华门—雨花门 S6 : ZHG-YHG	快速路 Expressways 0.69km/km^2 ② 干路 Primary roads 2.30km/km^2 ② 支路 Branch roads 3.39km/km^2 ②	10.79% ③ 36.06% ⑥ 53.14% ②	7.05 piece/km^2 0.49 piece/km^2 0 piece/km^2 2.73 piece/km

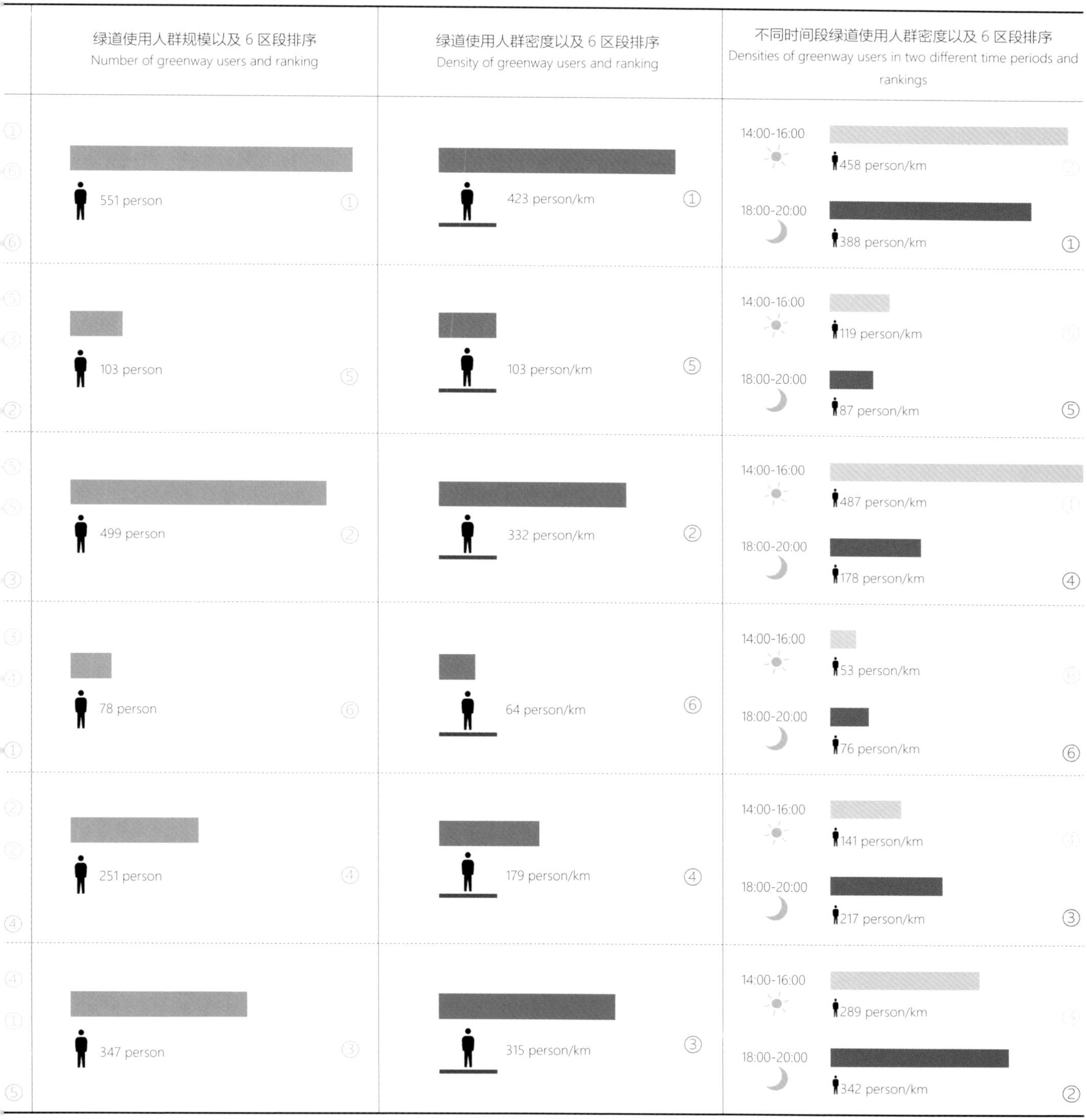

绿道使用人群规模以及 6 区段排序 Number of greenway users and ranking		绿道使用人群密度以及 6 区段排序 Density of greenway users and ranking		不同时间段绿道使用人群密度以及 6 区段排序 Densities of greenway users in two different time periods and rankings			
				14:00-16:00		18:00-20:00	
551 person	①	423 person/km	①	458 person/km	②	388 person/km	①
103 person	⑤	103 person/km	⑤	119 person/km	⑤	87 person/km	⑤
499 person	②	332 person/km	②	487 person/km	①	178 person/km	④
78 person	⑥	64 person/km	⑥	53 person/km	⑥	76 person/km	⑥
251 person	④	179 person/km	④	141 person/km	④	217 person/km	③
347 person	③	315 person/km	③	289 person/km	③	342 person/km	②

外部空间交通条件与服务绩效

6个区段中，使用人群密度最高的区段1（挹江门—华严岗门）其外部交通条件中支路占比在6个区段中排位第一，支路密度排位第三，快速路占比最低，同时公交站点密度排位也居首位。使用人群密度排位第二的区段3（草场门—清凉门），则干路占比最高，但支路占比和密度指标排位最后。使用人群密度排位最后的区段4（水西门—集庆门）和区段2（定淮门—草场门）快速路占比分列前两位，支路比重和密度则分别位居第四和第五位。

另一方面，使用人群密度最高的区段1（挹江门—华严岗门）的绿道出入口密度指标排位最后，而使用人群密度最低的区段4（水西门—集庆门）的绿道出入口密度指标则排位第一。可见在外部空间条件没有达到最佳状态时，绿道出入口的数量多少实际上很难对其绿道服务绩效产生决定性的影响。

同样，将公交和地铁站点密度指标与绿道使用密度指标对比，也很难发现有明显关联。实际上，由于明城墙绿道主要是沿线性历史遗存布局的绿道，其并没有直接穿越城市其他功能区，因此绿道游憩功能要远大于交通功能。绿道空间载体基本为明城墙沿线绿地，并未直接串联公交与地铁站点，因此不存在太多与公交、地铁站点的接驳问题，这也弱化了公交和地铁站点对绿道使用人群的导入作用。而另一方面，就目前明城墙绿道的外来游客或市民群体而言，乘坐公共交通工具来专门使用绿道的人群比重也较低，这也进一步削弱了绿道周边公交和地铁站点密度对其使用强度的影响。

External traffic conditions and service performance

Among the six segments, S1 YJG-HYGG has the highest density of users, the percentage of the branch roads among its external traffic conditions ranks first, and the density of its branch roads ranks third. The percentage of its expressways is the smallest. At the same time, the density of its bus stops also ranks first. The density of the users on S3 CCG-QLG ranks second. The percentage of its primary roads is the highest, but the percentage of its branch roads and its density index rank last. The densities of the users in S4 SXG-JQG and S2 DHG-CCG rank last. The percentages of their expressways are the greatest, with the percentages of their branch roads and densities ranking fourth and the fifth, respectively.

On the other hand, S1 YJG-HYGG has the highest density of users. The density index of the entrances in this greenway segment ranks last, while that on S4 SXG-JQG, which has the lowest user density ranks first. It can be seen that when the exterior space is not in best possible condition, it is difficult for the number of entrances to the greenway to influence the service performance of the greenway.

Similarly, after comparing the density index of bus stops and metro stations with the greenway's, there is no noticeable connection. In fact, since the Nanjing Ming Dynasty City Wall Greenway is mainly installed along the historic city sites and does not go directly through other functional zones of the city, the recreational function of the greenway is greater than its traffic function. The spatial carrier of the greenway is basically the green space along the Ming Dynasty City Wall without any direct connection to bus stops or metro stations, so there are no problems of connection to bus stops or metro stations, which also lessens the function of bus stops and metro stations to lead in greenway users. On the other hand, in terms of visitors and residents of the Nanjing Ming Dynasty City Wall Greenway at present, the percentage of those who take public transportation just to use the greenway is not big, which also further weakens the influence of the density of surrounding bus stops and metro stations on the service performance of the greenway.

绿道外部各级路网密度
Densities of road networks at different levels
绿道使用人群规模及密度
Number and density of greenway users
0.31
1.86
2.63
551
423
区段 1：挹江门—华严岗门
S1：YJG-HYGG
0.63
2.23
1.84
103
10
区段 2：定淮门—草场门
S2：DHG-CCG
0.35
2.12
1.46
499
332
区段 3：草场门—清凉门
S3：CCG-QLG
0.59
2.26
1.94
78
64
区段 4：水西门—集庆门
S4：SXG-JQG
0.90
3.34
4.22
251
179
区段 5：凤台路—中华门
S5：FTR-ZHG
0.69
2.30
3.39
347
315
区段 6：中华门—雨花门
S6：ZHG-YHG
快速路密度 (km/km²)
Density of Expressways
干路密度 (km/km²)
Density of primary roads
支路 (km/km²)
Density of branch ways
使用人群规模（单位：人）
Number of users (Unit: person)
使用人群密度（单位：人）
Density of users (Unit: person)

外部空间特征与服务绩效关联分析总结

Summary of correlation analysis of the characteristics of the exterior space and the service performance

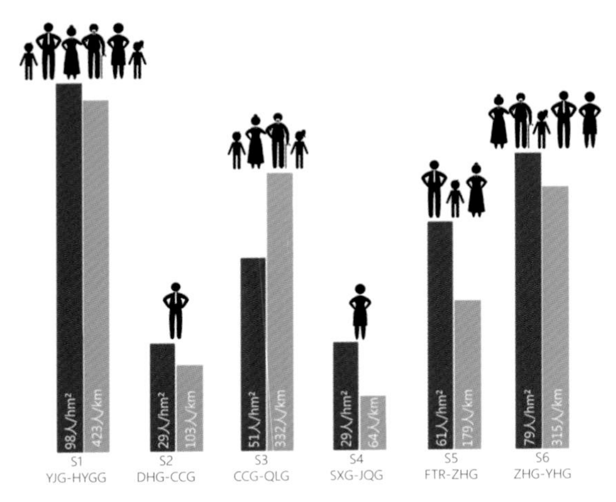

绿道使用人群密度（均值）

Density of greenway users (average)

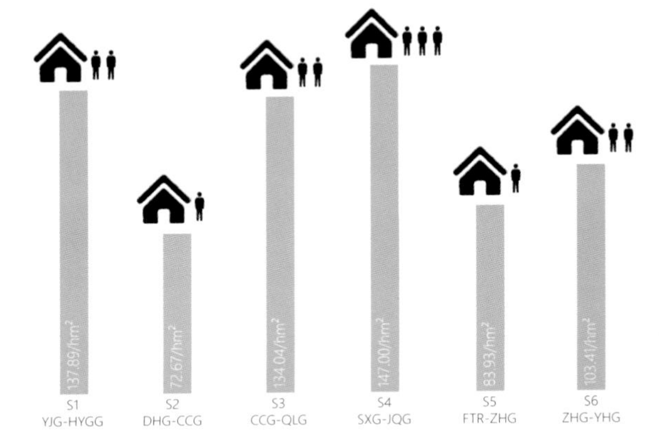

区域居住人口密度

Density of residents in the study district

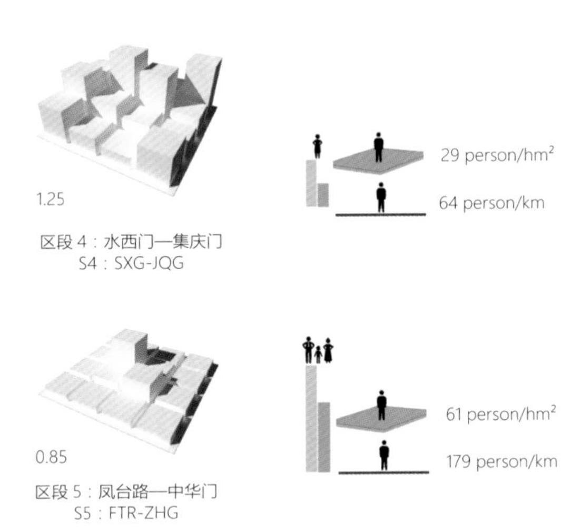

绿道所在区域容积率与使用人群密度

Plot ratio of the district and density of greenway users

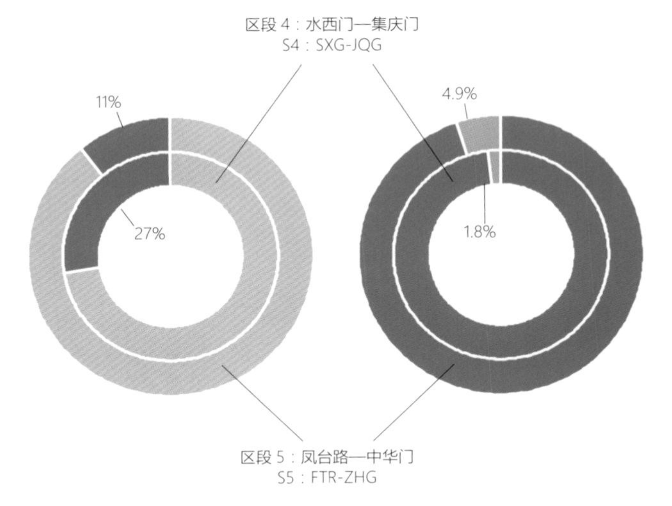

区域绿地占比，绿道用地面积在绿地中的占比

Percentage of green space，Percentage of the area of the greenway

绿道服务绩效与周边游憩资源的关系

单纯从外部空间特征指标看，区段 4（水西门—集庆门）绿道几乎具备了高服务绩效的所有因素，例如其周边居住用地占比排序第一并且呈两侧布局模式，居住人口密度排序也居首位，但其使用人群规模和总密度均排序最后，出现大幅反差。

为探索产生该反差的原因，我们对该区段其他数据做了进一步分析。其中一个特别之处引起了我们注意，即该区段所在区域中绿地面积占比也为 6 个区段最高，加之该区段绿道空间狭小，造成绿道面积在研究区域绿地面积中的占比指标在 6 个区段中排位最后。值得注意的是，该区段除空间紧张无较好的游憩资源外，也是 6 个区段中唯一没有紧邻明城墙布局区段，导致其对使用人群吸引力较之其他区段相对较弱。另外，在绿道附近布有莫愁湖公园（60 岁以上老人免票）、南湖公园（免票）等城市重要游憩资源，其对周边居民的游憩吸引力远超该绿道区段。由于这些公园均没有和绿道相连，进而造成区域内使用人群的大幅分流，加之秦淮河和东侧对岸滨河城市快速路凤台路对东面使用人群的进入阻隔效应明显。这些因素很大程度上共同造成了该区段使用人群规模、密度指标与外部空间条件指标的巨大反差。

反观区段 5（凤台路—中华门）虽然周边用地开发强度和居民密度均最低，但绿道面积在区域内绿地面积中的占比却是 6 个区段最高，并成为附近为数不多的公共活动空间，因而使用人群密度能够在 6 个区段中位居中游。而区段 1（挹江门—华严岗门）虽然北面有阅江楼风景区、绣球公园，南面有小桃园等城市核心游憩资源，但由于其均位于绿道沿线，因而实际能与绿道功能互补，同时绿道区段中部还有景观步行桥连接西面河心岛和对岸滨河绿地，这些因素均促进了绿道与周边景点及整个地区内的游客共享，成为该区段绿道使用人群密度指标处于高位的重要支撑。

The relationship between the service performance of the greenway and other surrounding recreational resources

Simply judging from the characteristic index of the exterior space of the greenway, we can see that S4 SXG-JQG possesses almost all the elements of high service performance. For example, the percentage of its surrounding residential area ranks first and its residential area is distributed on both sides of the greenway with a very short average distance to reach the greenway. The density of its residential population also ranks first, but the number of its users and the total density both rank last, showing a sharp contrast.

To explore the cause of such a sharp contrast, we further analyzed the other data from this segment. A special aspect catches our attention. The percentage of the area of green space of the district where this segment is located ranks the first among the six segments. In addition, the greenway space is narrow in this segment, so the proportion of the area of the greenway to the area of green space of the district in this study ranks last among the six segments. It is worth noting that in addition to a lack of space and good recreational resources, this segment is the only one that is not adjoined to the Ming Dynasty City Wall. As a result, this segment is not as attractive to users as the other segments. Besides, a lot of important urban recreational resources are distributed near the greenway, including Mochou Lake Park (free entrance for people aged over 60) and Nanhu Park (free entrance), both of which are far more attractive to surrounding residents than the greenway segment itself. Since none of these parks are connected to the greenway, the users of this district are greatly distributed and attracted to parks other than the greenway. In addition, the Qinhuai River and Fengtai Road, an urban expressway at the waterfront of the east side of the river, noticeably obstructs the access of potential users from the east of the greenway. All of these factors work together to cause the sharp contrast between the number and density of users and the conditions of the exterior space.

When we look back on S5 FTR-ZHG, we find that although the development intensity of the surrounding land and the density of residential population both rank last and the percentage of the area of greenway in the area of green space of the district ranks first among the six segments. So this segment becomes one of the few places for public activities and thus ranks in the middle among the six segments in terms of user density. In S1 YJG-HYGG, there are Yuejianglou Scenic Spot and Xiuqiu Park to the north and Little Peach Orchard to the south, as well as other such core urban recreational resources similar to that around S4 SXG-JQG. However, since most of them are located in the way of or close to the greenway, the functions of the greenway and surrounding recreational resources can actually support each other as supplements and work as an organic system. At the same time, the middle part of this greenway segment has landscaping footbridges that are connected to the tiny islands on the west and the green space at the waterfront across the river, providing convenient access for residents across the river to enter the greenway. All of these factors result in the user sharing between greenway and the surrounding scenic spots and continuously cause S1 YJG-HYGG to have a stably high user density.

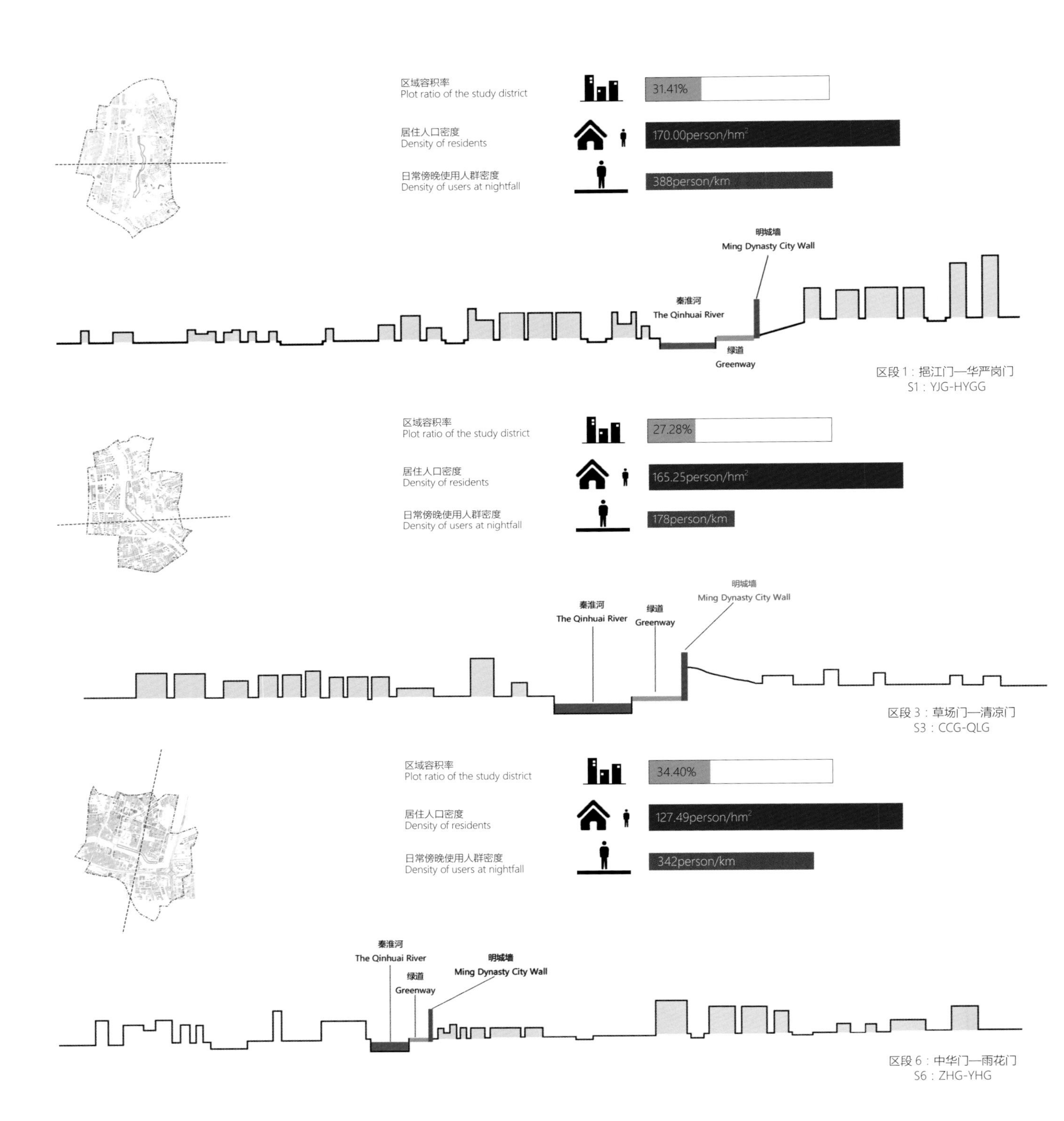

区域容积率
Plot ratio of the study district
31.41%
居住人口密度
Density of residents
170.00person/hm²
日常傍晚使用人群密度
Density of users at nightfall
388person/km
明城墙
Ming Dynasty City Wall
秦淮河
The Qinhuai River
绿道
Greenway
区段 1：挹江门—华严岗门
S1：YJG-HYGG
区域容积率
Plot ratio of the study district
27.28%
居住人口密度
Density of residents
165.25person/hm²
日常傍晚使用人群密度
Density of users at nightfall
178person/km
明城墙
Ming Dynasty City Wall
秦淮河
The Qinhuai River
绿道
Greenway
区段 3：草场门—清凉门
S3：CCG-QLG
区域容积率
Plot ratio of the study district
34.40%
居住人口密度
Density of residents
127.49person/hm²
日常傍晚使用人群密度
Density of users at nightfall
342person/km
秦淮河
The Qinhuai River
明城墙
Ming Dynasty City Wall
绿道
Greenway
区段 6：中华门—雨花门
S6：ZHG-YHG

护城河的空间阻隔效应

从理论上看，护城河（含秦淮河部分河道）既是绿道的景观游憩资源，同时也是绿道的空间阻隔要素，即对对岸居民日常使用绿道造成空间阻碍。为了验证该空间阻隔因素作用的大小，我们选取了几个典型区段来进行比较分析。

首先在居住用地三种布局形式中各选取一个区段进行比较，分别为区段1（挹江门—华严岗门）（两侧布局）、区段3（草场门—清凉门）（对岸布局）和区段6（中华门—雨花门）（紧邻布局）。如将周边居民分布影响进行叠加，并以使用人群主体为周边居民的日常傍晚时段使用人群规模指标进行比较，可发现由于居住用地分布形式差异，区段1（挹江门—华严岗门）在周边居民规模比区段6（中华门—雨花门）仅高出约6.3%的情况下，日常傍晚使用人数则高出35.8%，该结果显示区段1（挹江门—华严岗门）周边居民日常傍晚使用绿道的比重较之区段6（中华门—雨花门）可能要高很多。2个区段除了两端外，中部在城墙一侧均没有出入口，但由于区段1（挹江门—华严岗门）绿道一侧有较好的游憩资源，并且在中部有景观步行桥（鹊桥）联系两岸，实际弱化了护城河对对岸居民使用绿道的空间阻隔效应，而使此处周边开发强度较高、居民在绿道两侧密集分布、平均空间距离较短的可达性优势显现出来。

而区段1（挹江门—华严岗门）、区段6（中华门—雨花门）比区段3（草场门—清凉门）周边居民规模分别低约16.2%和21.2%的情况下，日常傍晚使用人群规模却分别比后者高出约89.1%和约39.3%，反映出前两者周边居民使用绿道的比重和便捷程度要远高于后者。与区段1（挹江门—华严岗门）相似，区段3（草场门—清凉门）绿道一侧同样拥有较好的游憩资源，并且中部有国防园步行桥连通两岸，虽能部分弱化护城河（该处为秦淮河河道）对对岸居民进入绿道的空间阻隔效应，但由于绝大部分居住用地均位于护城河对岸，将很可能使河道对对岸使用人群的空间阻隔效应累积并集中凸显。

The spatial obstruction effect of the moat

In theory, the moat (including partial river courses of the Qinhuai River) is not only a recreational resource but also an element that obstructs access for daily greenway users. It means that it obstructs in space the residents across the moat to use the greenway on a daily basis. To verify the effect of this spatial obstruction index, we select several typical segments for comparison, contrast, and analysis.

First, we select one segment from the three types of layout forms of surrounding residential land for comparison, namely, S1 YJG-HYGG (laid out on both sides), S3 CCG-QLG (laid out across the other side of the river), and S6 ZHG-YHG (laid out in close proximity). According to data, whether it is on a weekend afternoon or every day at nightfall, the ranking of user density is as follows: S1 YJG-HYGG ranks the first, S2 DHG-CCG last, and S6 ZHG-YHG second.

If we consider and overlay the influence of the distribution of the surrounding residential population, the number of users every day at nightfall is more suitable for analysis because the primary users in this time period are nearby residents. From the indicators, we can see that due to the difference in the pattern of spatial layout of residential land, the number of surrounding residents of S1 YJG-HYGG is only 6.3% higher than that of S6 ZHG-YHG but the number of users of the former every day at nightfall is 35.8% higher than the latter. This result shows that the percentage of the surrounding residents of S1 YJG-HYGG who use the greenway every day at nightfall is probably much larger than that of S6 ZHG-YHG. Neither segment has an entrance on the side of the City Wall in their middle sections. However, since S1 YJG-HYGG has very good recreational resources on the side of the greenway and a landscaping footbridge (Magpie Bridge) that connects both sides of the greenway, the spatial obstruction effect of the moat is weakened for those across the moat to use the greenway. The advantages of intensely development and the short average distance between the residential area and the greenway can be significantly improved at the same time.

The number of residents surrounding S1 YJG-HYGG and S6 ZHG-YHG is 16.2% and 21.2% lower than in S3 CCG-QLG, respectively. Under such circumstances, the number of users every day at nightfall is about 89.1% and 39.3% higher than the latter. It means that it is very possible that the percentages of surrounding residents of the former two segments who use the greenway are much higher than the latter. Similar to S1 YJG-HYGG, the greenway side of S3 CCG-QLG also has very good recreational resources, and the middle part of the segment equips the National Defense Theme Park footbridge that connects both sides of the river. Although it can weaken to a certain extent the spatial obstruction effect of the moat (Qinhuai River of this section) on nearby residents, most residential land is located across the opposite side the moat, so that the spatial obstruction effect of the river course on users probably accumulates and stands out.

人群构成因素对绿道服务绩效的影响

使用人群数据显示，区段 3（草场门—清凉门）（对岸布局）和区段 6（中华门—雨花门）（紧邻布局）的使用人群规模和密度指标在周末下午和日常傍晚均出现反转。该 2 个区段使用人群总密度指标几乎相同，外部空间条件中居住用地占比也相差无几，开发强度、周边人群规模和密度区段 6（中华门—雨花门）稍低。

在日常傍晚时段，外部居住用地紧邻布局的区段 6（中华门—雨花门）人群使用密度是居住用地对岸布局的区段 3（草场门—清凉门）的近 2 倍，护城河（秦淮河河道）空间阻隔效应较明显，但在周末下午人群使用密度却仅为区段 3（草场门—清凉门）的约 60%。

为了探索该反差产生的原因，我们对两个区段周末下午和日常傍晚使用人群来源进行了分析，并发现当绿道周边布有城市重要景点时（如区段 3 边上的国防园和石头城公园，区段 6 周边的中华门、老门东等），在周末下午的外来游客占比要大大高于日常傍晚，部分区段（如区段 3）甚至超过一半成为绿道主导使用人群。

由于外来游客进入绿道实际并不受护城河这一空间阻隔要素制约，如果在使用人群中占比较大，就能在很大程度上影响甚至决定该区段绿道周末下午的使用强度。护城河虽然对对岸人群的空间阻隔效依然存在，但对绿道使用强度的负面影响在一定程度上被弱化。

The influence of user composition on the service performance of the greenway

According to user data, the number of users and density index of S3 CCG-QLG (laid out across the other side of the river) and S6 ZHG-YHG (laid out in close proximity) are reversed on weekend afternoons and every day at nightfall. The average user density indexes in two time periods of the two segments are almost the same, and the percentages of external residential land of the two segments are also nearly the same. The development intensity, the number of surrounding users, and the density of surrounding users of S6 ZHG-YHG are slightly lower than the former.

Every day at nightfall, the user density of S6 ZHG-YHG, whose surrounding residential land is laid out in close proximity, is nearly twice that of S3 CCG-QLG, whose surrounding residential land is laid out across the other side of the river. In this segment, the spatial obstruction effect of the moat is fairly evident. However, on a weekend afternoon, the user density of S6 ZHG-YHG is only about 60% of S3 CCG-QLG's.

To explore the cause of such a sharp contrast, we analyzed the user sources of the two segments on weekend afternoons and every day at nightfall. We found that when key scenic spots of the city (for example, Defense Park and Stone City Wall Park beside S3 CCG-QLG, Zhonghua Gate and East Zhonghua Gate Historical Culture Block surrounding S6 ZHG-YHG) are located on the surrounding area of the greenway, the percentage of tourists on weekend afternoons is much larger than that every day at nightfall, with some segments (S3 CCG-QLG) even exceeding half of the total so that the dominant users become tourists instead of surrounding residents.

Since the access for tourists to the greenway is actually not limited by the spatial obstruction element of the moat, a big group of tourists can largely influence and even determine the service intensity of this greenway segment on a weekend afternoon. Although the spatial obstruction effect of the moat on the crowds across the river still exists, its negative influence on the service intensity of the greenway during this time period is weakened.

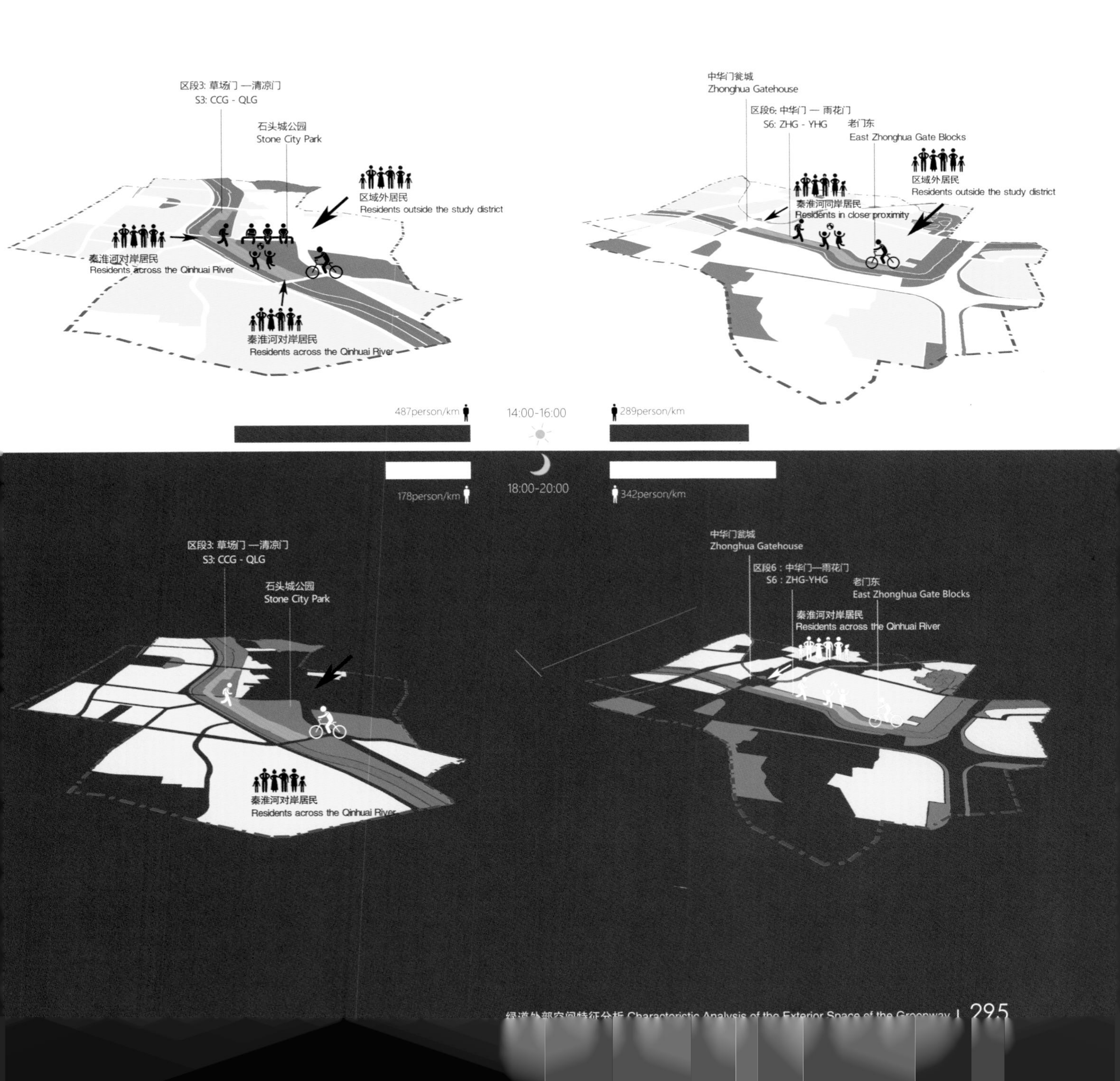
区段3: 草场门 —清凉门
S3: CCG - QLG
石头城公园
Stone City Park
区域外居民
Residents outside the study district
秦淮河对岸居民
Residents across the Qinhuai River
秦淮河对岸居民
Residents across the Qinhuai River
中华门瓮城
Zhonghua Gatehouse
区段6: 中华门 — 雨花门
S6: ZHG - YHG
老门东
East Zhonghua Gate Blocks
区域外居民
Residents outside the study district
秦淮河同岸居民
Residents in close proximity
487person/km
14:00-16:00
289person/km
178person/km
18:00-20:00
342person/km
区段3: 草场门 —清凉门
S3: CCG - QLG
石头城公园
Stone City Park
秦淮河对岸居民
Residents across the Qinhuai River
中华门瓮城
Zhonghua Gatehouse
区段6：中华门—雨花门
S6：ZHG-YHG
老门东
East Zhonghua Gate Blocks
秦淮河对岸居民
Residents across the Qinhuai River

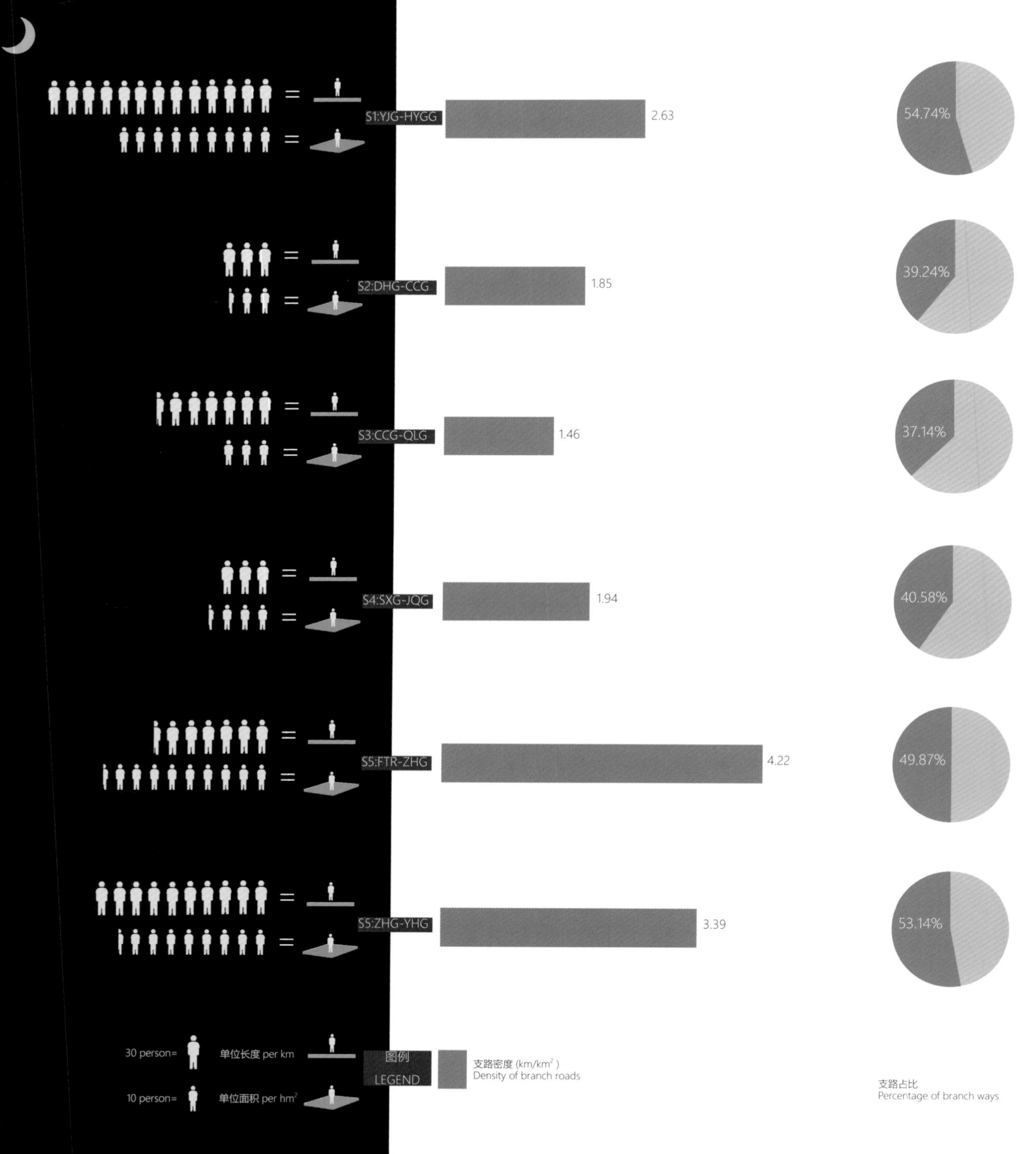
S1:YJG-HYGG
2.63
54.74%
S2:DHG-CCG
1.85
39.24%
S3:CCG-QLG
1.46
37.14%
S4:SXG-JQG
1.94
40.58%
S5:FTR-ZHG
4.22
49.87%
S5:ZHG-YHG
3.39
53.14%
30 person=
单位长度 per km
10 person=
单位面积 per hm²
图例
LEGEND
支路密度 (km/km²)
Density of branch roads
支路占比
Percentage of branch ways

支路密度对绿道服务绩效的影响

通常情况下，较高支路占比和密度反映了该区域内具有开展慢行交通方式的良好条件，有利于居民的日常游憩出行。我们通过比对以周边居民为使用主体的日常傍晚时段使用人群密度指标，发现排位前三的区段 1（挹江门—华严岗门）、区段 6（中华门—雨花门）和区段 5（凤台路—中华门）其支路密度和占比指标也均排位前三。区段 3（草场门—清凉门）虽然总体使用密度排位第二，但支路占比和密度指标排位最后，其傍晚使用人群密度指标也未进前三。

但值得注意的是，支路密度和占比排位前三的区段居住用地布局模式均为两岸布局和紧邻布局形式，护城河对绿道使用的空间阻隔效应相对较低。除区段 4（水西门—集庆门）情况特殊外，另外两个支路占比、支路密度以及使用人群密度较低的区段居住用地布局模式均为对岸布局型，其中护城河对绿道使用的阻隔效应较大。因此，鉴于护城河空间阻隔效应的共同作用，支路密度对绿道使用强度的影响大小仍需通过更进一步的深入研究才能明确。

The influence of the density of branch roads on the service performance of the greenway

Under normal circumstances, the high percentage and density of branch roads reflects that this district has good conditions for the development of slow traffic in favor of daily recreation and travel of residents. Through comparison of the density indexes of users every day at nightfall with surrounding residents as the main body of users, we found that S1 YJG-HYGG, S6 ZHG-YHG, and S5 FTR-ZHG rank first, second, and third, and that the densities and percentages of their branch roads also rank first, second, and third. Although S3 CCG-QLG ranks second in terms of total user density, the percentage and density indexes of its branch roads rank last, and the density index of its users at nightfall does not enter the top three.

It is worth noting that the layout patterns of residential lands of the segments whose densities and percentages of branch roads rank first, second, and third are all laid out on both sides and in close proximity to the greenway, and that the spatial obstruction effect of the moat on the use of the greenway is relatively weak. Except in the case of S4 SXG-JQG, the layout patterns of residential lands of the other two segments with relatively smaller percentages and lower densities of branch roads and lower user densities are both laid out across the other side of the river, and the spatial obstruction effect of the moat on the use of the greenway is strong. Therefore, in view of the combined action of the spatial obstruction effect of the moat, the influence of branch road density on the use of greenway cannot be easily determined in this study.

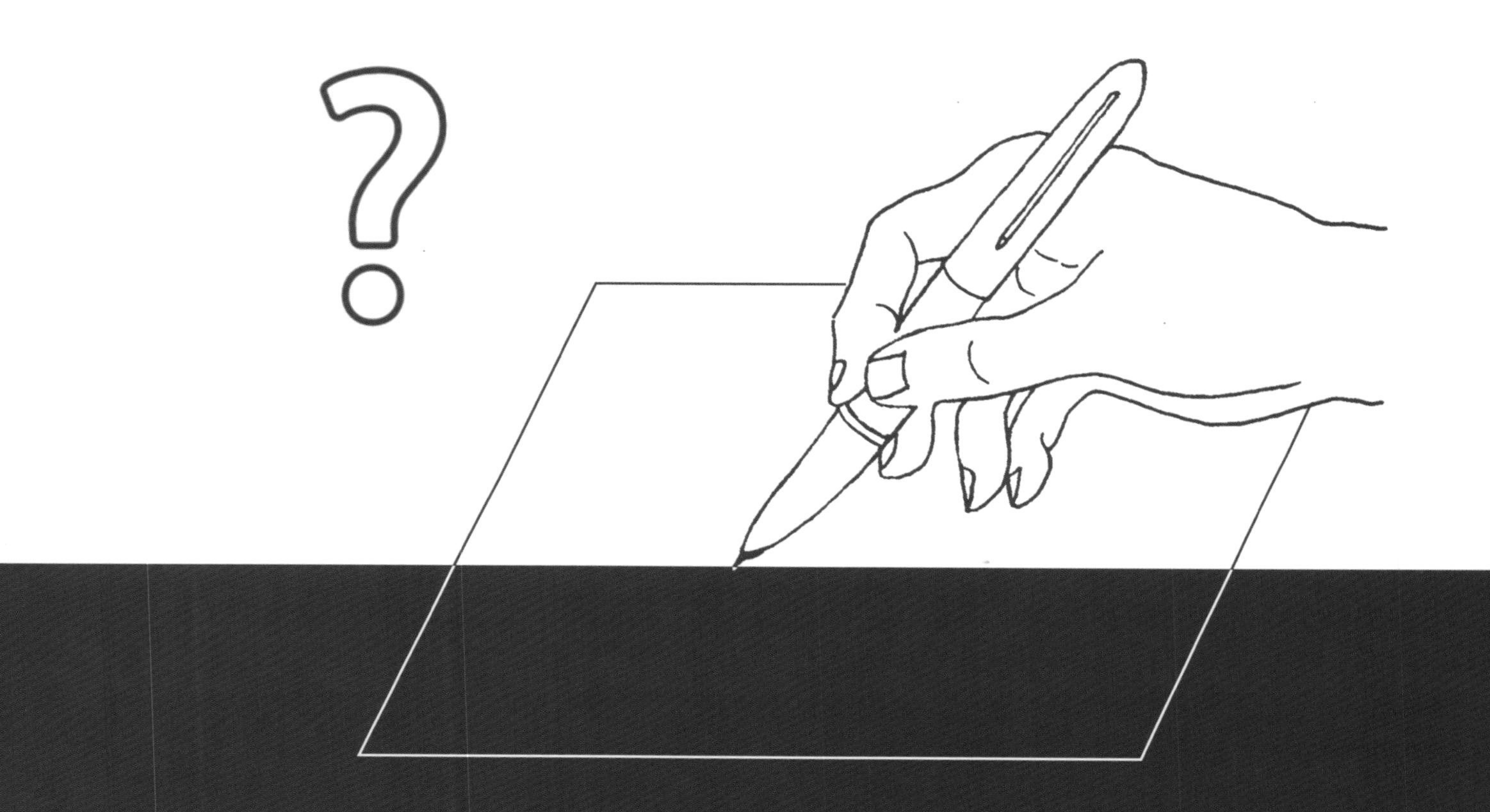

对规划设计的反思与讨论

Reflections and Discussions of Design and Planning

如何改良建成环境绿道的空间设计

How to improve space design of the greenway in a built environment

实际上通过南京明城墙绿道服务绩效和空间特征调查分析，除了能了解现状绿道内部空间对其服务绩效的影响外，更重要的是能够结合当前中国城市居民在绿道中的使用特征和规律，进一步发现和反思建成环境绿道内部空间设计的问题，并为未来绿道空间改良和优化设计提出针对性建议。因此，我们选取了调查分析中发现的 3 个绿道内部空间设计问题进行着重讨论：

In fact, through survey and analysis of the service performance and spatial characteristics of the Nanjing Ming Dynasty City Wall Greenway, in addition to understanding the influence of the current interior space of the greenway on the service performance of the greenway, more importantly we can combine the characteristics and patterns of how urban residents use greenways in China to further discover and reflect on problems regarding the design of the interior space of a greenway in a built environment and provide corresponding suggestions for the optimization of greenway space design in the future. Therefore, we select three problems related to the design of the interior space of the greenway, which we determined through survey and analysis, for discussion.

讨论一：空间序列的组织安排问题

Discussion I: Organization and arrangement of spatial sequence

绿道作为线性空间要素，很多情况下会让人自然想到其空间是均匀延展的，在课题研究的部分区段由于空间条件所限，它也不得不采取空间均匀延展的布局形式，如区段 2（定淮门—草场门）和区段 4（水西门—集庆门），但它能吸引停留休息、康体活动等类型活动人群数量和比重均大大低于空间有明显开合变化的区段 1（挹江门—华严岗门）以及区段 3（草场门—清凉门），部分数据甚至相差 5 倍以上。

As a linear spatial element, a greenway reminds people of an evenly extended space in many cases. For some segments in our study, limited by spatial conditions, the greenway cannot but adopt the layout form of evenly extended space such as in S2 DHG-CCG and S4 SXG-JQG, but the number and percentage of users that the greenway can attract to stop over and exercise is much smaller than in S1 YJG-HYGG and S3 CCG-QLG, whose spaces are noticeably open and closed, respectively, with more than five times the difference in terms of service performance.

水波的韵律

The rhythm of ripples

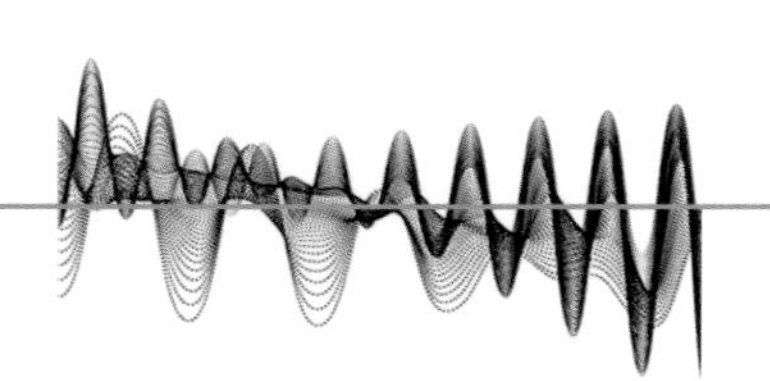

音乐的韵律

The rhythm of music

建筑空间的韵律

The rhythm of buildings

6 个绿道中空间开闭序列最清晰的是区段 1（挹江门—华严岗门）。该区段由于城墙的蜿蜒形态产生了一组较明显的开闭空间序列，而切片图上使用人群的分布形态也随着空间的开闭有韵律的起伏，其中较开敞的空间往往成为绿道沿线重要节点，也是休息停留和康体人群聚集的地段。从数据上看，该区段休息停留和康体活动人群总密度在日常傍晚和周末下午均排序第一，人群规模则分别排序第一和第二，成为 6 个区段中最能吸引人群停留的区段。

Among the six greenways, S1 YJG-HYGG is the segment that has the most featured spatial sequence by opening up and narrowing down the space rhythmically that is mostly generated by change of space along the zigzagging City Wall. The user distribution in space on the slice diagram also fluctuates rhythmically with the opening and closing of the space of the greenway, and the fairly wide-open space often becomes a very important node along the greenway, making it a place where people gather around to stop over or exercise. Judging from the data, we can see that the total user density of those who stop over and exercise on this segment ranks first both on weekend afternoons and every day at nightfall, and that the number of users in both time periods ranks first and the second, respectively, which makes this segment the most attractive to users.

在绿道调查和使用分析过程中，可以清楚地发现绿道有节奏的空间开闭变化能打破单一均质的线性空间，并给使用人群创造游憩兴奋点以及停留活动的机会。即便是在空间相对均质，空间开闭变化较少的区段 2（定淮门—草场门）和区段 4（水西门—集庆门），也能发现但凡通过穿插开敞硬地或绿地来打破均质空间形态的地段，通常会带来停留人群的大幅上升，并成为区段为数不多的人群使用高光地段。

During the survey and analysis of the greenway, we can clearly find out that the rhythmic space change of the opening and closing of the greenway can break the dull and homogenous linear space and create opportunities for users to stop over and carry out other recreational activities. Even in S2 DHG-CCG and S4 SXG-JQG, which have fairly homogenous space and few changes of open and closed space, we can also find out that as long as a segment breaks a homogenous spatial form through the insertion of a series of wide-open hard sites or green spaces, the number of people who are willing to stop over is usually increased drastically, which makes these areas highlights in the segment.

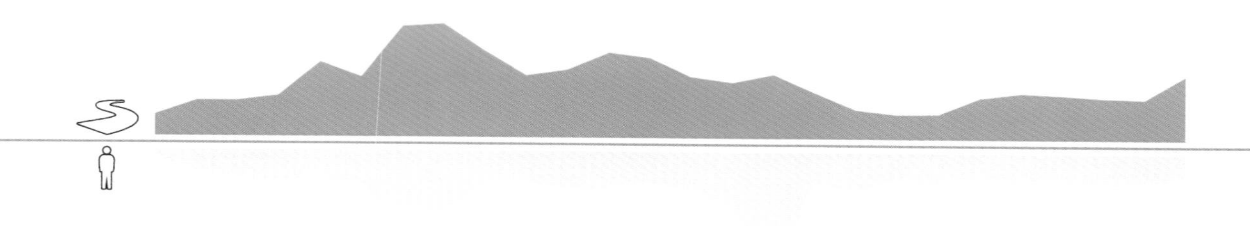

绿道空间韵律 & 人群分布起伏

Spatial change of the greenway (upper) & the fluctuation of the number of users (lower)

讨论二：空间设计的刚性与弹性平衡问题

通常情况下，注重设计刚性，意味着清晰界定内部空间、明确服务人群和活动类型，在提升设计针对性的同时，排他性也很明显。而提升设计的弹性，实际上在弱化空间特殊性和针对性的同时，也增加了空间被各类人群或不同活动类型使用的可能性，即提升了空间的包容性。如何妥善平衡绿道沿线空间设计中的刚性和弹性问题将对绿道使用和服务绩效产生重要影响。

从本课题调查和分析中我们发现，目前南京明城墙绿道沿线的空间设计中大多刚性有余，而弹性不足。书中曾就同一绿道区段在不同时段内使用人群规模最高地段的空间转移问题进行过讨论，实际上反映了部分区段主要活动节点空间无法适应另一时段由于活动类型改变而产生的空间需求差异问题。例如，区段 6（中华门—雨花门）切片 5 节点场地被花坛、座椅划分为多个小空间，恰好能满足周末下午周边中、老年人小群体棋牌娱乐和围观的活动需求，但到了傍晚，同样是以周边中老年人为主的广场舞等集体健身活动则难以在此类空间中开展，导致该时段使用者寥寥无几。而诸如此类绿道内部空间设计和使用中刚性有余、弹性不足的问题，在南京明城墙绿道沿线屡见不鲜。因此，要提升此类空间的服务绩效，则需提升绿道沿线场地空间设计和使用中的弹性，从而有效兼顾各时段、各类人群的不同活动需求。

以集中硬质场地上的座位设置为例，目前南京明城墙沿线的座椅座位总数和密度并不低，但由于座位均是固定座位并且部分座位摆放位置不当，在我们调查中经常可看到场地上许多座位空置的同时，大量人群却坐在座椅边的台阶、花坛边缘、护栏等设施上，这其实显示出座位的供需发生错位，即供非所求。尤其是场地中间的固定座椅，除了空间品质和使用体验不佳外，还往往会切割场地空间，妨碍场地被整合集中使用，其

Discussion II: The balance between rigidity and flexibility of space design

Under normal circumstances, the focus on the rigidity of design means clear definition and division of interior space and the definition of service group and activity type, which is considered to enforce the pertinence of the design, as well as exclusiveness. The improvement of the flexibility of design actually not only weakens the particularity and pertinence of the space but also increases the possibility that the space be used by all kinds of people or for all kinds of activities, which strengthens the inclusiveness of the space. How to properly balance the rigidity and flexibility of space design along the greenway will have significant influence on the use of greenway and its service performance.

From the survey and analysis of this study, we found that the rigidity in space design along the Nanjing Ming Dynasty City Wall Greenway is far greater than its flexibilities. This book has discussed spatial transference of the districts with the biggest number of users in different time periods at the same segment, which actually reflects that many node spaces of the recreational activities along the greenway cannot adapt to the difference in space requirement generated and raised by the change of recreational activities. For example, the main node site of slice 5 of S6 ZHG-YHG is divided into multiple small spaces by flowerbeds and bench seats, which happens to satisfy the needs of small groups of surrounding middle-aged and senior people who would like to go to the greenway to play cards and chess for entertainment or just watch on a weekend afternoon. At nightfall, it is very hard for the same groups of people to exercise in groups like square dancing in this type of space, as a result of which few of them use the greenway continuously in this following time period. Similar problems can be found in many other places along the Nanjing Ming Dynasty City Wall Greenway. Therefore, to improve the service performance of this type of space, it is necessary to enhance the flexibilities of site design and use along the greenway, so as to effectively take into account different activity requirements of different groups of people in different time periods.

Take for example the installation of bench seats on hard sites. At present, the number of bench seats along the Nanjing Ming Dynasty City Wall is not small, and their density is not low either. However, since the bench seats are fixed to the ground and some of them are not properly installed, there are often a lot of empty bench seats. A great many people would rather sit on the steps next to the bench seats, the edges of the flowerbeds, and the guardrails and other such facilities. This actually reflects unmatched supply and demand of bench seats; that is to say, what is supplied does not match what is demanded. In addition to destroying the space quality and user experience, the fixed bench seats in the middle of the site often divide the space into pieces, thus preventing the site from being integrally and centrally used. The most typical is the node sites of slice 5 of S6 ZHG-YHG.

1 2
3 4

中最典型的就是区段 6（中华门—雨花门）切片 5 节点场地。

事实上，绿道沿线集中硬质场地作为使用人群集散的空间，应更多考虑在场地边缘，结合台阶、花坛边缘或草坪营造大量弹性灵活的可坐空间，既远比在场地上安放固定座椅要经济和高效，并且在大多情况下也更受使用人群欢迎。另外，设置一定数量可移动座椅让使用人群自己决定在什么位置，以什么样的方式停留休息也能很大程度激发场地的服务绩效潜能，这一点已经在国外许多案例中印证，如纽约市布莱恩特公园。但基于目前中国国情而言，出于维护管理等方面的原因，该举措要推行的难度仍然较大。此外，除了优化场地座椅布局和设计外，在设计中避免异形或狭长场地、保障或改善场地夜间照明条件、添置可移动设施、提供免费 Wi-Fi 等措施均能为场地弹性使用，提升绿道综合服务绩效创造条件。

In fact, since the centralized hard sites along the greenway function as the space that gathers and distributes users, consideration should be given to the creation of a great deal of seating space on the edge of the sites in combination with steps, the edge of flowerbedsand grasslands. This method is not only more economical and efficient than the installation of fixed bench seats on the sites but also more popular among users in most cases. Besides, a certain amount of portable and movable bench seats can be installed, so that users can decide for themselves where and how they prefer to stop over and sit down. This can unleash the potential of the service performance of the sites to a great extent, which has been proven by a lot of cases in other countries such as Bryant Park in New York City in U.S. However, based on the current national conditions of China, for the reason of maintenance and management, it is still very difficult to promote such an approach. Besides, in addition to the optimization of the layout and design of bench seats, flexible use and service performance of the site can also be enhanced by many other approaches, such as avoiding the sites with a narrow and long shape in design, guaranteeing and improving the conditions of night illumination, adding portable facilities, and providing users with free Wi-Fi.

讨论三：空间感知与人群引导问题

在我们调查中，很多绿道沿线空间服务绩效低下的重要原因并非没有潜在使用者，而是潜在使用者无法感知或不能明确感知到游憩服务空间的存在。美国城市社会学家W. H. Whyte 曾经说过“在空间使用中，视线非常重要，人们只会使用有着开阔视线的城市公共空间”。在南京明城墙绿道中，视线的通透性同样影响着沿线空间的使用和服务绩效，其中反映最明显的是对护城河滨水地段的使用。由于南京明城墙绿道全线均位于护城河滨水地段，其往往会根据驳岸的坡度进行竖向设计，形成缓坡型驳岸（区段1）、两级台地型驳岸（区段4、区段5、区段6）、三级台地型驳岸（区段2、区段3）三种驳岸类型，并会在滨水高程较低处设置亲水步道，而在高程较高接近城墙地段设置主游径。但在周末下午时段，许多区段使用人群均集中在主游径上，而较少使用游憩体验相对更好的亲水步道，其中核心的因素就是视线阻塞和通行不顺。

以区段6（中华门—雨花门）为例，在东侧连接江宁路的城市出入口地段，由于服务建筑以及周边茂密乔灌木的遮挡和压迫造成该入口进入绿道的人群基本很难意识到亲水步道的存在，同时也不易发现连接亲水步道的台阶，因而很大程度上阻碍了人群对亲水步道的使用。同时，该区段亲水步道内侧沿线均布有较浓密的乔灌木，既遮挡了在主游径上人群视线，让亲水步道难以被主游径上的人群发现，并且还成为两级游径竖向交通连接的障碍。而在区段3（草场门—清凉门），虽然驳岸采用的是三级台地形式，但由于各级台地间视线非常开敞，且连接通道较丰富，因此各级台地和步道的使用相对较均衡，与区段6（中华门—雨花门）形成鲜明对比。此外，能否对使用者的视线进行有效安排，还将对绿道沿线草坪绿地使用、硬质场地使用以及绿道出入口人群导入等方面产生关键性影响。

Discussion III: Spatial perception and the leading of crowd

In our study, we find that an important reason why a lot of spaces along the greenway have low service performance is not that they lack potential users but that potential users cannot clearly perceive the existence of recreational spaces.

W. H. Whyte, an American urban sociologist, once said, “Sight lines are important. If people do not see a space, they will not use it.” In terms of the Nanjing Ming Dynasty City Wall Greenway, the permeability of sight line also influences the use and service performance of the spaces along the greenway, which is reflected most by the use of waterfront area of the moat. Since the Nanjing Ming Dynasty City Wall Greenway is located at the waterfront of the moat, the greenway is vertically designed in some spots according to the slope of the revetment so as to form a gentle sloped revetment (S1 YJG-HYGG), two-level platform revetments (S4 SXG-JQG, S5 FTR-ZHG, and S6 ZHG-YHG), and three-level platform revetments (S2 DHG-CCG and S3 CCG-QLG). Hydrophilic trails are installed at the lower level of the waterfront elevation, while the main trails are installed at the higher level of the elevation that is close to the City Wall. However, on a weekend afternoon, it can be found in this survey that a lot of users often gather around on the main trails, instead of using the hydrophilic trails that provide better recreational experience. The core factor lies in the blocking of sight line and the roughness of the passage for people to reach.

Take S6 ZHG-YHG for example. At the urban entrance connected to Jiangning Road on the east side, blocked and constricted by service buildings and surrounding leafy trees and shrubs, those who enter the greenway by this entrance may find it hard to perceive the existence of hydrophilic trails, and they may also find it difficult to discover the steps connecting to the hydrophilic trails, which thus prevents many people from using the hydrophilic trails. At the same time, the areas along the inside of the hydrophilic trails of the segment are all laid out with leafy trees and shrubs, which not only block users' vision on the main trails and makes it difficult for them to discover the hydrophilic trails, but also become obstacles to the vertical traffic connection of two levels of trails (left figure on facing page 88 of 5.21 edition). Although S3 CCG-QLG adopts the three-level platform revetment, since the sight lines of platforms at different levels are wide-open and have a variety of connecting channels, the use of platforms and trails at different levels are balanced, which is in sharp contrast to S6 ZHG-YHG. Besides, whether the sight lines of users can be effectively arranged will exert strong influence on the use of lawns and green spaces along the greenway, the use of hard sites, and the leading of crowds at entrances of the greenway.

区段 6（中华门—雨花门）鸟瞰图
Birds-eye view of S6 (ZHG-YHG)

区段 3（草场门—清凉门）鸟瞰图
Birds-eye view of S3(CCG-QLG)

区段 6（中华门—雨花门）人视图
Human vision of S6 (ZHG-YHG)

区段 3（草场门—清凉门）人视图
Human vision of S3 (CCG-QLG)

如何优化建成环境绿道的规划选线

How to optimize the greenway planning in a built environment

由于南京明城墙绿道主要以线性的城墙遗址为空间载体，其线路主要沿城墙展开，并不存在太多的规划选线问题。但研究明城墙绿道周边空间环境对其服务绩效的作用方式和影响规律，实际上可以为在其他建成环境绿道规划选线中服务绩效的保障提供些许有价值的参考。

讨论一：游憩服务供需两端的串联平衡问题

建成环境绿道除了自身具备一定的游憩服务能力外，还应在城市游憩服务的供需端之间建立便捷联系，因此其连接对象可分为游憩服务供给类节点（如城市公园、广场以及商业休闲、历史文化、娱乐体育中心等）和游憩服务需求类节点（如居住区）。

本研究涉及的南京明城墙绿道 6 个区段中，区段 4（水西门—集庆门）所在区域居住用地占比和居民密度均位于 6 个区段首位，绿道沿线均是游憩服务需求较强地段，但是实际数据显示该区段使用人群规模和密度指标却居于末位。其中一个重要原因就是区段线路并未有效串联周边诸如莫愁湖公园、南湖公园等重要的城市游憩节点，导致绿道与此类节点成为竞争关系，使用人群被大量分流。而在研究预调研过程中所调查的南京滨江绿道却刚好相反，该绿道沿线有着大量优质的游憩资源，但由于周边大量用地尚未开发，居民密度较低，人气低迷，同样也导致绿道服务绩效低下，大量游憩资源被闲置浪费。

而对于游憩服务供需节点整合较好的典型区段则是区段 1（挹江门—华严岗门），其周边居民密度排位第二，游憩服务需求旺盛，同时绿道线路有效连接了阅江楼风景区、绣球公园，小桃园等城市重要游憩节点，在游憩服务供需两端取得了较好

Since the Nanjing Ming Dynasty City Wall Greenway mainly relies on linear historic sites of the City Wall as spatial carriers, its route is mainly developed along the City Wall, without too much decision-making in terms of greenway route arrangement in planning. However, the study of the mode of action and influence of the surrounding spatial environment of the Nanjing Ming Dynasty City Wall Greenway on its service performance can actually provide valuable reference for guaranteeing service performance in the greenway planning, especially for the work of greenway route selection in built environments.

Discussion 1 : The connection and balance of supply and demand of recreational services

In addition to the ability of providing recreational services, the greenway in built environment should also set up a convenient connection between the supply and demand sides of urban recreational services. Therefore, the objects of connection can be divided into supply nodes of recreational services (e.g. city parks, plazas, commercial and leisure centers, historic and cultural centers, entertainment and sports centers) and demand nodes of recreational services (e.g. residential areas).

Among the six segments of the Nanjing Ming Dynasty City Wall Greenway in this study, the percentage of the residential land in the district where S4 SXG-JQG is located and the density of its residential population both rank the first. The areas along the greenway all have the strongest demand of recreational services, but actual data show that the user numbers and user density indexes in this segment both rank last. One of the important reasons is that the route of this segment has yet to effectively connect in a series the other surrounding urban recreational nodes including Mochou Lake Park, Nanhu Park and others, so that the greenway and these types of nodes become competitors, and users are greatly distributed amongst them. Nanjing Yangtze Riverside Greenway in the preliminary survey of this study is quite the opposite. The area along this greenway has a lot of quality recreational resources, but since its surrounding land has yet to be developed, the density of residential population is low As a result, the service performance of this greenway is also very poor and a lot of recreational resources remain unused.

S1 YJG-HYGG is a typical segment that integrates well the supply and demand nodes of recreational services. The density of its surrounding residential population ranks second, and recreational services are in strong demand in the district surrounding this

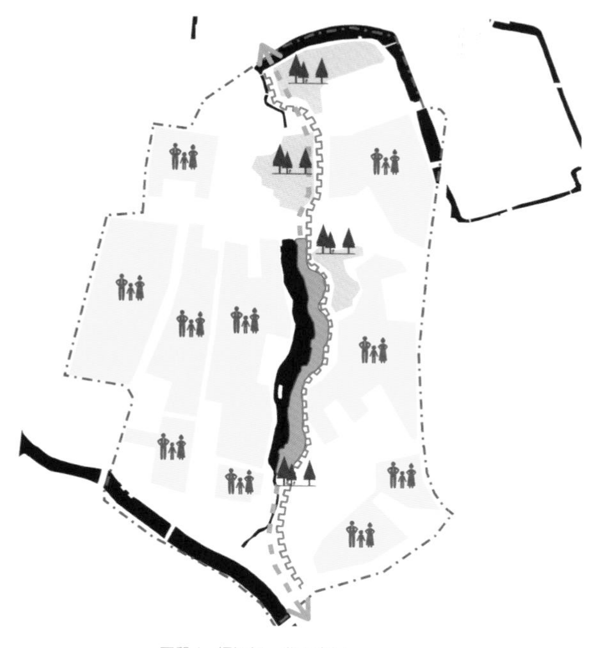

区段 1：挹江门—华严岗门
S1：YJG-HYGG

区段 3：草场门—清凉门
S3：CCG-QLG

区段 4：水西门—集庆门
S4：SXG-JQG

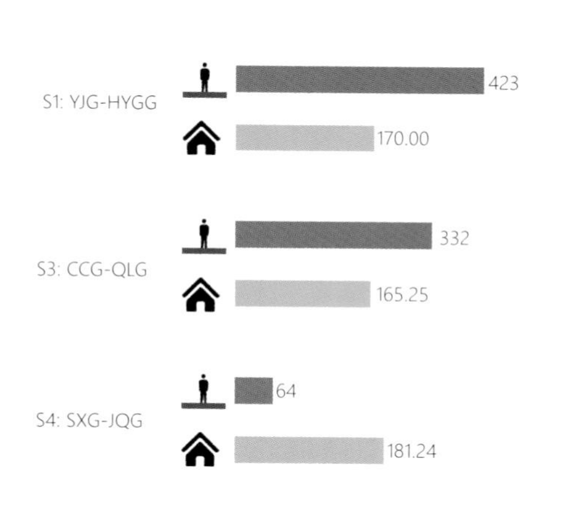

使用人群密度（单位：人 /km）
Density of users (Unit : person/km)

居住人口密度（单位：人 /hm²）
Density of residents (Unit:person/hm²)

游憩服务供给点
Supply points of recreational services

游憩服务需求点
Demand points of recreational services

的平衡。数据结果显示，其使用人群密度总体指标排位第一，并且在周末下午（排位第二）和日常傍晚（排位第一）两个时段均稳居前列。

上述绿道或绿道区段的服务绩效差异进一步反映出绿道选线过程中，在游憩服务供需端间建立平衡对保障其服务绩效的重要性。但在绿道规划选线实际操作中，要实现游憩服务供需端的统筹平衡并非易事，并需解决两大关键问题。其中，首要问题是对游憩服务供需节点串联价值的衡量评价与比较问题，即当有多个游憩服务供需节点存在并无法全部串联时，如何对其进行分析对比，确定优先串联次序是大多数建成环境绿道规划选线无法回避的问题。其次，受制于建成环境已有空间格局，很多情况下对一定区域内游憩服务供需节点的串联和整合很难通过单条绿道完成，而需多条绿道甚至绿道网的综合统筹。在此情况下，如何分配各个绿道的主要任务，并形成高效互补，整体功效最优的空间服务体系成为绿道规划选线过程中的另一难题。

这些难题均需通过针对性研究进行更加深入的探讨。实际上在该课题开展的同时，研究团队同时也正在结合此类问题展开研究，并以《东营市中心城绿道网综合规划及示范线设计实施导则》项目为载体，探索通过定量方法建立建成环境下的绿道选线潜力评价模型，对游憩服务供需端的价值进行分项和整合评价，形成选线潜力数据，为绿道规划选线提供有力参照。另一方面，通过将建成环境绿道做进一步分级（城市级、组团级和社区级），并设定每一级绿道在网络中的核心功能，协调整体网络系统中各绿道的功能和相互关联。目前该研究尚在进行中，具体的成果将在进行系统化梳理和验证后进行分享。

segment. Meanwhile, the route of this segment effectively connects key urban recreational nodes including Yuejianglou Scenic Spot, Xiuqiu Park, Little Peach Orchard and others and balances well the supply and demand sides of recreational services. Data shows that the overall index of user desnity ranks the first. Besides, the density ranks first on both the afternoon weekends (the second) and every day at nightfall (the first).

The difference in the service performance of the abovementioned greenway or greenway segments further reflects the importance of building a balance between the supply and demand of recreational services so as to guarantee the service performance during route selection of the greenway. However, during the actual route selection of the greenway, it is not easy to achieve an overall balance between the supply and demand sides of recreational services. Two key issues need to be solved. The first and foremost is how to measure, evaluate, and compare the connection value of supply and demand nodes of recreational services. When there are multiple supply and demand nodes of recreational services, which cannot be connected in series, how to analyze and compare them and how to define the priorities of connection in series become the unavoidable questions for the route selection in the planning of most greenways in built environments. Second, limited by the existing spatial pattern of a built environment, in most cases, it is very hard to connect in series and integrate the supply and demand nodes of recreational services in certain districts through a single greenway. It takes multiple greenways and even a greenway network to comprehensively plan as a whole. In this case, how to allocate the main tasks of every greenway and achieve a highly efficient and complementary spatial service system of the best overall effect becomes another difficulty during the process of route selection of the greenway.

We have in-depth discussions on these difficulties through targeted study. In fact, in addition to the development of this study, our team is also combining these types of questions to carry out a study. We use the project of Comprehensive Planning of the Greenway Network of Dongying Central City and Guidelines on the Design and Implementation of Pilot Routes as the carrier to explore the potential evaluation model of route selection of greenways in built environments through quantitative methods, conduct itemized and integrated evaluation of the value of the supply and demand sides of recreational services, formulate comprehensive potential values for route selection, and provide strong reference for the planning and route selection of the greenway. On the other hand, through further breaking up greenways in built environments into municipal level, district level, and community level, and setting up core functions of the greenway at every level in the greenway network, we adjust the functions of all the greenways in the entire greenway network system and their mutual relationships. This study is ongoing. Specific results will be shared after they are systematically sorted out and verified.

讨论二：空间载体的适宜性问题

建成环境绿道主要空间载体主要为城市道路及沿线绿地（沿路型绿道），河道水系及沿线绿地（滨水型绿道）、线性历史遗址及沿线绿地（遗址型绿道）。南京明城墙绿道实际上主要为遗址型和滨水型绿道，而且均是在原有绿地或设施基础上改造而成。从南京明城墙绿道已建成部分使用情况和空间品质来看，各个绿道空间载体适宜性存在较大差异。

需要明确的是，绿道作为提供休闲游憩服务的载体，其并非由单一的游径要素组成，而是多个系统要素共同构成。绿道自上而下可分为自然系统和人工系统，而人工系统又可进一步拆分为游径系统、服务设施系统、市政设施系统、标识系统等多个子系统，所有的系统要素均需一定空间来承载。但由于建成环境通常空间紧张，因而能够用以进行绿道改造和建设的空间十分受限。因此，在建成环境绿道规划中，需要在确保绿道线路最优的情况下，进一步挑选出适宜用于绿道改造建设的空间载体。

Discussion 2 : Suitability of spatial carrier

The main spatial carriers of greenways in a built environment mainly include urban roads and green spaces along the roads (greenway along roads), river system and green spaces along the rivers (greenway at the waterfront), linear historic sites and green spaces along the sites (greenway around historic sites). In fact, the Nanjing Ming Dynasty City Wall Greenway mainly consists of greenway around historic sites and at the waterfront, both of which are remolded based on the original green spaces or facilities. Judging from the service condition and space quality of part of the Nanjing Ming Dynasty City Wall Greenway that has already been built, we find that the spatial carriers of different greenway segments differ a lot in suitability.

To be clear, as a carrier of recreational services, a greenway consists of multiple system elements, instead of just a single trail system. From top to bottom, a greenway can be classified into natural systems and manmade systems. Then, manmade systems can be further divided into trail systems, service facilities systems, municipal facilities systems, sign systems, and other such systems. All system elements need certain spatial carriers. However, since built environments usually have limited space, there is very limited space that can be remolded and built into a greenway. Therefore, for the planning of a greenway in a built environment, it is necessary to further select spatial carriers that are suitable for remolding the greenway under the circumstance of guaranteeing the quality of the route of the greenway.

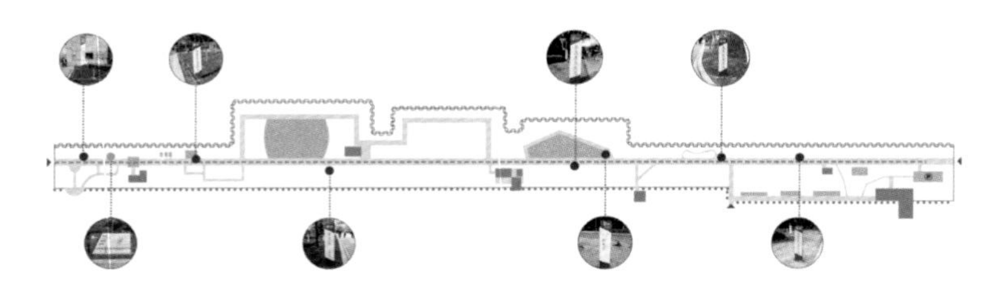

区段 1：挹江门—华严岗门
S1：YJG-HYGG

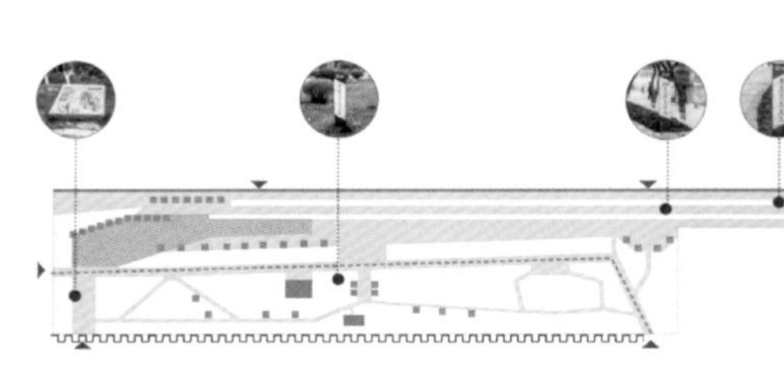

区段 3：草场门—清凉门
S3：CCG-QLG

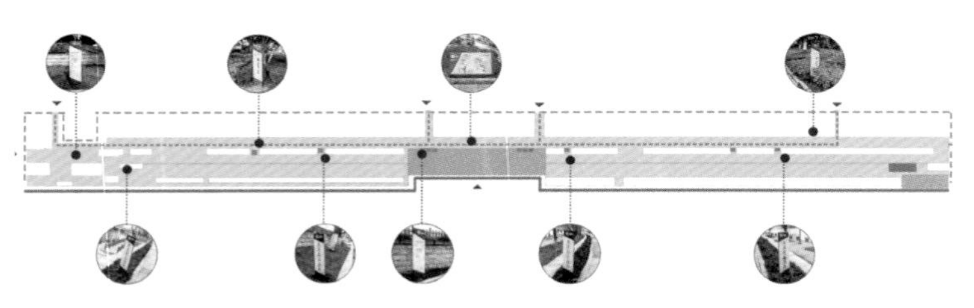

区段 2：定淮门—草场门
S2：DHG-CCG

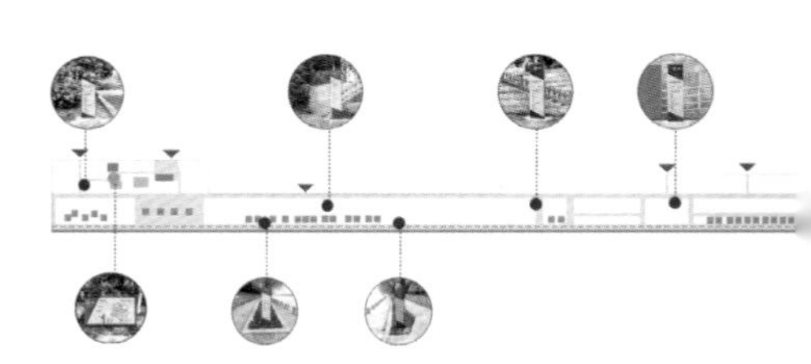

区段 4：水西门—集庆门
S4：SXG-JQG

虽然南京市明城墙绿道已有过半建成并已投入使用，但在调研过程中，我们发现建成使用区段的绿道系统要素中，实际建设完成度最高的是游径系统和标识系统，而绿道沿线的服务设施系统、市政设施系统等完成度并不高。而导致该问题最主要的原因之一实际上就是环明城墙沿线并非每个区段都适宜被用以进行绿道改造。例如，区段 4（水西门—集庆门）的大部分绿道宽度均不超过 15m，且空间狭窄和均质，扣除主游径和亲水平台后空间所剩无几，因此几乎没有多余的空间来配置绿道的集中服务点、自行车停放点等服务设施，该区段沿线也没有突出的游憩资源，实际上并不是理想的绿道选线空间载体，也正是这些因素造成了其游憩吸引力较低，远不及附近的莫愁湖公园和南湖公园等其他城市游憩节点，最终导致其服务绩效低下。而区段 1（挹江门—华严岗门）和区段 3（草场门—清凉门）空间相对宽敞，沿线也有大量场地、设施和服务建筑可以被有效整合进绿道的要素系统，并且线路上布有多个重要的城市游憩节点，均具有较高游憩吸引力，因此较适宜作为绿道选线的空间载体。

Although over half of the Nanjing Ming Dynasty City Wall Greenway has already been completed and put to use, we found that among the system elements of the greenway segments that have already been completed and put to use, the trail system and sign system are the ones with the highest degree of completion, in contrast to the service facilities system and municipal facilities system that are not completed. One of the primary reasons behind this issue is that not every segment along the Ming Dynasty City Wall is suitable for being remolded into a greenway. For example, most parts of the greenway in S4 SXG-JQG are less than 15m wide, with narrow and homogeneous space. After the main trails and hydrophilic platforms are taken out, this segment does not have any space left, so that there is almost no extra space for the installation of information and service centers, service points for bicycle storage and other such service facilities on the greenway. Nor does this segment have any special recreational resources. In fact, it is not the ideal spatial carrier of route selection of the greenway. All of these factors lead to a low utilization ratio in this segment and result in much poorer service performance, compared to nearby urban recreational nodes such as Mochou Lake Park and Nanhu Park. In contrast, the spaces in S1 YJG-HYGG and S3 CCG-QLG are much wider. A great many existing sites, facilities, and service buildings along this greenway segment can be effectively integrated into the element system of the greenway. In addition, the area along this greenway segment is distributed with multiple urban recreational nodes that are very attractive in terms of recreation. Therefore, the areas in these conditions can be considered as the priorities in the route selection of the greenway.

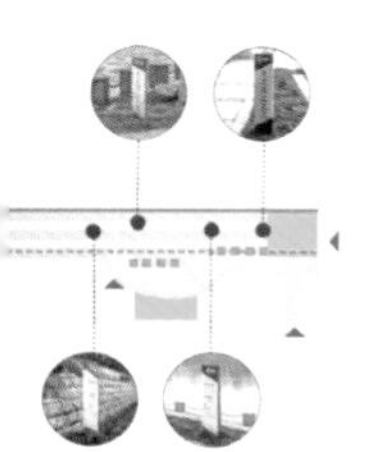

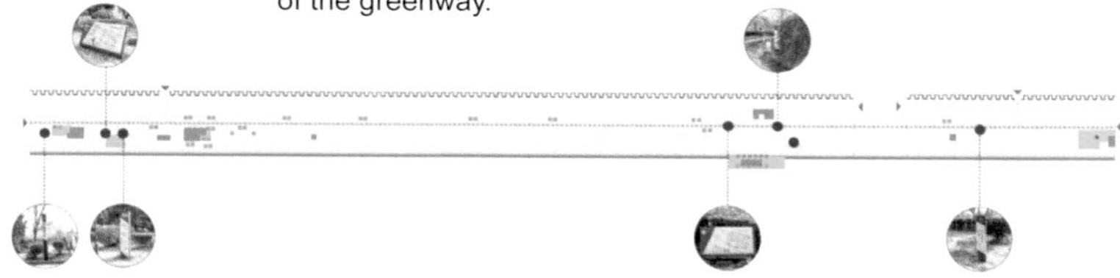

区段 5：凤台路—中华门
S5：FTR-ZHG

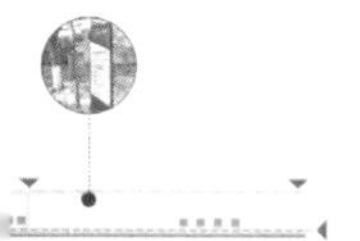

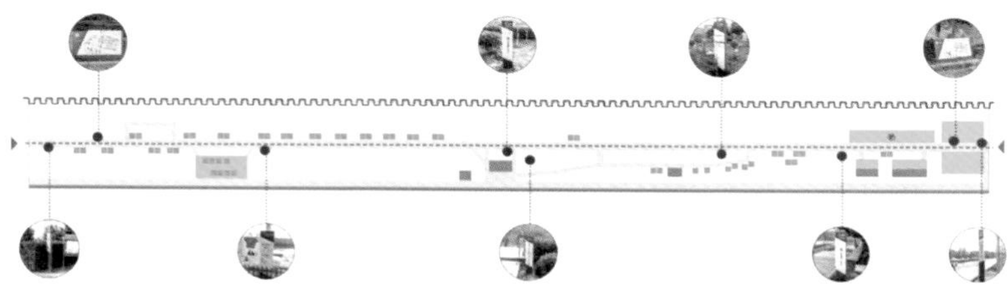

区段 6：中华门—雨花门
S6：ZHG-YHG

讨论三：空间阻隔要素的规避与整合问题

在南京明城墙绿道研究过程中，我们专门针对护城河这一要素的空间阻隔效应进行了分析，并且对其作用特征进行了初步研究。而在绿道实际选线过程中，城市中还有大量诸如此类的空间阻隔要素，例如铁路、高速路、快速路、高压走廊、河道水系甚至城墙自身等，并需要在选线中规避或整合。

其中，部分空间阻隔要素自身具有一定危险性，并将对绿道正常使用形成干扰，例如高压走廊、铁路等，此类空间阻隔要素在绿道选线过程中应该坚决回避。还有一类要素具有强烈的空间阻隔效应，但对绿道正常使用干扰性较低，例如高速路、快速路等，此类要素如果在选线过程中无法规避，则可结合要素两侧的游憩服务需求强弱进行优选，并可通过人行天桥或地道来联系要素两侧，弱化其空间阻隔效应。第三类要素则较特殊，其自身既是良好的景观游憩资源，同时在空间上会有一定的阻隔性，例如本书中分析的秦淮河。这是因为城市河道除了在交通上具有阻隔性之外，河道两岸沿线的滨水绿带在游憩使用上也具有竞争性。在该课题的调查中我们发现，如没有特殊的游憩资源吸引情况下，即便在有步行桥等设施连通两岸的情况下，居民仍旧倾向使用近岸一侧的滨水绿地进行日常游憩活动。这也是为何周边大量居住用地位于河道对岸的区段 2（定淮门—草场门）、区段 3（草场门—清凉门）在日常傍晚时段，服务绩效低下的重要原因。针对该类型要素，如果要素两侧的居民分布出现单侧明显聚集的情况，绿道应优先结合居民聚集一侧地段进行选线，同时将游憩资源的建设投入重心朝该侧地段针对性倾斜。

Discussion 3： The avoidance and integration of spatial obstruction elements

During the study of the Nanjing Ming Dynasty City Wall Greenway, we analyzed the spatial obstruction effect of the moat and have made a preliminarily discussion on the characteristics of the way it acts. After summarizing the information of the space in the city, we find out that the city also has a lot of other such spatial obstruction elements including railways, highways, expressways, high-tension corridors, river system, and even the City Wall itself, which need to be avoided or integrated during the process of route selection.

Among them, some spatial obstruction elements are dangerous and interfere with normal use of the greenway such as high-tension corridors, railways and so on. This type of spatial obstruction elements should be firmly avoided during the process of route selection of the greenway. Other types of elements have a strong spatial obstruction effect but with low interference with normal use of the greenway, including highways, expressways, and so on. If these types of elements cannot be avoided during the process of route selection of the greenway, they can be optimized in combination with whether the demand of recreational services on both sides of the elements is weak or strong, and their spatial obstruction effect can be weakened by adding pedestrian overcrossings or underpasses to connect both sides of these kind of elements. The third type of elements is very special. They are not only very good landscape recreational resources but also function as spatial obstruction like the Qinhuai River analyzed in this book. Not only can city rivers obstruct traffic, but the waterfront green space along the rivers are also competitive in terms of recreational use. In the survey of this study, we found that under the circumstance of no special recreational resources even if footbridges and other such facilities connect both sides of a river, residents still prefer to travel and use the waterfront green space on the same riverside for their daily recreational activities, which is nearer to their homes. This is why S2 DHG-CCG and S3 CCG-QLG, in which large areas of surrounding residential land are located across the rivers, have poor service performance every day at nightfall. In the light of this type of element, if the residents are imbalanced by gathering around on one side, priority should be given to the area where the residents gather around most in the route selection of the greenway, and recreational resources should be provided and put into use specifically on this side.

主要参考文献 | MAIN REFERENCES

程茂吉 . 紧凑城市理论在南京城市总体规划修编中的运用 [J]. 城市规划 , 2012, 36(2): 43-50.

Cheng Maoji. Application of Compact City Theory in Nanjing Master Plan Revising[J]. City Planning Review, 2012, 36(2): 43-50.

崔俊贵 , 田心 . 二维视角下的城乡规划法 [J]. 国际关系学院学报 , 2009(3): 87-91.

Cui Jungui, Tian Xin. The Urban-Rural Planning Law on Two-dimensional View [J]. Journal of University of International Relations, 2009(3): 87-91.

丁林可 , 田燕 . 工业用地集约利用评价指标体系初探 [J]. 国土资源科技管理 , 2007, 24(5): 18-21.

Ding Linke, Tian Yan. A Discussion on Evaluation Indicator System for Intensive Utilization of Industrial Land [J].Scientific and Technological Management of Land and Resources,2007, 24(5): 18-21.

董云波 . 老工业城市迸发新活力 —— 山东省淄博市大力推进节约集约用地的做法 [J]. 国土资源通讯 , 2013(11): 40-41,44.

DONG Yunbo. Old Industrial City Burst New Life: Zibo in Shandong Province Vigorously Promote the Economical and Intensive Land Use Practices [J]. National Land & Resources Information, 2013 (11): 40-41,44.

方创琳 , 祁巍锋 . 紧凑城市理念与测度研究进展及思考 [J]. 城市规划学刊 , 2007(4): 65-73.

Fang Chuanglin, Qi Weifeng. Research Progress and Thinking of Compact City and Its Measurement Methods[J].Urban Planning Forum, 2007(4): 65-73.

耿宏兵 . 紧凑但不拥挤 —— 对紧凑城市理论在我国应用的思考 [J]. 城市规划 , 2008(6): 48-54.

Geng Hongbing. Compact Without Crowd: Application of Compact City Theory in China[J]. City Planning Review, 2008(6): 48-54.

郭贯成 , 任宝林 , 吴群 . 基于 ArcGIS 的江苏省金坛市工业用地集约利用评价研究 [J]. 中国土地科学 , 2009, 23(8): 24-30.

Guo Guancheng, Ren Baolin, Wu Qun. Study on Assessment on Intensive Use of Industrial Land Based on ArcGIS in Jintan City[J]. Jiangsu Province China Land Science, 2009, 23(8): 24-30.

国土资源部 . 国土资源部：严控土地使用标准力促节约集约用地 [J]. 资源与人居环境 , 2012(10): 20-21.

Resources Environment Inhabitant Ministry of Land and Resources. Ministry of Land and Resources：Strict land-use Standards to Promoting Intensive Land Use [J]. Resources Environment Inhabitant, 2012(10): 20-21.

韩笋生 , 秦波 . 借鉴 " 紧凑城市 " 理念 , 实现我国城市的可持续发展 [J]. 国外城市规划 , 2004, 19(6): 5.

Han Sunsheng, Qin Bo. The Compact City and Sustainable Urban Development in China[J]. Urban Planning, 2004, 19(6): 5.

黄大全 , 洪丽璇 , 梁进社 . 福建省工业用地效率分析与集约利用评价 [J]. 地理学报 , 2009, 64(4): 479-486.

Huang Daquan, Hong Lixuan, Liang Jinshe. Analysis and Evaluation of Industrial Land Efficiency and Intensive Use in Fujian Province[J].Acta Geographica Sinica, 2009, 64(4): 479-486.

杰克 · 艾亨 , 周啸 . 论绿道规划原理与方法 [J]. 风景园林 , 2011(5): 104-107.

Jack Ahern (US), Zhou Xiao. Greenway Planning Theory and Methods[J].Landscape Architecture, 2011(5): 104-107.

金云峰，周聪惠.《城乡规划法》颁布对我国绿地系统规划编制的影响 [J]. 城市规划学刊，2009(5): 49-56.
Jin Yunfeng, Zhou Conghui. A Research on the Influence of the Enactment of Urban-Rural Planning Law on Green Space System Planning [J].Urban Planning Forum, 2009(5): 49-56.
金云峰，周聪惠. 城市绿地系统规划要素组织架构研究 [J]. 城市规划学刊，2013(3): 86-92.
Jin Yunfeng, Zhou Conghui. A Research on the Framework of Urban Green Space System Planning[J]. Urban Planning Forum, 2013(3): 86-92.
金云峰，周聪惠. 绿道规划理论实践及其在我国城市规划整合中的对策研究 [J]. 现代城市研究，2012(3): 4-12.
Jin Yunfeng, Zhou Conghui. Research on the Greenway Planning Theory Evolution and Its Practice in China[J]. Modern Urban Research, 2012(3): 4-12.
李琳."紧凑"与"集约"的并置比较——再探中国城市土地可持续利用研究的新思路 [J]. 城市规划，2006, 30(10): 19-24.
Li Lin. A Comparative Study of "Compact" and "Intensive"[J]. City Planning Review, 2006, 30(10): 19-24.
李琳. 紧凑城市中"紧凑"概念释义 [J]. 城市规划学刊，2008(3): 41-45.
Li Lin. A conceptual Analysis on "Compact" in Compact City[J]. Urban Planning Forum, 2008(3): 41-45.
李伟芳，吴迅锋，杨晓平. 宁波市工业用地节约和集约利用问题研究 [J]. 中国土地科学，2008, 22(5): 23-27.
Li Weifang, Wu Xunfeng, Yang Xiaoping. Study on Economical and Intensive Use of Industrial Land in Ningbo[J]. China Land Science, 2008, 22(5): 23-27.
理查德·罗杰斯，郑小东. 紧凑城市 [J]. 世界建筑，2009(11): 118-121.
Richard Rogers, Zheng Xiaodong. The Compact City[J]. World Architecture, 2009(11): 118-121.
刘滨谊，贺炜，刘颂. 基于绿地与城市空间耦合理论的城市绿地空间评价与规划研究 [J]. 中国园林，2012, 28(5): 42-46.
Liu Binyi, He Wei, Liu Song. Study of the Evaluation and Planning of Urban Green Space Based on the Coupling Theory of Green Space and City Space[J]. Chinese Landscape Architecture , 2012, 28(5): 42-46.
刘滨谊. 城乡绿道的演进及其在城镇绿化中的关键作用 [J]. 风景园林，2012(3): 62-65.
Liu Binyi. The Evolution of Urban and Rural Greenways and Their Key Role in Urban Greenery[J].Landscape Architecture, 2012(3):62-65.
刘冬娟，董捷. 基于灰色系统理论的工业用地集约度测度 [J]. 国土资源科技管理，2008, 25(4): 101-104.
Liu Dongjuan, Dong Jie. A Study of Land Use Intensity Measurement of Industrial Land Based on Grey System Theory[J].Scientific and Technological Management of Land and Resources, 2008, 25(4): 101-104.
刘颂，刘滨谊. 城市绿地空间与城市发展的耦合研究——以无锡市区为例 [J]. 中国园林，2010, 26(3): 14-18.
Liu Song, Liu Binyi. Coupling Analysis Between Urban Green Space and Urban Development: A Case Study of Wuxi[J]. Chinese Landscape Architecture, 2010, 26(3): 14-18.

刘文俭，张传翔 . 土地资源的节约集约利用与城市经济的持续稳定发展 [J]. 现代城市研究 , 2006, 21(5): 53-58.

Liu Wenjian, Zhang Chuanxiang. Utilizing land Resources by Economizing Intensive Ways for Urban Economy Development Continuously and Steadily[J]. Modern Urban Research, 2006, 21(5): 53-58.

吕斌，祁磊 . 紧凑城市理论对我国城市化的启示 [J]. 城市规划学刊 , 2008(4): 61-63.

Lü Bin, Qi Lei. Compact City: A Sustainable Way of Urbanization[J]. Urban Planning Forum, 2008(4): 61-63.

马奕鸣 . 紧凑城市理论的产生与发展 [J]. 现代城市研究 , 2007, 22(4): 10-16.

Ma Yiming. The Birth and Development of Compact City[J].Modern Urban Research, 2007, 22(4): 10-16.

彭晖 . 紧凑城市的再思考 —— 紧凑城市在我国应用中应当关注的问题 [J]. 国际城市规划 , 2008, 23(5): 83-87.

Peng Hui. Rethinking of Compact City: Key Issues in the Application of Compact City Theory in China[J]. Urban Planning International, 2008, 23(5): 83-87.

涂志华，王兴平 . 城市建设用地集约性评价指标体系研究 —— 基于规划编制和规划管理的视角 [J]. 城市规划学刊 , 2012(4): 86-91.

Tu Zhihua, Wang Xingping. A Research on the Evaluation of Urban Development Land-Use Intensity: A Views from Planning and Management[J]. Urban Planning Forum, 2012(4): 86-91.

王慧 . 新城市主义的理念与实践、理想与现实 [J]. 国外城市规划 , 2002(3): 35-38.

Wang Hui. The Concept and Practice, the Ideal and Reality of New Urbanism [J]. Urban Planning Overseas, 2002(3): 35-38.

王晶，曾坚，苏毅 . 可持续性 " 纤维 " 绿廊在紧凑城区规划中的应用 —— 以大野秀敏 2050 年东京概念规划方案为例 [J]. 城市规划学刊 , 2009(4):68-73.

Wang Jing, Zeng Jian, Su Yi. Application of Sustainable Fibre Greenway in Compact Urban Planning: A Case Study on Ohno Hidetoshi's Concept Planning for Tokyo 2050[J]. Urban Planning Forum, 2009(4): 68-73.

韦亚平，赵民，汪劲柏 . 紧凑城市发展与土地利用绩效的测度：" 屠能 - 阿隆索 " 模型的扩展与应用 [J]. 城市规划学刊 , 2008(3): 32-40.

Wei Yaping, Zhao Min, Wang Jinbai. Compact Development & Land Use Performance Measurements: An Applicable Expansion of "Thunne-Alonso" Model [J]. Urban Planning Forum, 2008(3): 32-40.

夏燕榕，曲福田，姜海，等 . 基于集约评价的城市开发区规模计量研究 —— 以南京市省级以上开发区为例 [J]. 中国人口 · 资源与环境 , 2010,20(2): 37-42.

Xia Yanrong, Qu Futian, Jiang Hai, et al. Quantitative Research on the Scale of Urban Development Zone Based on Evaluation of Intensive Use: A Case Study in Development Zones above the Province Level in Nanjing[J]. China Population Resources and Environment, 2010, 20(2): 37-42.

谢敏，郝晋珉，丁忠义，等 . 城市土地集约利用内涵及其评价指标体系研究 [J]. 中国农业大学学报 , 2006, 11(5): 117-120.

Xie Min, Hao Jinmin, Ding Zhongyi, et al. Study on Connotation of Intensified Urban Land Use and Its Evaluation Index[J]. Journal of China Agricultural University, 2006, 11(5): 117-120.

杨俊宴，史北祥，杨扬 . 城市中心区土地集约利用的评价模型：基于 50 个样本的定量分析 [J]. 东南大学学报（自然科学版）, 2013, 43(4): 877-884.

Yang Junyan, Shi Beixiang , Yang Yang. Research on Evaluation Model of Intensive Use of Land in Urban Central District: Quantitative Analysis Based on 50 Samples[J]. Journal of Southeast University(Natural Science Edition), 2013, 43(4): 877-884.

姚遥，魏晓冬，冯伟，等．丽水市中心城区自行车交通系统规划 [J]. 城市规划 , 2015, 39(z1): 102-106.

Yao Yao,Wei Xiaodong, Feng Wei, et al. Bicycle Traffic System Planning of Lishui Urban Area[J].City Planning Review, 2015, 39(z1):102-106.

岳宜宝．紧凑城市的可持续性与评价方法评述 [J]. 国际城市规划 , 2009, 24(6): 95-101.

Yue Yibao. The Reviews of Sustainability of Compact City and Evaluation Method[J]. Urban Planning International, 2009, 24(6): 95-101.

张润朋，周春山，明立波．紧凑城市与绿色交通体系构建 [J]. 规划师 , 2010, 26(9): 11-15.

Zhang Runpeng, Zhou Chunshan, Ming Libo. Compact City and Green Transportation Development[J]. Planners, 2010, 26(9): 11-15.

张天洁，李泽．高密度城市的多目标绿道网络 —— 新加坡公园连接道系统 [J]. 城市规划 , 2013, 37(5): 67-73.

Zhang Tianjie, Li Ze. Multi-Objective Greenway Network in High-Density Cities: The Park Connector Network in Singapore [J]. City Planning Review , 2013, 37(5): 67-73.

赵兵，王健，范月．江苏省节约型园林绿化扩展性研究与实践 [J]. 中国园林 , 2010, 26(11): 68-71.

Zhao Bing, Wang Jian, Fan Yue. Extended Research and Practice of Conservation-oriented Landscape in Jiangsu Province[J]. Chinese Landscape Architecture, 2010, 26(11): 68-71.

周聪惠，金云峰．“精细化”理念下的城市绿地复合型分类框架建构与规划应用 [J]. 城市发展研究 , 2014, 21(11): 118-124.

Zhou Conghui , Jin Yunfeng. Establishment and Planning Practice of the Multiple Classification Framework of Urban Green Space Led by the Delicacy Theory[J].Urban Development Studies, 2014, 21(11): 118-124.

周聪惠，金云峰．城市绿地系统中线状要素的规划控制途径研究 [J]. 规划师 , 2014(5): 96-102.

Zhou Conghui, Jin Yunfeng. Linear Greenery Control In Urban Green System Planning [J]. planners, 2014(5): 96-102.

周聪惠．城市内部绿道布局特征及量化控制指标体系建构研究 [J]. 中国城市林业 , 2014(1): 56-61.

Zhou Conghui. Research on Characteristics of Urban Interior Greenway and Its Quantitative Control System[J]. Journal of Chinese Urban Forestry, 2014(1): 56-61.

周聪惠．基于选线潜力定量评价的中心城绿道布局方法 [J]. 中国园林 , 2016, 32(10): 104-109.

Zhou Conghui. Layout Planning Method of Greenways in Central Urban Area Based on Route Selection Potential Quantitative Evaluation [J]. Chinese Landscape Architecture, 2016, 32(10): 104-109.

周聪惠．精细化理念下的公园绿地集约型布局优化调控方法 [J]. 现代城市研究 , 2015(10): 47-54.

Zhou Conghui. The Public Park Intensive Layout Optimization and Regulation Methods Oriented by Delicacy Management Theory[J]. Modern Urban Research, 2015(10): 47-54.

Ahern J. Greenways as a planning strategy[J]. Landscape and Urban Planning, 1995, 33(1-3): 131-155.

Andresen T, De Aguiar F B, Curado M J. The Alto Douro Wine Region Greenway[J]. Landscape and Urban Planning, 2004, 68(2-3):289-303.

Asakawa S, Yoshida K, Yabe K. Perceptions of Urban Stream Corridors Within the Greenway System of Sapporo, Japan[J]. Landscape and Urban Planning, 2004, 68(2-3): 167-182.

Bueno J A, Tsihrintzis V A, Alvarez L. South Florida Greenways: A Conceptual Framework for the Ecological Re-connectivity of the Region[J]. Landscape and Urban Planning, 1995, 33(1-3).

Conine A, Xiang W-N, Young J, et al. Planning for Multi-purpose Greenways in Concord, North Carolina[J]. Landscape and Urban Planning, 2004, 68(2-3): 271-287.

Coutts C, Miles R. Greenways as Green Magnets: The Relationship between the Race of Greenway Users and Race in Proximal Neighborhoods[J]. Journal of Leisure Research, 2011, 43(3): 317-333.

Coutts C. Multiple Case Studies of the Influence of Land-Use Type on the Distribution of Uses Along Urban River Greenways [J]. Journal of Urban Planning & Development, 2009, 135(1): 31-38.

De Sousa C A. Turning Brown Fields into Green Space in the City of Toronto[J]. Landscape and Urban Planning, 2003, 62(4): 181-198.

Erickson D L. The Relationship of Historic City form and Contemporary Greenway Implementation: A Comparison of Milwaukee, Wisconsin (USA) and Ottawa, Ontario (Canada)[J]. Landscape and Urban Planning, 2004, 68(2-3): 199-221.

Fábos J G, Ryan R L. An Introduction to Greenway Planning Around the World[J]. Landscape and Urban Planning, 2006, 76(1-4): 1-6.

Fábos J G, Ryan R L. International Greenway Planning: An Introduction[J]. Landscape and Urban Planning, 2004, 68(2-3): 143-146.

Fábos J G. Greenway Planning in the United States: Its Origins and Recent Case Studies[J]. Landscape and Urban Planning, 2004, 68(2-3): 321-342.

Fabos J G. Introduction and Overview: The Greenway Movement, Uses and Potentials of Greenways [J]. Landscape and Urban Planning,1995, 33(1-3): 1-13.

Freilich R H, Sitkowski R J, Mennillo S D. From Sprawl to Sustainability: Smart Growth, New Urbanism, Green Development, and Renewable Energy [M]. American Bar Association, 2010.

Fuller R A, Gaston K J. The Scaling of Green Space Coverage in European Cities [J]. Biology Letters, 2009, 5(3): 352-355.

Giordano L D C, Riedel P S. Multi-criteria Spatial Decision Analysis for Demarcation of Greenway: A Case Study of the City of Rio Claro, Sao Paulo, Brazil[J]. Landscape and Urban Planning, 2008, 84(3-4): 301-311.

Gobster P H. Perception and Use of a Metropolitan Greenway System for Recreation[J]. Landscape and Urban Planning, 1995, 33(1):401-413.

Hellmund P C, Smith D. Designing Greenways : Sustainable Landscapes for Nature and People[M]. Washington, DC, USA: Island Press, 2006.

Hoover A P, Shannon M A. Building Greenway Policies Within a Participatory Democracy Framework[J]. Landscape and Urban Planning,1995, 33(1-3): 433-459.

Jim C Y. Green-space Preservation and Allocation for Sustainable Greening of Compact Cities [J]. Cities, 2004, 21(4): 311-320.

Kent R L, Elliott C L. Scenic Routes Linking and Protecting Natural and Cultural Landscape Features: A Greenway Skeleton[J].

Landscape and Urban Planning, 1995, 33(1-3): 341-355.

Linehan J, Gross M, Finn J. Greenway Planning: Developing a Landscape Ecological Network Approach[J]. Landscape and Urban Planning, 1995, 33(1-3): 179-193.

Little C E. Greenways for America[M]. Baltimore: Johns Hopkins University Press, 1995.

Miller W, Collins M G, Steiner F R, et al. An Approach for Greenway Suitability Analysis[J]. Landscape and Urban Planning, 1998, 42(2-4): 91-105.

Moseley D, Marzano M, Chetcuti J, et al. Green Networks for People: Application of a Functional Approach to Support the Planning and Management of Green Space[J]. Landscape and Urban Planning, 2013, 116(0): 1-12.

Ohno H. Fibre City Tokyo: Design for the Demographic Change of Tokyo[C]. Internationale Bauausstellung Stadtumbau Sachsen- Anhalt 2010, 2010: 1-6.

Rottle N D. Factors in the Landscape-based Greenway: A Mountains to Sound Case Study[J]. Landscape and Urban Planning, 2006,76(1-4): 134-171.

Ryan R L, Fábos J G, Allan J J. Understanding Opportunities and Challenges for Collaborative Greenway Planning in New England[J].Landscape and Urban Planning, 2006, 76(1-4): 172-191.

Schrader C C. Rural Greenway Planning: The Role of Stream Land Perception in Landowner Acceptance of Land Management Strategies[J]. Landscape and Urban Planning, 1995, 33(1-3): 375-390.

Shafer C S, Lee B K, Turner S. A Tale of Three Greenway Trails: User Perceptions Related to Quality of Life[J]. Landscape and Urban Planning, 2000, 49(3-4): 163-178.

Shannon S, Smardon R, Knudson M. Using Visual Assessment as a Foundation for Greenway Planning in the St. Lawrence River Valley[J]. Landscape and Urban Planning, 1995, 33(1–3): 357-371.

Sinclair K E, Hess G R, Moorman C E, et al. Mammalian Nest Predators Respond to Greenway Width, Landscape Context and Habitat Structure[J]. Landscape and Urban Planning, 2005, 71(2-4): 277-293.

Tan K W. A greenway Network for Singapore[J]. Landscape and Urban Planning, 2006, 76(1-4): 45-66.

Teng M, Wu C, Zhou Z, et al. Multipurpose Greenway Planning for Changing Cities: A Framework Integrating Priorities and a Least-cost Path Model[J]. Landscape and Urban Planning, 2011, 103(1): 1-14.

Turner T. Greenway Planning in Britain: Recent Work and Future Plans[J]. Landscape and Urban Planning, 2006, 76(1-4): 240-251.

Zakaria El Adli Imam K. Role of Urban Greenway Systems in Planning Residential Communities: A Case Study from Egypt[J]. Landscape and Urban Planning, 2006, 76(1-4): 192-209.

后记 | AFTERWORD

研究团队成员前后历经 20 个月的时间，出动无人机航拍器进行数据采集超过 100 架次，跨越寒暑进行了一轮又一轮枯燥的数据和信息采集，当中夹杂的临时补充调查次数更是不计其数，可谓是一次不折不扣辛苦艰难的调查过程。整个研究经历了计划制订、预调研、研究对象锁定、多轮数据采集、问题分析和补充调查等多个阶段，技术路线和研究方法也经历了不断调适和再优化的过程，最终形成当前成果并整理成书。

在研究中，我们发现建成环境中对绿道服务绩效产生的影响因素不计其数，每个因素的作用方式也各不相同，定量方法虽然提供了一种研究复杂因素及其作用方式的直观途径，但要想对各个因素进行全面梳理和深入的系统分析仍是难上加难，尤其是如何针对各影响因素的作用方式找到针对的定量衡量方法，如何将不同要素的共同影响和作用机制进行定量整合分析，如何将绿道内部空间布局特征进行定量描述等方面还有很长的路要走。因此，我深知目前的研究方法、过程和结果难免会有诸多不成熟之处有待进一步优化和完善，也恳请批评指正。

谨在本书出版之际，要衷心感谢在 20 个月时间里全程参与和承担课题调研、研究分析、成果整理等辛苦工作的吴韵、胡樱、容梓昊、戴文嘉等研究团队核心成员，也需要感谢在本人任教和博士后在站研究期间一直提供指导、帮助和支持的成玉宁教授以及为本书出版提出宝贵意见的哥伦比亚大学城市设

After twenty months, our team mobilized drones for data collection more than 100 times and conducted multiple rounds of monotonous data and information collection across all seasons. There were many temporary supplementary surveys, which can be described as a toilsome and difficult process. The whole study had many phases, including planning, preliminary survey and analysis, the locking of study object, rounds of data collection, problem analysis, supplementary surveys, and so on. Our technical route and study methodology also experienced a process of continuous adaptation and re-optimization, so as to form the current results, which were ultimately compiled into this book.

During study, we found countless influencing factors of service performance of greenways in built environments, each of which functioned in a different way. Although quantitative methods could provide a straightforward way for us to study complex factors and their functions, we found it extremely difficult to put in order and make in-depth systematic analysis of all the factors in a comprehensive way, which was especially true for how to find out targeted quantitative measuring methods aimed at the modes of action of all different influencing factors, how to quantitatively integrate and analyze common effects and mechanisms of action of different factors, and how to make a quantitative description of the internal spatial layout characteristics of the greenway. In terms of these aspects, we still have a long way to go. Therefore, I know very well that the current study methodology, process, and results may unavoidably be under developing, and await further improvement. Therefore, we cordially invite criticisms and corrections.

At the time of the publication of this book, I would like to express my sincere thanks to core members of the team including Wu Yun, Hu Ying, Rong Zihao, Dai Wenjia and others, who fully participated in and shouldered the responsibility of all the hard work including survey, analysis, and result collation and so on in those twenty months. I would also like to thank Professor Cheng Yuning, who has continuously offered me a great deal of help and support on

计实验室主任 Richard Plunz 教授。另外，王建国院士、阳建强教授、张宏教授也在本研究开展的过程中提出过宝贵建议，徐宁老师、殷铭老师和郝一涵也为本研究的开展提供过很多帮助和支持，高寒玉同学也在课题研究的前期做了很多工作，另外，田恬老师的翻译以及哥伦比亚大学城市设计实验室助理 Steff Cnarvez 的英文译审工作也为本书能以双语形式出版提供了重要保障。在此一并感谢。

正是得益于各位前辈和同行专家的指导和帮助，该研究才能得以完成，并顺利成书出版。同时，也感谢一直以来在生活和研究工作中给予我巨大帮助和支持的家人和朋友们，还要感谢东南大学出版社的徐步政、孙惠玉编辑为本书顺利出版所付出的努力！

周聪惠

2017 年 7 月 30 日

东南大学

my research since I became a faculty at Southeast University, and Professor Richard Plunz, Director of the Urban Design Lab at Columbia University, who has given valuable suggestions and encouragement to the publication of this book. In addition, Professor Wang Jianguo, Professor Yang Jianqiang, and Professor Zhang Hong also made precious comments during the process of this study. I would also like to offer my sincere gratitude to Dr. Xu Ning, Dr. Yin Ming, and Hao Yihan who also provided a great deal of help and support, Gao Hanyu who did a lot of work in the early stages of this study, Tian Tian who is the main translator of this book, and Steff Charvez, assistant of the Urban Design Lab at Columbia University, who contributed a lot to the English edit.

It is thanks to the help and support of the aforesaid predecessors and experts in the same field that this study can be completed and that this book can be published. Meanwhile, I would like to express heartfelt thanks to my family and friends who have always given me great help and support during my life and research. Last but not least, I would like to thank my editors Xu Buzheng and Sun Huiyu at Southeast University Press for making efforts for the successful publication of this book!

Zhou Conghui

July 30, 2017

Southeast University

作者简介

About the Main Author

周聪惠，1982 年生于湖南省衡阳市。现任东南大学建筑学院景观学系副教授，美国景观建筑师协会（ASLA）国际会员。本科、硕士和博士均就读于同济大学建筑与城市规划学院景观学系，博士阶段赴美国哥伦比亚大学城市设计系联合培养，罗马大学建筑系访问学者。近年来主持国家自然科学基金课题及省部级科研课题 3 项，主持完成景观规划设计及城市规划设计实践项目 10 余项。著有学术专著 2 部，参与国外学术专著编写 2 部，并分别在美国和意大利出版，在国内外重要期刊和会议上发表城市绿色基础设施规划、城市污染场地治理等方面的学术论文共 20 余篇，科研、设计和教学成果均多次获奖。

Zhou Conghui is an Associate Professor of Department of Landscape Architecture of School of Architecture of Southeast University and an international member of American Society of Landscape Architects (ASLA). He got a Ph.D degree in Landscape Architecture from Tongji University. During the pursuit of the doctor's degree, he studied at the Urban Design Program of Columbia University for a one-year's joint-training fully funded by CSC program and went to Department of Architecture at Sapienza University in Rome as a visiting scholar. In recent years, his research on the planning, establishment, and optimization of green infrastructure in built environment has been funded by National Natural Science Foundation of China and other ministerial and provincial funding. He presided over and completed more than 10 green infrastructure planning, landscape design, and urban planning and design projects in Chongqing, Shandong Province and Jiangsu Province. He was the main author of 2 academic monographs, and participant of the writing and compilation of 2 academic monographs published in US and Italy, respectively. He has published more than 20 academic papers on the research of green infrastructure planning and contaminated site remediation in academic journals and conferences both home and abroad and has won a series of awards for academic, designing and teaching achievements.